SEMICONDUCTOR DEVICES
Physics and Technology

SEMICONDUCTOR DEVICES
Physics and Technology

S.M. SZE
AT&T Bell Laboratories
Murray Hill, New Jersey

JOHN WILEY & SONS
New York Chichester Brisbane Toronto Singapore

Library of Congress Cataloging in Publication Data:

Sze, S. M., 1936-
  Semiconductor devices, physics and technology.

  Includes index.
  1. Semiconductors.   I. Title.
TK7871.85.S9883   1985   621.381'52   85-3217
ISBN 0-471-87424-8

Printed in the United States of America

20 19 18

To Therese, Raymond and Julia

Preface

This book is an introduction to the physical principles of semiconductor devices and their fabrication technology. It is intended as a textbook for undergraduate students in applied physics, electrical engineering, and materials science; it can also serve as a reference for practicing engineers and scientists who need an update on device and technology developments.

The text is organized into three parts. The first part, Chapters 1 and 2, describes the basic properties of semiconductors and their conduction processes, with special emphasis on the two most important semiconductors: silicon (Si) and gallium arsenide (GaAs). The second part, Chapters 3 through 7, considers the physics and characteristics of semiconductor devices. We begin with the $p-n$ junction, which is the building block of most semiconductor devices; we proceed to bipolar and unipolar devices and then cover special microwave and photonic devices. The third part, Chapters 8 through 12, deals with processing technology from crystal growth to lithographic pattern transfer. We present the theoretical and practical aspects of the major steps in device fabrication with an emphasis on integrated devices. Although each chapter is more or less independent of the other chapters, it is recommended that instructors follow the sequence of topics from Chapter 1 to Chapter 12 for a logical and coherent presentation. The problems at the end of each chapter form an integral part of the development of the topics.

Many people have assisted me in writing the book. I would first like to express my deep appreciation to the management of AT&T Bell Laboratories for their support of this project. I have benefited significantly from suggestions made by the reviewers Drs. A.C. Adams, J. Agraz-Guerena, J.R. Brews, J.H. Bruning, D.I. Caplan, A.Y. Cho, C.M. Drum, W. Fichtner, D.B. Fraser, W.D. Johnston, L.E. Katz, T.P. Lee, M.P. Lepselter, S. Luryi, W.T. Lynch, D.A. McGillis, C.J. Mogab, K.K. Ng, M.B. Panish, L.C. Parrillo, C.W. Pearce, T.P. Pearsall, T.E. Seidel, T.T. Sheng, G.E. Smith, G.W. Taylor, and J.C.C. Tsai of AT&T Bell Laboratories; and Professors C.Y. Chang, the National Cheng Kung University, C.R. Crowell, University of Southern California, H. Melchior, Swiss Federal Institute of Technology, and H.W. Thim, Technical University of Vienna.

I am further indebted to Mr. E. Labate for the literature search; to Mr. N. Erdos, Mr. J.A. Foley, Ms. P.F. Foley, Mr. W.M. Heskes, Ms. A.C. Johnson, and Ms. N.J.

Miller for technical editing of the manuscript; to Ms. J. Maye, Ms. J.T. McCarthy, Ms. A.M. McDonough, and the members of the Text Processing Center who typed the final manuscript; and to Mr. R.T. Anderson and the members of the drafting department who furnished the hundreds of technical illustrations used in the book. In each case where an illustration was used from another published source, I have received permission from the copyright holder. Even though all illustrations were then adopted and redrawn, I appreciate being granted these permissions.

I wish to thank Mr. M.G. Floyd and Mr. G.V. Novotny of John Wiley and Sons, who encouraged me to undertake this project. My sincere appreciation is also extended to Mr. Ranjit S. Mand, a doctoral student at the University of Bradford, for his contribution in improving the content of the text and for preparation of the problem solutions. Finally, I am grateful to my wife and children for their assistance in many ways, including typing the entire first draft and preparing the illustrations, appendixes, and index.

Murray Hill, New Jersey
January 1985   *S. M. Sze*

Contents

SEMICONDUCTOR DEVICES
Physics and Technology

1

Energy Bands and Carrier Concentration

In this chapter we consider some basic properties of semiconductors. We begin with a discussion of crystal structure, which is the arrangement of atoms in a semiconductor. We then present the concepts of valence bonds and energy bands, which relate to conduction in semiconductors. Finally, we discuss the concept of carrier concentration under thermal equilibrium. These concepts will be used throughout this book.

1.1 SEMICONDUCTOR MATERIALS

Solid-state materials can be grouped into three classes—insulators, semiconductors, and conductors. Figure 1 shows the electrical conductivities σ (and the corresponding resistivities $\rho \equiv 1/\sigma)^*$ associated with some important

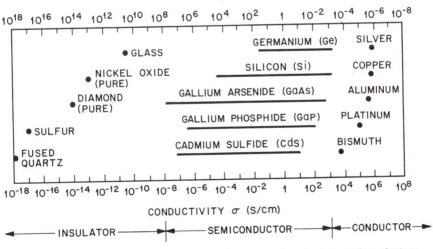

Fig. 1 Typical range of conductivities for insulators, semiconductors, and conductors.

---

* A list of symbols is given in Appendix A.

materials in each of the three classes. Insulators such as fused quartz and glass have very low conductivities, in the order of 10^{-18} to 10^{-8} S/cm; and conductors such as aluminum and silver have high conductivities, typically from 10^4 to 10^6 S/cm.[†] Semiconductors have conductivities between those of insulators and those of conductors. The conductivity of a semiconductor is generally sensitive to temperature, illumination, magnetic field, and minute amount of impurity atoms. This sensitivity in conductivity makes the semiconductor one of the most important materials for electronic applications.

The study of semiconductor materials began in the early nineteenth century.[1] Over the years many semiconductors have been investigated. Table 1 shows a portion of the periodic table related to semiconductors. The element semiconductors, those composed of single species of atoms, such as silicon (Si) and germanium (Ge), can be found in Column IV. However, numerous compound semiconductors are composed of two or more elements. For example, gallium arsenide (GaAs) is a III–V compound that is a combination of gallium (Ga) from Column III and arsenic (As) from Column V. Table 2 lists some of the element and compound semiconductors.

Prior to the invention of the bipolar transistor in 1947, semiconductors were used only as two-terminal devices, such as rectifiers and photodiodes. In the early 1950s, germanium was the major semiconductor material. However, germanium proved unsuitable in many applications because germanium devices exhibited high leakage currents at only moderately elevated temperatures. In addition, germanium oxide is water soluble and unsuited for device fabrication. Since the early 1960s silicon has become a practical substitute and has now virtually supplanted germanium as a material for semiconductor fabrication. The main reasons we now use silicon are that silicon devices exhibit much lower leakage currents, and high-quality silicon dioxide can be grown thermally. There is also an economic consideration. Device grade silicon costs much less than any other semiconductor material. Silicon in the form of silica

Table 1   Portion of the Periodic Table Related to Semiconductors

Period	Column II	III	IV	V	VI
2		B Boron	C Carbon	N Nitrogen	
3	Mg Magnesium	Al Aluminum	Si Silicon	P Phosphorus	S Sulfur
4	Zn Zinc	Ga Gallium	Ge Germanium	As Arsenic	Se Selenium
5	Cd Cadmium	In Indium	Sn Tin	Sb Antimony	Te Tellurium
6	Hg Mercury		Pb Lead		

[†] The international system of units is presented in Appendix B.

Table 2 Element and Compound Semiconductors

Element	IV–IV Compounds	III–V Compounds	II–VI Compounds	IV–VI Compounds
Si	SiC	AlAs	CdS	PbS
Ge		AlSb	CdSe	PbTe
		BN	CdTe	
		GaAs	ZnS	
		GaP	ZnSe	
		GaSb	ZnTe	
		InAs		
		InP		
		InSb		

and silicates comprises 25% of the Earth's crust, and silicon is second only to oxygen in abundance. At present, silicon is one of the most studied elements in the periodic table; and silicon technology is by far the most advanced among all semiconductor technologies.

Many of the compound semiconductors have electrical and optical properties that are absent in silicon. These semiconductors, especially gallium arsenide (GaAs), are used mainly for microwave and photonic applications. Although we do not know as much about the technology of compound semiconductors as we do about that of silicon, compound semiconductor technology has advanced partly because of the advances in silicon technology. In this book we are concerned mainly with device physics and processing technology of silicon and gallium arsenide.

1.2 CRYSTAL STRUCTURE

The semiconductor materials we will study are single crystals, that is, the atoms are arranged in a three-dimensional periodic fashion. The periodic arrangement of atoms in a crystal is called a *lattice*. In a crystal, an atom never strays far from a single, fixed position. The thermal vibrations associated with the atom are centered about this position. For a given semiconductor, there is a *unit cell* that is representative of the entire lattice; by repeating the unit cell throughout the crystal, one can generate the entire lattice.

Figure 2 shows some basic cubic-crystal unit cells. Figure 2a shows a simple cubic crystal; each corner of the cubic lattice is occupied by an atom that has six equidistant nearest neighboring atoms. The dimension *a* is called the *lattice constant*. Only polonium is crystallized in the simple cubic lattice. Figure 2b is a body-centered cubic (bcc) crystal, where in addition to the eight corner atoms, an atom is located at the center of the cube. In a bcc lattice, each atom has eight nearest-neighboring atoms. Crystals exhibiting bcc lattices include those of sodium and tungsten. Figure 2c shows a face-centered

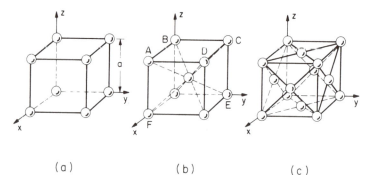

(a)   (b)   (c)

Fig. 2 Three cubic-crystal unit cells. (a) Simple cubic. (b) Body-centered cubic. (c) Face-centered cubic.

cubic (fcc) crystal that has one atom at each of the six cubic faces in addition to the eight corner atoms. In an fcc lattice, each atom has 12 nearest-neighboring atoms. A large number of elements exhibit the fcc lattice form, including aluminum, copper, gold, and platinum.

Problem

If we pack hard spheres in a bcc lattice such that the atom in the center just touches the atoms at the corners of the cube, find the fraction of the bcc unit cell volume filled with hard spheres.

Solution

Each corner sphere in a bcc unit cell is shared among eight neighboring cells; thus, each unit cell contains one eighth of a sphere at each of the eight corners for a total of one sphere. In addition, each unit cell contains one central sphere. We have

Spheres (atoms) per unit cell = 1 (corner) + 1 (center) = 2

Nearest-neighbor distance (along the diagonal AE in Fig. 2b) = $a\sqrt{3}/2$

Radius of each sphere = $a\sqrt{3}/4$

Volume of each sphere = $\dfrac{4}{3}\pi(a\sqrt{3}/4)^3 = \dfrac{\pi a^3\sqrt{3}}{16}$

Maximum fraction of unit cell filled

$$= \frac{\text{number of spheres} \times \text{volume of each sphere}}{\text{total volume of unit cell}}$$

$$= \frac{2 \times \pi a^3\sqrt{3}/16}{a^3} = \pi\sqrt{3}/8 = 0.68 .$$

Therefore, 68% of the bcc unit cell volume is filled with hard spheres, and 32% of the volume is empty.

The element semiconductors, silicon and germanium, have a *diamond lattice* structure as shown in Fig. 3a. This structure also belongs to the cubic-crystal family and can be seen as two interpenetrating fcc sublattices with one sublattice displaced from the other by one quarter of the distance along a diagonal of the cube (i.e., a displacement of $a\sqrt{3}/4$). All atoms are identical

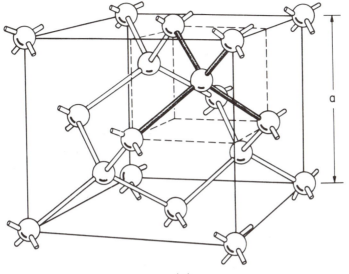

(a)

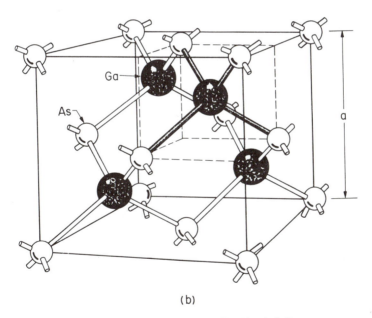

(b)

Fig. 3 (a) Diamond lattice. (b) Zincblende lattice.

in a diamond lattice, and each atom in the diamond lattice is surrounded by four equidistant nearest neighbors that lie at the corners of a tetrahedron (refer to the spheres connected by darkened bars in Fig. 3a). Most of the III–V compound semiconductors (e.g., GaAs) have a *zincblende lattice*, shown in Fig. 3b, which is identical to a diamond lattice except that one fcc sublattice has Column III atoms (Ga) and the other has Column V atoms (As). Appendix F gives a summary of the lattice constants and other properties of important semiconductors.

Problem

At 300 K the lattice constant for silicon is 5.43 Å. Calculate the number of silicon atoms per cubic centimeter and the density of silicon at room temperature.

Solution

There are eight atoms per unit cell. Therefore,

$$\frac{8}{a^3} = \frac{8}{(5.43 \times 10^{-8})^3} = 5 \times 10^{22} \text{ atoms/cm}^3$$

$$\text{Density} = \frac{\text{no. of atoms/cm}^3 \times \text{atomic weight}}{\text{Avogadro constant}}$$

$$= \frac{5 \times 10^{22} \text{ (atoms/cm}^3) \times 28.09 \text{ (g/mole)}}{6.02 \times 10^{23} \text{ (atoms/mole)}} = 2.33 \text{ g/cm}^3.$$

In Fig. 2b we note that there are four atoms in the ABCD plane and five atoms in the ACEF plane (four atoms from the corners and one from the center) and that the atomic spacings are different for the two planes. Therefore, the crystal properties along different planes are different, and the electrical and other device characteristics are dependent on the crystal orientation. A convenient method of defining the various planes in a crystal is to use *Miller indices*.[2] These indices are obtained using the following steps:

(1)   Find the intercepts of the plane on the three Cartesian coordinates in terms of the lattice constant.

(2)   Take the reciprocals of these numbers and reduce them to the smallest three integers having the same ratio.

(3)ʹ Enclose the result in parentheses (*hkl*) as the Miller indices for a single plane.

Example

As shown in Fig. 4, the plane has intercepts at *a*, 2*a*, and 2*a* along the three coordinates. Taking the reciprocals of these intercepts, we get 1, ½, and ½. The smallest three integers having the same ratio are 2, 1, and 1 (obtained by multiplying each fraction by 2). Thus, the plane is referred to as a (211)-plane.

Figure 5 shows the Miller indices of important planes in a cubic crystal. Some other conventions are given as follows:

($\bar{h}kl$): For a plane that intercepts the *x*-axis on the negative side of the origin, such as ($\bar{1}$00).

{*hkl*}: For planes of equivalent symmetry—such as {100} for (100), (010), (001), ($\bar{1}$00), (0$\bar{1}$0), and (00$\bar{1}$) in cubic symmetry.

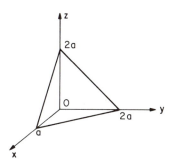

Fig. 4   A (211)-crystal plane.

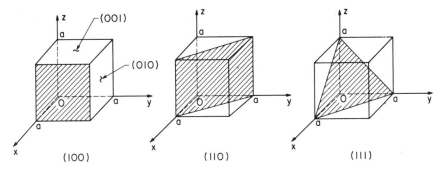

Fig. 5   Miller indices of some important planes in a cubic crystal.

[*hkl*]: For a crystal direction, such as [100] for the *x*-axis. Thus, the [100]-direction is perpendicular to (100)-plane, and the [111]-direction is perpendicular to the (111)-plane.

<*hkl*>: For a full set of equivalent directions—such as <100> for [100], [010], [001], [$\bar{1}$00], [0$\bar{1}$0], and [00$\bar{1}$].

1.3   VALENCE BONDS

As discussed in Section 1.2, each atom in a diamond lattice is surrounded by four nearest neighbors. Figure 6*a* shows the tetrahedron configuration of a diamond lattice. A simplified two-dimensional bonding diagram for the tetrahedron is shown in Fig. 6*b*. Each atom has four electrons in the outer orbit, and each atom shares these valence electrons with its four neighbors. This sharing of electrons is known as *covalent bonding*; each electron pair constitutes a covalent bond. Covalent bonding occurs between atoms of the same element or between atoms of different elements that have similar outer-shell electron configurations. Each electron spends an equal amount of time with each nucleus. However, both electrons spend most of their time between the two nuclei. The force of attraction for the electrons by both nuclei holds the

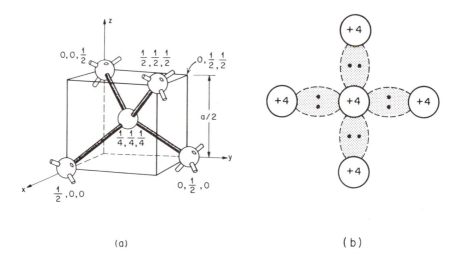

Fig. 6 (*a*) A tetrahedron bond. (*b*) Schematic two-dimensional representation of a tetrahedron bond.

two atoms together. For a zincblende lattice such as gallium arsenide, the major bonding force is from the covalent bonds. However, gallium arsenide has a slight ionic bonding force that is an electrostatic attractive force between each Ga^- ion and its four neighboring As^+ ions, or between each As^+ ion and its four neighboring Ga^- ions.

At low temperatures, the electrons are bound in their respective tetrahedron lattice; consequently, they are not available for conduction. At higher temperatures, thermal vibrations may break the covalent bonds. When a bond is broken, a free electron results that can participate in current conduction. Figure 7*a* shows the situation when a valence electron becomes a free electron. An electron deficiency is left in the covalent bond. This deficiency may be filled by one of the neighboring electrons, which results in a shift of the deficiency location, as from location A to location B in Fig. 7*b*. We may therefore consider this deficiency as a particle similar to an electron. This fictitious particle is called a *hole*. It carries a positive charge and moves, under the influence of an applied electric field, in the direction opposite to that of an electron. The concept of a hole is analogous to that of a bubble in a liquid. Although it is actually the liquid that moves, it is much easier to talk about the motion of the bubble in the opposite direction.

1.4 ENERGY BANDS

For an isolated atom, the electrons of the atom can have only discrete energy levels. For example, the energy levels for an isolated hydrogen atom are given by the Bohr model[3]:

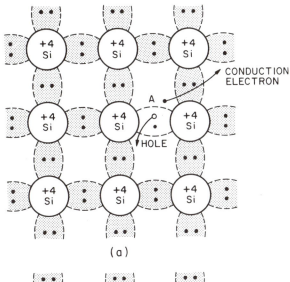

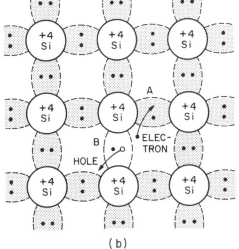

Fig. 7 The basic bond pictures of intrinsic silicon. (a) A broken bond at position A, resulting in a conduction electron and a hole. (b) A broken bond at position B.

$$E_H = \frac{-m_0 q^4}{8\epsilon_0^2 h^2 n^2} = \frac{-13.6}{n^2} \text{ eV} \qquad (1)$$

where m_0 is the free-electron mass, q is the electronic charge, ϵ_0 is the free-space permittivity, h is the Planck constant, and n is a positive integer called the principal quantum number. The discrete energies are -13.6 eV for the ground level ($n = 1$), -3.4 eV for the first excited level ($n = 2$), etc.

We now consider two identical atoms. When they are far apart, the allowed energy levels for a given principal quantum number (e.g., $n = 1$) consist of one doubly degenerate level, that is, each atom has exactly the same energy (e.g., -13.6 eV for $n = 1$). As the two atoms approach one another, the doubly degenerate energy level will split into two levels by the interaction between the atoms. When we bring N atoms together to form a crystal, the N-fold degenerate energy level will split into N separate but closely spaced levels due to atomic interaction. This results in an essentially continuous band of energy.

The detailed energy band structures of crystalline solids have been calculated using quantum mechanics. Figure 8 is a schematic diagram of the formation of a diamond lattice crystal from isolated silicon atoms. Each isolated atom has its discrete energy levels (two levels are shown on the far right of the diagram). As the interatomic spacing decreases, each degenerate energy level splits to form a band. Further decrease in spacing causes the bands originating from different discrete levels to lose their identities and merge together, forming a single band. When the distance between atoms approaches the equilibrium interatomic spacing of the diamond lattice (with a lattice constant of 5.43 Å for silicon), this band splits again into two bands. These bands are separated by a region which designates energies that the electrons in the solid cannot possess. This region is called the forbidden gap, or *bandgap* E_g. The upper band is called the *conduction band*, while the lower band is called the *valence band*, as shown on the far left of Fig. 8.

Figure 9 shows the energy band diagrams of three classes of solids— insulators, semiconductors, and conductors. In an insulator such as silicon dioxide (SiO_2), the valence electrons form strong bonds between neighboring

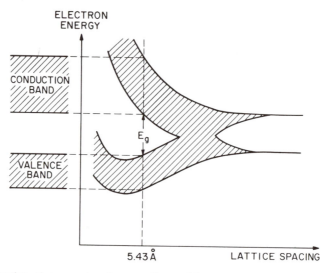

Fig. 8 Formation of energy bands as a diamond lattice crystal is formed by bringing together isolated silicon atoms.

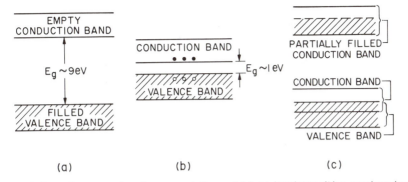

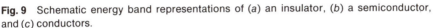

Fig. 9 Schematic energy band representations of (*a*) an insulator, (*b*) a semiconductor, and (*c*) conductors.

atoms. These bonds are difficult to break, and consequently there are no free electrons to participate in current conduction. As shown in the energy band diagram Fig. 9a, there is a large bandgap. Note that all energy levels in the valence band are occupied by electrons and all energy levels in the conduction band are empty. Thermal energy or an applied electric field cannot raise the uppermost electron in the valence band to the conduction band. Therefore, silicon dioxide is an insulator, it cannot conduct current.

As we discussed in Section 1.3, bonds between neighboring atoms in a semiconductor are only moderately strong. Therefore, thermal vibrations will break some bonds. When a bond is broken, a free electron along with a free hole result. Figure 9b shows that the bandgap of a semiconductor is not as large as that of an insulator (e.g., Si with a bandgap of 1.12 eV). Because of this, some electrons will be able to move from the valence band to the conduction band, leaving holes in the valence band. When an electric field is applied, both the electrons in the conduction band and the holes in the valence band will gain kinetic energy and conduct electricity.

In conductors such as metals, Fig. 9c, the conduction band either is partially filled or overlaps the valence band so that there is no bandgap. As a consequence, the uppermost electrons in the partially filled band or electrons at the top of the valence band can move to the next-higher available energy level when they gain kinetic energy (e.g., from an applied electric field). Therefore, current conduction can readily occur in conductors.

The energy band diagrams shown in Fig. 9 indicate electron energies. When the energy of an electron is increased, the electron moves to a higher position in the band diagram. On the other hand, when the energy of a hole is increased, the hole moves downward in the valence band. (This is because a hole has a charge opposite that of an electron.)

The relation of the electron and hole energies is illustrated in Fig. 10. As we discussed before, the separation between the energy of the lowest conduction band and that of the highest valence band is called the bandgap E_g, which is the most important parameter in semiconductor physics. We desig-

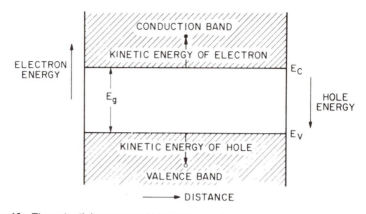

Fig. 10   The potential energy and kinetic energy in an energy band representation.

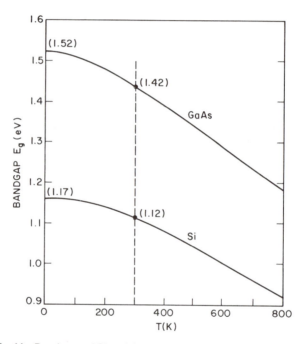

Fig. 11   Bandgaps of Si and GaAs as a function of temperature.[4]

nate E_C as the bottom of the conduction band; E_C corresponds to the potential energy of an electron, that is, the energy of a conduction electron that is at rest. The kinetic energy of an electron is measured upward from E_C. Similarly, we designate E_V as the top of the valence band; E_V corresponds to the potential energy of a hole. The kinetic energy of a hole is measured downward from E_V.

At room temperature and under normal atmosphere, the values of the bandgap are 1.12 eV for silicon and 1.42 eV for gallium arsenide. Figure 11 shows the measured bandgap as a function of temperature. The bandgap approaches 1.17 eV for silicon and 1.52 eV for gallium arsenide at 0 K. The variation of bandgaps with temperature can be expressed as[4]

$$E_g(T) = 1.17 - \frac{(4.73 \times 10^{-4})\, T^2}{(T + 636)} \qquad \text{for Si} \qquad (2a)$$

and

$$E_g(T) = 1.52 - \frac{(5.4 \times 10^{-4})\, T^2}{(T + 204)} \qquad \text{for GaAs}. \qquad (2b)$$

The temperature coefficient dE_g/dT is negative for both silicon and gallium arsenide (i.e., the bandgap decreases with increasing temperature).

For a free electron, the kinetic energy E is given by

$$E = \frac{p^2}{2m_0} \qquad (3)$$

where p is the particle momentum and m_0 is the free-electron mass. An electron in the conduction band is similar to a free electron in that it is relatively free to move about in a semiconductor. However, because of the periodic potential of the nuclei, the *effective mass* of a conduction electron is different from the mass of a free electron. The energy-momentum relationship of a conduction electron can be written as

$$E = \frac{\bar{p}^2}{2m_n} \qquad (4)$$

where $\bar{p}$ is the *crystal momentum* and m_n is the electron effective mass (the subscript n refers to the negative charge on an electron). A similar expression can be written for holes (with effective mass m_p, where the subscript p refers to the positive charge on a hole). The crystal momentum is analogous to the particle momentum. The effective-mass concept is very useful because it enables us to treat holes and electrons essentially as classical charged particles.

The energy band diagram of Fig. 10 is a simplified representation of a rather complex energy band structure. Figure 12 shows slightly more complicated energy band diagrams for silicon and gallium arsenide in which the energy is plotted against the crystal momentum for two crystal directions.

We note that there is a bandgap E_g between the bottom of the conduction band and the top of the valence band. For silicon, Fig. 12a, the maximum in the valence band occurs at $\bar{p} = 0$; however, the minimum in the conduction band occurs along the [100] direction at $\bar{p} = \bar{p}_c$. This situation shows the main difference between the particle momentum and the crystal momentum. The particle momentum for a free electron is zero when the kinetic energy is zero. However, an electron at the conduction band minimum (with zero kinetic energy) can have crystal momentum different from zero (e.g., $\bar{p} = \bar{p}_c$ as shown in Fig. 12a). Therefore, in silicon, when an electron makes a transition from the valence band to the conduction band, it requires not only an

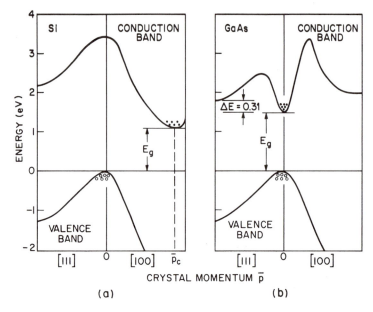

Fig. 12   Energy band structures of Si and GaAs. Circles (○) indicate holes in the valence bands and dots (•) indicate electrons in the conduction bands.

energy change ($\geqslant E_g$) but also some change in the crystal momentum. For gallium arsenide, Fig. 12b, the maximum in the valence band and the minimum in the conduction band occur at the same crystal momentum ($\bar{p} = 0$). Thus, an electron making a transition from the valence band to the conduction band can do so without a change in $\bar{p}$ value.

Gallium arsenide is called a *direct semiconductor*, because a transition from the valence band to the conduction band does not require a change in crystal momentum for the electron. Silicon is called an *indirect semiconductor*, because a change of crystal momentum is required in a transition. This difference between direct and indirect band structures is very important for light-emitting diodes and semiconductor lasers. These devices require direct semiconductors for efficient generation of photons (refer to Chapter 7).

The energy-momentum relationships near the minimum of the conduction band or near the maximum of the valence band are indeed parabolic as expressed by Eq. 4. With a known $E-\bar{p}$ relationship, one can obtain the effective mass from the second derivative of E with respect to $\bar{p}$:

$$m_n = \left[\frac{d^2E}{d\bar{p}^2}\right]^{-1}. \tag{5}$$

Therefore, the narrower the parabola, the smaller the effective mass. For example, for gallium arsenide with a very narrow conduction band parabola, the electron effective mass is $0.07m_0$ whereas for silicon with a wider conduction band parabola, the electron effective mass is $0.19m_0$. (The electron effective mass in silicon is dependent on crystal direction; the value of $0.19m_0$ is for the effective mass perpendicular to the [100]-direction.)

1.5 DENSITY OF STATES

When electrons move back and forth along the x-direction in a semiconductor material, the movements can be described by standing-wave oscillations. The wavelength λ of a standing wave is related to the length of the semiconductor L by

$$\frac{L}{\lambda} = n_x \tag{6}$$

where n_x is an integer.[3] The wavelength can be expressed as

$$\lambda = \frac{h}{\bar{p}_x} \tag{7}$$

where h is the Planck constant and $\bar{p}_x$ is the crystal momentum in the x-direction. Substituting Eq. 7 into Eq. 6 gives

$$L\bar{p}_x = hn_x . \tag{8}$$

The incremental momentum $d\bar{p}_x$ required for a unity increase in n_x is

$$L\, d\bar{p}_x = h . \tag{9}$$

For a three-dimensional cube of side L, we have

$$L^3\, d\bar{p}_x\, d\bar{p}_y\, d\bar{p}_z = h^3 . \tag{10}$$

The volume $d\bar{p}_x\, d\bar{p}_y\, d\bar{p}_z$ in the momentum space for a unit cube ($L = 1$) is thus equal to h^3. Each incremental change in n corresponds to a unique set of integers ($n_x,\ n_y,\ n_z$), which in turn corresponds to an allowed energy state. Thus, the volume in momentum space for an energy state is h^3. Figure 13 shows the momentum space in spherical coordinates. The volume between

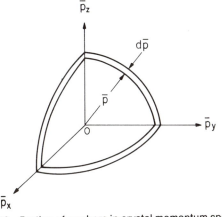

Fig. 13 Portion of a sphere in crystal momentum space.

two concentric spheres (from $\bar{p}$ to $\bar{p} + d\bar{p}$) is $4\pi\bar{p}^2\, d\bar{p}$. The number of energy states contained in this volume is then $2(4\pi\bar{p}^2\, d\bar{p})/h^3$, where the factor 2 accounts for the electron spins. From Eq. 4 we can substitute E for $\bar{p}$ and obtain

$$N(E)\, dE = \frac{8\pi\bar{p}^2\, d\bar{p}}{h^3} = 4\pi\left[\frac{2m_n}{h^2}\right]^{3/2} E^{1/2}\, dE \tag{11}$$

and

$$N(E) = 4\pi\left[\frac{2m_n}{h^2}\right]^{3/2} E^{1/2} \tag{12}$$

where $N(E)$ is called the *density of states*, that is, the density of allowed energy states per unit volume.

1.6 INTRINSIC CARRIER CONCENTRATION

At finite temperatures, continuous thermal agitation results in the excitation of electrons from the valence band to the conduction band and leaves an equal number of holes in the valence band. An *intrinsic semiconductor* is one that contains relatively small amounts of impurities compared to the thermally generated electrons and holes. To obtain the electron density (i.e., the number of electrons per unit volume) in an intrinsic semiconductor, we first evaluate the electron density in an incremental energy range dE. This density $n(E)$ is given by the product of the density of allowed energy states per unit volume $N(E)$ and by the probability of occupying that energy range $F(E)$. Thus, the electron density in the conduction band is given by integrating $N(E)\, F(E)\, dE$ from the bottom of the conduction band (E_C initially taken to be $E = 0$ for simplicity) to the top of the conduction band E_{top}:

$$n = \int_0^{E_{\text{top}}} n(E)\, dE = \int_0^{E_{\text{top}}} N(E)\, F(E)\, dE \ . \tag{13}$$

The probability that an electronic state with energy E is occupied by an electron is given by the *Fermi–Dirac distribution function* (which is also referred to as the Fermi distribution function)

$$F(E) = \frac{1}{1 + e^{(E - E_F)/kT}} \tag{14}$$

where k is the Boltzmann constant, T is the absolute temperature in degrees Kelvin, and E_F is the *Fermi level*. The Fermi level is the energy at which the probability of occupation by an electron is exactly one half. The Fermi distribution is illustrated in Fig. 14 for different temperatures. Note that $F(E)$ is symmetrical around the Fermi level E_F. For energies that are $3kT$ above or below the Fermi level, the exponential term in Eq. 14 becomes larger than 20 or smaller than 0.05, respectively. The Fermi distribution function can thus be

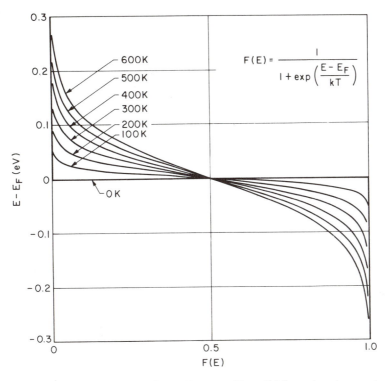

Fig. 14 Fermi distribution function $F(E)$ versus $(E - E_F)$ for various temperatures.

approximated by simpler expressions:

$$F(E) \simeq e^{-(E - E_F)/kT} \quad \text{for} \quad (E - E_F) > 3kT \quad (15a)$$

and

$$F(E) \simeq 1 - e^{-(E_F - E)/kT} \quad \text{for} \quad (E - E_F) < -3kT . \quad (15b)$$

Equation 15b can be regarded as the probability that a hole occupied a state located at energy E.

Figure 15 shows schematically from left to right the band diagram, the density of states (which varies as $\sqrt{E}$), the Fermi distribution function, and the carrier concentrations for an intrinsic semiconductor. The carrier concentrations can be obtained graphically from Fig. 15 using Eq. 13; that is, the product of $N(E)$ in Fig. 15b and $F(E)$ in Fig. 15c gives the $n(E)$-versus-E curve (upper curve) in Fig. 15d. The upper shaded area in Fig. 15d corresponds to the electron density.

In the conduction band there are a large number of allowed states. However, for intrinsic semiconductors there will be only a few electrons in the conduction band, hence the probability of an electron occupying one of these states is small. There also are a large number of allowed states in the valence

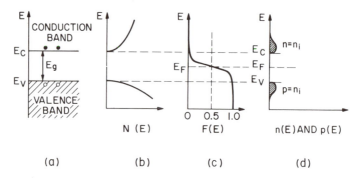

Fig. 15 Intrinsic semiconductor. (*a*) Schematic band diagram. (*b*) Density of states. (*c*) Fermi distribution function. (*d*) Carrier concentration.

band. By contrast most of these are occupied by electrons. Thus, the probability of an electron occupying one of these states in the valence band is nearly unity. There will be only a few unoccupied electron states, that is, holes, in the valence band. As can be seen, the Fermi level is located near the middle of the bandgap (i.e., E_F is many kT below E_C). Because $F(E)$ is an exponentially decreasing function of E (Eq. 15a), the value of E_{top} can be replaced by infinity. Substituting Eqs. 12 and 15a into Eq. 13 yields

$$n = 4\pi \left[\frac{2m_n}{h^2} \right]^{3/2} \int_0^\infty E^{1/2} \exp \left[- \frac{E - E_F}{kT} \right] dE . \tag{16}$$

If we let $x \equiv E/kT$, Eq. 16 becomes

$$n = 4\pi \left[\frac{2m_n}{h^2} \right]^{3/2} (kT)^{3/2} \exp \left[\frac{E_F}{kT} \right] \int_0^\infty x^{1/2} e^{-x} dx . \tag{17}$$

The integral in Eq. 17 is of the standard form and equals $\sqrt{\pi}/2$. Therefore, Eq. 17 becomes

$$n = 2 \left[\frac{2\pi m_n kT}{h^2} \right]^{3/2} \exp \left[\frac{E_F}{kT} \right] . \tag{18}$$

If we refer to the bottom of the conduction band as E_C instead of $E = 0$, we obtain for the electron density in the conduction band

$$n = 2 \left[\frac{2\pi m_n kT}{h^2} \right]^{3/2} \exp \left[- \frac{E_C - E_F}{kT} \right]$$

$$= N_C \exp \left[- \frac{E_C - E_F}{kT} \right] \tag{19}$$

and

$$N_C \equiv 2 \left[\frac{2\pi m_n kT}{h^2} \right]^{3/2} \tag{20}$$

where N_C is the *effective density of states in the conduction band.* At room temperature (300 K), N_C is 2.8×10^{19} cm^{-3} for silicon and 4.7×10^{17} cm^{-3} for gallium arsenide.

Similarly, we can obtain the hole density p in the valence band:

$$
p = 2 \left[\frac{2\pi m_p kT}{h^2} \right]^{3/2} \exp \left[-\frac{E_F - E_V}{kT} \right]
$$

$$
= N_V \exp \left[-\frac{E_F - E_V}{kT} \right] \tag{21}
$$

and

$$
N_V \equiv 2 \left[\frac{2\pi m_p kT}{h^2} \right]^{3/2} \tag{22}
$$

where N_V is the *effective density of states in the valence band.* At room temperature, N_V is 1.04×10^{19} cm^{-3} for silicon and 7.0×10^{18} cm^{-3} for gallium arsenide.

For an intrinsic semiconductor as defined previously, the number of electrons per unit volume in the conduction band is equal to the number of holes per unit volume in the valence band, that is, $n = p = n_i$, where n_i is the *intrinsic carrier density.* This relation of electrons and holes is depicted in Fig. 15d. Note that the shaded area in the conduction band is the same as that in the valence band.

The Fermi level for an intrinsic semiconductor is obtained by equating Eqs. 19 and 21:

$$
E_F = E_i = \frac{E_C + E_V}{2} + \frac{kT}{2} \ln \left[\frac{N_V}{N_C} \right]
$$

$$
= \frac{E_C + E_V}{2} + \frac{3kT}{4} \ln \left[\frac{m_p}{m_n} \right]. \tag{23}
$$

At room temperature, the second term is much smaller than the bandgap. Hence, the *intrinsic Fermi level* E_i, that is, the Fermi level of an intrinsic semiconductor, generally lies very close to the middle of the bandgap.

The intrinsic carrier density is obtained from Eqs. 19, 21, and 23:

$$
np = n_i^2 \tag{24}
$$

$$
n_i^2 = N_C N_V \exp \left[-\frac{E_g}{kT} \right] \tag{25}
$$

and

$$
n_i = \sqrt{N_C N_V} \exp \left[-\frac{E_g}{2kT} \right] \tag{26}
$$

where $E_g \equiv (E_C - E_V)$. Equation 24 is called the *mass action law,* which is valid for both intrinsic and extrinsic (i.e., doped with impurities) semi-conductors under a thermal equilibrium condition. In an extrinsic semi-conductor, the increase of one type of carrier tends to reduce the number of the other type through recombination (described in Chapter 2); thus, the product of the two types of carriers will remain constant at a given temperature.

Figure 16 shows the temperature dependence of n_i for silicon and gallium arsenide.[4] At room temperature, n_i is 1.45×10^{10} cm^{-3} for silicon and 1.79×10^6 cm^{-3} for gallium arsenide. As expected, the larger the bandgap, the smaller the intrinsic carrier density.

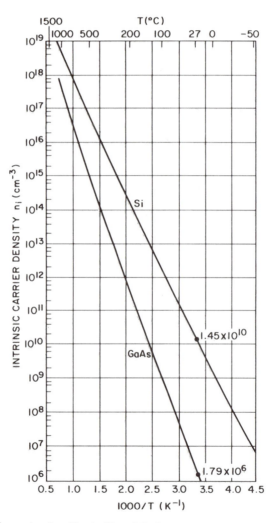

Fig. 16 Intrinsic carrier densities in Si and GaAs as a function of the reciprocal of temperature.[4]

1.7 DONORS AND ACCEPTORS

When a semiconductor is doped with impurities, the semiconductor becomes *extrinsic* and impurity energy levels are introduced. Figure 17*a* shows schematically that a silicon atom is replaced (or substituted) by an arsenic atom with five valence electrons. The arsenic atom forms covalent bonds with its four neighboring silicon atoms. The fifth electron becomes a conduction electron that is "donated" to the conduction band. The silicon becomes *n*-type because of the addition of the negative charge carrier, and the

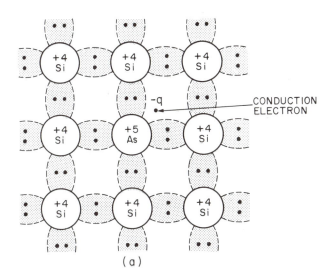

(*a*)

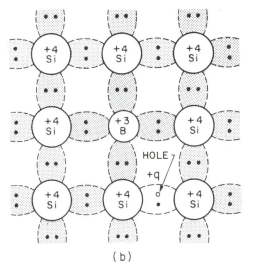

(b)

Fig. 17 Schematic bond pictures for (*a*) *n*-type Si with donor (arsenic) and (*b*) *p*-type Si with acceptor (boron).

arsenic atom is called a *donor*. Similarly, Fig. 17*b* shows that when a boron atom with three valence electrons substitutes for a silicon atom, an additional electron is "accepted" to form four covalent bonds around the boron, and a positively charged "hole" is created in the valence band. This is a *p*-type semiconductor, and the boron is an *acceptor*.

To calculate the impurity energy levels, the simplest approach is to use the hydrogen atom model. We can estimate the *ionization energy* for the donor E_D by replacing m_0 with the electron effective mass m_n and replacing ϵ_0 with the permittivity of the semiconductor ϵ_s in Eq. 1:

$$E_D = \left[\frac{\epsilon_0}{\epsilon_s}\right]^2 \left[\frac{m_n}{m_0}\right] E_H .$$ (27)

The ionization energy for donors (measured from the conduction band edge) as calculated from Eq. 27 is 0.025 eV for silicon and 0.007 eV for gallium arsenide. The hydrogen atom calculation for the ionization level of acceptors is similar to that for donors. We consider the unfilled valence band as a filled band plus a hole in the central force field of a negatively charged acceptor. The calculated ionization energy (measured from the valence band edge) is 0.05 eV for both silicon and gallium arsenide.

This simple hydrogen atom model cannot account for the details of ionization energy, particularly for the deep impurity levels in semiconductors (i.e., with ionization energies $\gtrsim 3kT$). However, the calculated values do predict the correct order of magnitude of the true ionization energies for shallow impurity levels. Figure 18 shows the measured ionization energies[5] for various impurities in silicon and gallium arsenide. Note that it is possible for a single atom to have many levels; for example, oxygen in silicon has two donor levels and two acceptor levels in the forbidden energy gap.

For shallow donors in silicon and gallium arsenide, there usually is enough thermal energy to supply the energy E_D to ionize all donor impurities at room temperature and thus provide an equal number of electrons in the conduction band. This condition is called *complete ionization*. Under a complete ionization condition, we can write the electron density as

$$n = N_D$$ (28)

where N_D is the donor concentration. Figure 19*a* illustrates complete ionization where the donor level E_D is measured with respect to the bottom of the conduction band and equal concentrations of electrons (which are mobile) and donor ions (which are immobile) are shown. From Eqs. 19 and 28, we obtain the Fermi level in terms of the effective density of states N_C and the donor concentration N_D:

$$E_C - E_F = kT \ln \left[\frac{N_C}{N_D}\right] .$$ (29)

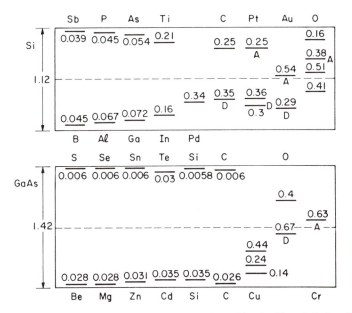

Fig. 18 Measured ionization energies for various impurities in Si and GaAs. The levels below the gap center are measured from the top of the valence band and are acceptor levels unless indicated by D for donor level. The levels above the gap center are measured from the bottom of the conduction band and are donor levels unless indicated by A for acceptor level.[5]

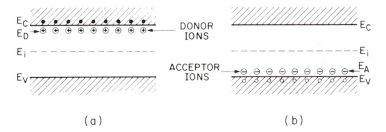

(a)   (b)

Fig. 19 Schematic energy band representation of extrinsic semiconductors with (a) donor ions and (b) acceptor ions.

Similarly, for shallow acceptors as shown in Fig. 19b, if there is complete ionization, the concentration of holes is

$$p = N_A \qquad (30)$$

where N_A is the acceptor concentration. We can obtain the corresponding Fermi level from Eqs. 21 and 30:

$$E_F - E_V = kT \ln \left[\frac{N_V}{N_A} \right] \qquad (31)$$

From Eq. 29 it is clear that the higher the donor concentration, the smaller the energy difference $(E_C - E_F)$, that is, the Fermi level will move closer to the bottom of the conduction band. Similarly, for higher acceptor concentrations, the Fermi level will move closer to the top of the valence band. Figure 20 shows the graphic procedure to obtain the carrier concentration. This is similar to that shown in Fig. 15. However, the Fermi level is closer to the bottom of the conduction band, and the electron concentration (upper shaded area) is much larger than the hole concentration (lower shaded area). Because of the mass action law (Eq. 24), the product of n and p is the same as that for the intrinsic case (i.e., $np = n_i^2$).

It is useful to express electron and hole densities in terms of the intrinsic carrier concentration n_i and the intrinsic Fermi level E_i, since E_i is frequently used as a reference level when discussing extrinsic semiconductors. From Eqs. 19 and 21 we obtain

$$
\begin{aligned}
n &= N_C \exp\left[-\frac{E_C - E_F}{kT}\right] \\
&= N_C \exp\left[-\frac{E_C - E_i}{kT}\right] \exp\left[\frac{E_F - E_i}{kT}\right] \\
&= n_i \exp\left[\frac{E_F - E_i}{kT}\right]
\end{aligned}
\tag{32}
$$

and

$$
\begin{aligned}
p &= N_V \exp\left[-\frac{E_F - E_V}{kT}\right] \\
&= n_i \exp\left[\frac{E_i - E_F}{kT}\right]
\end{aligned}
\tag{33}
$$

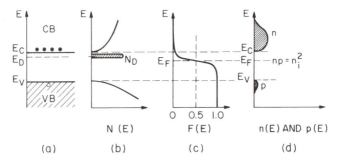

Fig. 20 *n*-type semiconductor. (*a*) Schematic band diagram. (*b*) Density of states. (*c*) Fermi distribution function. (*d*) Carrier concentration. Note that $np = n_i^2$.

Problem

A silicon ingot is doped with 10^{16} arsenic atoms/cm^3. Find the carrier concentrations and the Fermi level at room temperature (300 K).

Solution

At 300 K, we can assume complete ionization of impurity atoms. We have

$$n \simeq N_D = 10^{16} \text{ cm}^{-3} .$$

From Eq. 24,

$$p \simeq \frac{n_i^2}{N_D} = \frac{(1.45 \times 10^{10})^2}{10^{16}} = 2.1 \times 10^4 \text{ cm}^{-3} .$$

The Fermi level measured from the bottom of the conduction band is given by Eq. 29:

$$E_C - E_F = kT \ln \left[\frac{N_C}{N_D} \right]$$

$$= 0.0259 \ln \left[\frac{2.8 \times 10^{19}}{10^{16}} \right] = 0.206 \text{ eV} .$$

The Fermi level measured from the intrinsic Fermi level is given by Eq. 32:

$$E_F - E_i = kT \ln \left[\frac{n}{n_i} \right] \simeq kT \ln \left[\frac{N_D}{n_i} \right]$$

$$= 0.0259 \ln \left[\frac{10^{16}}{1.45 \times 10^{10}} \right] = 0.354 \text{ eV} .$$

These results are shown graphically in Fig. 21.

If both donor and acceptor impurities are present simultaneously, the impurity that is present in a greater concentration determines the type of conductivity in the semiconductor. The Fermi level must adjust itself to preserve

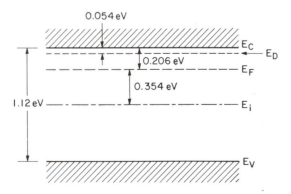

Fig. 21   Band diagram showing Fermi level E_F and intrinsic Fermi level E_i.

charge neutrality, that is, the total negative charges (electrons and ionized acceptors) must equal the total positive charges (holes and ionized donors):

$$n + N_A = p + N_D . \tag{34}$$

Solving Eqs. 24 and 34 yields the equilibrium electron and hole concentrations in an *n*-type semiconductor:

$$n_n = \frac{1}{2} \left[N_D - N_A + \sqrt{(N_D - N_A)^2 + 4n_i^2} \right] \tag{35}$$

$$p_n = \frac{n_i^2}{n_n} . \tag{36}$$

The subscript *n* refers to the *n*-type semiconductor. Because the electron is the dominant carrier, it is called the *majority carrier*. The hole in the *n*-type semiconductor is called the *minority carrier*. Similarly, we obtain the concentration of holes (majority carrier) and electrons (minority carrier) in a *p*-type semiconductor:

$$p_p = \frac{1}{2} \left[N_A - N_D + \sqrt{(N_A - N_D)^2 + 4n_i^2} \right] \tag{37}$$

and

$$n_p = \frac{n_i^2}{p_p} . \tag{38}$$

The subscript *p* refers to the *p*-type semiconductor.

Generally, the magnitude of the net impurity concentration $| N_D - N_A |$ is larger than the intrinsic carrier concentration n_i; therefore, the above relationships can be simplified to

$$n_n \simeq N_D - N_A \quad \text{if} \quad N_D > N_A \tag{39}$$

$$p_p \simeq N_A - N_D \quad \text{if} \quad N_A > N_D . \tag{40}$$

From Eqs. 35 to 38 together with Eqs. 19 and 21, we can calculate the position of the Fermi level as a function of temperature for a given acceptor or donor concentration. Figure 22 shows a plot of these calculations for silicon[6] and gallium arsenide. We have incorporated in the figure the variation of the bandgap with temperature. Note that as the temperature increases, the Fermi level approaches the intrinsic level, that is, the semiconductor becomes intrinsic.

Figure 23 shows electron density as a function of temperature for a donor concentration of $N_D = 10^{15} \, \text{cm}^{-3}$. At low temperatures, the thermal energy in the crystal is not sufficient to ionize all the donor impurities present. Some electrons are "frozen" at the donor level and the electron density is less than the donor concentration. As the temperature is increased, the condition of complete ionization is reached, (i.e., $n_n = N_D$). As the temperature is further increased, the electron concentration remains essentially the same over a wide

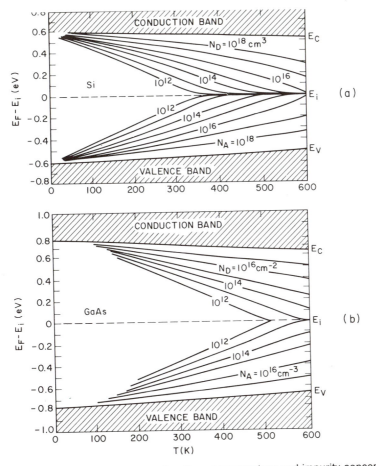

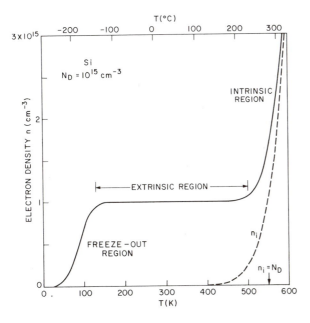

Fig. 22 Fermi level for Si and GaAs as a function of temperature and impurity concentration. The dependence of the bandgap on temperature is shown in the figure.[6]

Fig. 23 Electron density as a function of temperature for a Si sample with a donor concentration of 10^{15} cm^{-3}.

temperature range. This is the extrinsic region. However, as the temperature is increased even further, we reach a point where the intrinsic carrier concentration becomes comparable to the donor concentration. Beyond this point, the semiconductor becomes intrinsic. The temperature at which the semiconductor becomes intrinsic depends upon the impurity concentrations and can be obtained from Fig. 16 by setting the impurity concentration equal to n_i.

REFERENCES

1   R. A. Smith, *Semiconductors*, 2nd ed., Cambridge University Press, London, 1979.

2   C. Kittel, *Introduction to Solid State Physics*, Wiley, New York, 1976.

3   D. Halliday and R. Resnick, *Fundamentals of Physics*, 2nd ed., Wiley, New York, 1981.

4   C. D. Thurmond, "The Standard Thermodynamic Function of the Formation of Electrons and Holes in Ge, Si, GaAs, and GaP," *J. Electrochem. Soc.*, **122**, 1133 (1975).

5   S. M. Sze, *Physics of Semiconductors Devices*, 2nd ed., Wiley, New York, 1981.

6   A. S. Grove, *Physics and Technology of Semiconductor Devices*, Wiley, New York, 1967.

PROBLEMS

1   (a) What is the distance between nearest neighbors in silicon?
(b) Find the number of atoms per square centimeter in silicon in the (100), (110), and (111) planes.

2   Find the maximum fraction of the unit cell volume which can be filled by identical hard spheres in the simple cubic, face-centered cubic, and diamond lattices.

3   If a plane has intercepts at $2a$, $3a$, and $4a$ along the three Cartesian coordinates where a is the lattice constant, find the Miller indices of the plane.

4   (a) Calculate the density of GaAs (its lattice constant is given in Appendix F).
(b) A gallium arsenide sample is doped with tin. If the tin displaces gallium atoms in the lattice, are donors or acceptors formed? Why? Is the semiconductor n- or p-type?

5   Derive Eq. 21. *Hint*: In the valence band, the probability of occupancy of a state by a hole is $[1 - F(E)]$.

6   At room temperature (300 K) the effective density of states in the valence band is $1.04 \times 10^{19} \, cm^{-3}$ for silicon and $7 \times 10^{18} \, cm^{-3}$ for gallium arsenide, find the corresponding effective masses of holes. Compare these masses with the free-electron mass.

7   Calculate the location of E_i in silicon at liquid nitrogen temperature (77 K), at room temperature (300 K), and at 100°C (let $m_p = 0.5m_0$ and

$m_n = 0.3m_0$). Is it reasonable to assume that E_i is in the center of the forbidden gap?

8 Draw a simple energy band diagram for silicon doped with 10^{16} arsenic atoms/cm^3 at 77 K, 300 K, and 600 K. Show the Fermi level and use the intrinsic Fermi level as the energy reference.

9 Find the electron and hole concentrations and Fermi level in silicon at 300 K (a) for 1×10^{15} boron atoms/cm^3 and (b) for 3×10^{16} boron atoms/cm^3 and 2.9×10^{16} arsenic atoms/cm^3.

10 Calculate the Fermi level of silicon doped with 10^{15}, 10^{17}, and 10^{19} phosphorus atoms/cm^3 at room temperature, assuming complete ionization. From the calculated Fermi level, check if the assumption of complete ionization is justified for each doping.

2

Carrier Transport Phenomena

In this chapter we consider various transport phenomena that arise from the motion of charge carriers (electrons and holes) in semiconductors under the influence of an electric field and a carrier concentration gradient. We shall discuss the concept of injection of excess carriers, which gives rise to a none-quilibrium condition, that is, the carrier concentration product pn is different from its equilibrium value n_i^2. Return to an equilibrium condition through the generation–recombination processes will be considered next. We shall then derive the basic governing equations for semiconductor device operation, which include the current density equation and the continuity equation. The chapter closes with a brief discussion of high-field effects, which result in velocity saturation and impact ionization.

2.1 CARRIER DRIFT

2.1.1 Mobility

Consider an n-type semiconductor sample with uniform donor concentration in thermal equilibrium. As discussed in Chapter 1, the conduction electrons in the semiconductor conduction band are essentially free particles, since they are not associated with any particular lattice or donor site. The influence of crystal lattices is incorporated in the effective mass of conduction electrons, which differs somewhat from the mass of free electrons. Under thermal equilibrium, the average thermal energy of a conduction electron can be obtained from the theorem for equipartition of energy: $\frac{1}{2} kT$ units of energy per degree of freedom, where k is Boltzmann's constant and T is the absolute temperature.[1] The electrons in a semiconductor have three degrees of freedom; they can move about in a three-dimensional space. Therefore, the kinetic energy of the electrons is given by

$$\frac{1}{2} m_n v_{th}^2 = \frac{3}{2} kT \tag{1}$$

where m_n is the effective mass of electrons and v_{th} is the average thermal velocity.[1] At room temperature (300 K) the thermal velocity in Eq. 1 is about 10^7 cm/s for silicon and gallium arsenide.

The electrons in the semiconductor are therefore moving rapidly in all directions. The thermal motion of an individual electron may be visualized as a succession of random scattering from collisions with lattice atoms, impurity atoms, and other scattering centers, as illustrated in Fig. 1a. The random motion of electrons leads to a zero net displacement of an electron over a sufficiently long period of time. The average distance between collisions is called the *mean free path*, and the average time between collisions is called the *mean free time* τ_c. For a typical value of 10^{-5} cm for the mean free path, τ_c is about 1 ps (i.e., $10^{-5}/v_{th} \simeq 10^{-12}$ s).

When a small electric field $\mathscr{E}$ is applied to the semiconductor sample, each electron will experience a force $-q\mathscr{E}$ from the field and will be accelerated along the field (in the opposite direction to the field) during the time between collisions. Therefore, an additional velocity component will be superimposed upon the thermal motion of electrons. This additional component is called the *drift velocity*. The combined displacement of an electron due to the random thermal motion and the drift component is illustrated in Fig. 1b. Note that there is a net displacement of the electron in the direction opposite to the applied field.

We can obtain the drift velocity v_n by equating the momentum (force × time) applied to an electron during the free flight between collisions to the momentum gained by the electron in the same period. The equality is valid because in a steady state all momentum gained between collisions is lost to the lattice in the collision. The momentum applied to an electron is given by $-q\mathscr{E}\tau_c$, and the momentum gained is $m_n v_n$. We have

$$-q\mathscr{E}\tau_c = m_n v_n \tag{2}$$

or

$$v_n = -\left[\frac{q\tau_c}{m_n}\right]\mathscr{E} \tag{2a}$$

Equation 2a states that the electron drift velocity is proportional to the applied electric field. The proportionality factor depends on the mean free time and

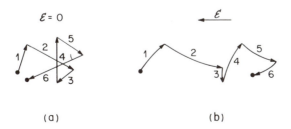

Fig. 1 Schematic path of an electron in a semiconductor. (a) Random thermal motion. (b) Combined motion due to random thermal motion and an applied electric field.

the effective mass. The proportionality factor is called the *electron mobility* μ_n in units of cm^2/V-s, or

$$\mu_n \equiv \frac{q\tau_c}{m_n}.$$
(3)

Thus,

$$v_n = -\mu_n \mathscr{E}$$
(4)

Mobility is an important parameter for carrier transport because it describes how strongly the motion of an electron is influenced by an applied electric field. A similar expression can be written for holes in the valence band:

$$v_p = \mu_p \mathscr{E}$$
(5)

where v_p is the hole drift velocity and μ_p is the hole mobility. The negative sign is removed in Eq. 5 because holes drift in the same direction as the electric field.

In Eq. 3 the mobility is related directly to the mean free time between collisions, which in turn is determined by the various scattering mechanisms. The two most important mechanisms are lattice scattering and impurity scattering. Lattice scattering results from thermal vibrations of the lattice atoms at any temperature above absolute zero. These vibrations disturb lattice periodic potential and allow energy to be transferred between the carriers and the lattice. Since lattice vibration increases with increasing temperature, lattice scattering becomes dominant at high temperatures; hence the mobility decreases with increasing temperature. Theoretical analysis[2] shows that the mobility due to lattice scattering μ_L will decrease in proportion to $T^{-3/2}$.

Impurity scattering results when a charge carrier travels past an ionized dopant impurity (donor or acceptor). The charge carrier path will be deflected due to Coulomb force interaction. The probability of impurity scattering depends on the total concentration of ionized impurities, that is, the sum of the concentration of negatively and positively charged ions. However, unlike lattice scattering, impurity scattering becomes less significant at higher temperatures. At higher temperatures, the carriers move faster; they remain near the impurity atom for a shorter time and are therefore less effectively scattered. The mobility due to impurity scattering μ_I can theoretically be shown to vary as $T^{3/2}/N_T$, where N_T is the total impurity concentration.[3]

The probability of a collision taking place in unit time, $1/\tau_c$, is the sum of the probabilities of collisions due to the various scattering mechanisms:

$$\frac{1}{\tau_c} = \frac{1}{\tau_{c,\ lattice}} + \frac{1}{\tau_{c,\ impurity}}$$
(6)

or

$$\frac{1}{\mu} = \frac{1}{\mu_L} + \frac{1}{\mu_I}.$$
(6a)

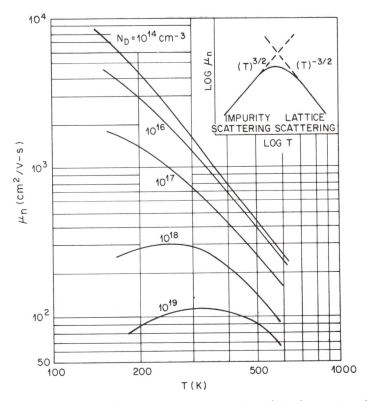

Fig. 2 Electron mobility in silicon versus temperature for various donor concentrations. Insert shows the theoretical temperature dependence of electron mobility.[4]

Figure 2 shows the measured electron mobility as a function of temperature for silicon with five different donor concentrations.[4] The insert shows the theoretical temperature dependence of mobility due to both lattice and impurity scatterings. For lightly doped samples (e.g., the sample with doping of 10^{14} cm^{-3}), the lattice scattering dominates, and the mobility decreases as the temperature increases. For heavily doped samples, the effect of impurity scattering is most pronounced at low temperatures. The mobility increases as the temperature increases, as can be seen for the sample with doping of 10^{19} cm^{-3}. For a given temperature, the mobility decreases with increasing impurity concentration, because of enhanced impurity scatterings.

Figure 3 shows the measured mobilities in silicon and gallium arsenide as a function of impurity concentration at room temperature.[4, 5] Mobility reaches a maximum value at low impurity concentrations; this corresponds to the lattice-scattering limitation. Both electron and hole mobilities decrease with increasing impurity concentration and eventually approach a minimum value at high concentrations. Note also that the mobility of electrons is greater than that of holes. Greater electron mobility is due mainly to the smaller effective mass of electrons.

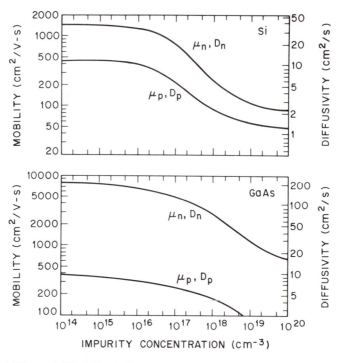

Fig. 3 Mobilities and diffusivities in Si and GaAs at 300 K as a function of impurity concentration.[4, 5]

2.1.2 Resistivity

We shall now consider conduction in a homogeneous semiconductor material. Figure 4a shows an n-type semiconductor and its band diagram at thermal equilibrium. Figure 4b shows the corresponding band diagram when a biasing voltage is applied to the right-hand terminal. (We assume that the contacts at the left-hand and right-hand terminals are ohmic, that is, there is negligible voltage drop at each of the contacts. The behavior of ohmic contacts will be considered in Chapter 5.) As mentioned previously, when an electric field $\mathscr{E}$ is applied to a semiconductor, each electron will experience a force $-q\mathscr{E}$ from the field. The force is equal to the negative gradient of potential energy; that is,

$$-q\mathscr{E} = - \text{ (gradient of electron potential energy) .} \qquad (7)$$

In Section 1.4 we discussed that the bottom of the conduction band E_C corresponds to the potential energy of an electron. Since we are interested in the gradient of the potential energy, we can use any part of the band diagram that is parallel to E_C (e.g., E_F, E_i, or E_V as shown in Fig. 4b). It is convenient to use E_i because we shall use E_i when we consider p–n junctions in

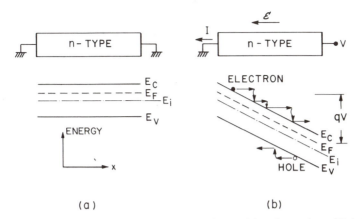

Fig. 4 Conduction process in an *n*-type semiconductor (*a*) at thermal equilibrium, and (*b*) under a biasing condition.

Chapter 3. Therefore, from Eq. 7 we have

$$\mathscr{E} = \frac{1}{q} \frac{dE_i}{dx} . \qquad (8)$$

We can define a related quantity ψ as the *electrostatic potential* whose negative gradient equals the electric field:

$$\mathscr{E} \equiv - \frac{d\psi}{dx} . \qquad (9)$$

Comparison of Eqs. 8 and 9 gives

$$\psi = - \frac{E_i}{q} \qquad (10)$$

which provides a relationship between the electrostatic potential and the potential energy of an electron. For a homogeneous semiconductor shown in Fig. 4*b*, the potential energy and E_i decrease linearly with distance; thus, the electric field is a constant in the negative *x*-direction. Its magnitude equals the applied voltage divided by the sample length.

The electrons in the conduction band move to the right side as shown in Fig. 4*b*. When an electron undergoes a collision, it loses some or all of its kinetic energy to the lattice and drops toward its thermal-equilibrium position. After the electron has lost some or all its kinetic energy, it will again begin to move toward the right, and the same process will be repeated many times. Conduction by holes can be visualized in a similar manner but in the opposite direction.

The transport of carriers under the influence of an applied electric field produces a current called the *drift current*. Consider a semiconductor sample shown in Fig. 5, which has a cross-sectional area A, a length L, and a carrier concentration of n electrons/cm³. Suppose we now apply an electric field $\mathscr{E}$ to

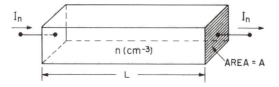

Fig. 5 Current conduction in a uniformly doped semiconductor bar with length L and cross-sectional area A.

the sample. The electron current density J_n flowing in the sample can be found by summing the product of the charge $(-q)$ on each electron times the electron's velocity over all electrons per unit volume n:

$$J_n = \frac{I_n}{A} = \sum_{i=0}^{n} (-qv_i) = -qnv_n = qn\mu_n\mathscr{E} \qquad (11)$$

where I_n is the electron current. We have employed Eq. 4 for the relationship between v_n and $\mathscr{E}$.

A similar argument applies to holes. By taking the charge on the hole to be positive, we have

$$J_p = qpv_p = qp\mu_p\mathscr{E}. \qquad (12)$$

The total current flowing in the semiconductor sample due to the applied field $\mathscr{E}$ can be written as the sum of the electron and hole current components:

$$J = J_n + J_p = (qn\mu_n + qp\mu_p)\mathscr{E}. \qquad (13)$$

The quantity in parentheses is known as *conductivity*:

$$\sigma = (qn\mu_n + qp\mu_p). \qquad (14)$$

The electron and hole contributions to conductivity are simply additive.

The corresponding resistivity of the semiconductor, which is the reciprocal of σ, is given by

$$\rho \equiv \frac{1}{\sigma} = \frac{1}{q(n\mu_n + p\mu_p)}. \qquad (15)$$

Generally, in extrinsic semiconductors, only one of the components in Eq. 13 or 14 is significant because of the many orders-of-magnitude difference between the two carrier densities. Therefore, Eq. 15 reduces to

$$\rho = \frac{1}{qn\mu_n} \qquad (15a)$$

for an *n*-type semiconductor and to

$$\rho = \frac{1}{qp\mu_p} \qquad (15b)$$

for a *p*-type semiconductor.

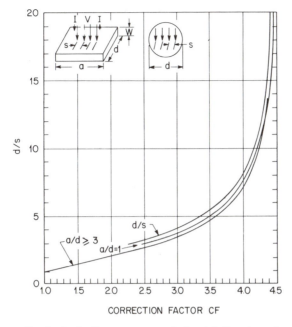

CORRECTION FACTOR CF

Fig. 6 Correction factor for the measurement of resistivity using a four-point probe.[4]

The most common method for measuring resistivity is the four-point probe method shown in the insert of Fig. 6. The probes are equally spaced. A small current from a constant-current source is passed through the outer two probes and the voltage is measured between the inner two probes. For a thin semiconductor sample with thickness W much smaller than either a or d, the resistivity is given by

$$\rho = \frac{V}{I} \cdot W \cdot CF \quad \Omega\text{-cm} \tag{16}$$

where CF is the correction factor shown in Fig. 6. At the limit when $d \gg s$, where s is the probe spacing, the correction factor becomes $\pi/\ln 2 = 4.54$.

Figure 7 shows the measured resistivity for silicon and gallium arsenide (at 300K) as a function of the impurity concentration. At this temperature and for low impurity concentrations, all donor or acceptor impurities that have shallow energy levels will be ionized. Under these conditions the carrier concentration is equal to the impurity concentration. From these curves we can obtain the impurity concentration of a semiconductor if the resistivity is known, or vice versa.

Problem

Find the room temperature resistivity of an n-type silicon doped with 10^{16} phosphorus atoms/cm^3.

Solution

At room temperature we assume that all donors are ionized; thus,

$$n \simeq N_D = 10^{16} \quad cm^{-3}.$$

From Fig. 7 we find $\rho = 0.48$ Ω-cm. We can also calculate the resistivity from Eq. 15a

$$\rho = \frac{1}{qn\mu_n} = \frac{1}{1.6 \times 10^{-19} \times 10^{16} \times 1300} = 0.48 \ \Omega\text{- cm}.$$

The mobility μ_n is obtained from Fig. 3.

2.1.3 The Hall Effect

The carrier concentration in a semiconductor may be different from the impurity concentration, because the ionized impurity density depends on the temperature and the impurity energy level. To measure the carrier concentration directly, the most commonly used method is the Hall effect. Hall measurement is also one of the most convincing methods to show the existence of holes as charge carriers, because the measurement can give directly the carrier type. Figure 8 shows an electric field applied along the *x*-axis and a magnetic field applied along the *z*-axis. Consider a *p*-type semiconductor sample. The Lorentz force $q\mathbf{v} \times \mathbf{B}$ ($= qv_x B_z$) due to the magnetic field will exert an average upward force on the holes flowing in the *x*-direction. The upward-directed current causes an accumulation of holes at the top of the sample that gives rise to a downward-directed electric field $\mathscr{E}_y$. Since there is no net current flow along the *y*-direction in the steady state, the electric field along

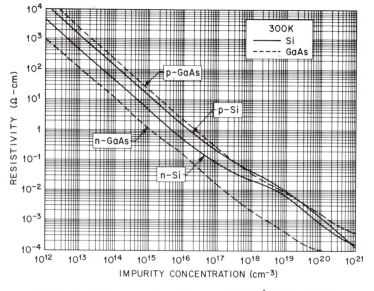

Fig. 7 Resistivity versus impurity concentration[4] for Si and GaAs.

the y-axis exactly balances the Lorentz force; that is,

$$q\mathscr{E}_y = qv_x B_z \tag{17}$$

or

$$\mathscr{E}_y = v_x B_z \; . \tag{18}$$

Once the electric field $\mathscr{E}_y$ becomes equal to $v_x B_z$, no net force along the y-direction is experienced by the holes as they drift in the x-direction.

The establishment of the electric field is known as the *Hall effect*. The electric field in Eq. 18 is called the *Hall field*, and the terminal voltage $V_H = \mathscr{E}_y W$ (Fig. 8) is called the *Hall voltage*. Using Eq. 12 for the hole drift velocity, the Hall field $\mathscr{E}_y$ in Eq. 18 becomes

$$\mathscr{E}_y = \left[\frac{J_p}{qp}\right]B_z = R_H J_p B_z \tag{19}$$

where

$$R_H \equiv \frac{1}{qp} \; . \tag{20}$$

The Hall field $\mathscr{E}_y$ is proportional to the product of the current density and the magnetic field. The proportionality constant R_H is the *Hall coefficient*. A similar result can be obtained for an n-type semiconductor, except that the Hall coefficient is negative:

$$R_H = -\frac{1}{qn} \; . \tag{21}$$

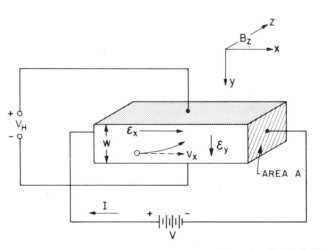

Fig. 8 Basic setup to measure carrier concentration using the Hall effect.

A measurement of the Hall voltage for a known current and magnetic field yields

$$p = \frac{1}{qR_H} = \frac{J_p B_z}{q\mathscr{E}_y} = \frac{(I/A)B_z}{q(V_H/W)} = \frac{IB_z W}{qV_H A} \tag{22}$$

where all the quantities in the right-hand side of the equation can be measured. Thus, the carrier concentration and carrier type can be obtained directly from the Hall measurement.

Problem

A sample of Si is doped with 10^{16} phosphorous atoms/cm^3. Find the Hall voltage in a sample with $W = 500$ μm, $A = 2.5 \times 10^{-3}$ cm^2, $I = 1$ mA, and $B_z = 10^{-4}$ Wb/cm^2.

Solution

The Hall coefficient is

$$R_H = -\frac{1}{qn} = -\frac{1}{1.6 \times 10^{-19} \times 10^{16}} = -625 \text{ cm}^3/\text{C}.$$

The Hall voltage is

$$V_H = \mathscr{E}_y W = \left[R_H \frac{I}{A} B_z \right] W$$

$$= \left[-625 \cdot \frac{10^{-3}}{2.5 \times 10^{-3}} \cdot 10^{-4} \right] 500 \times 10^{-4}$$

$$= -1.25 \text{ mV}.$$

2.2 CARRIER DIFFUSION

2.2.1 Diffusion Process

In the preceding section we considered the drift current, that is, the transport of carriers when an electric field is applied. Another important current component can exist if there is a spatial variation of carrier concentration in the semiconductor material, that is, the carriers tend to move from a region of high concentration to a region of low concentration. This current component is called diffusion current.

To understand the diffusion process, let us assume an electron density that varies in the x-direction, as shown in Fig. 9. The semiconductor is at uniform temperature, so that the average thermal energy of electrons does not vary with x, only the density $n(x)$ varies. We shall consider the number of electrons crossing the plane at $x = 0$ per unit time and per unit area. Because of finite temperature, the electrons have random thermal motions with a thermal velocity v_{th} and a mean free path l. (Note that $l = v_{th}\tau_c$, where τ_c is the mean free time.) The electrons at $x = -l$, one mean free path away on the left side, have equal chances of moving left or right; and in a mean free time

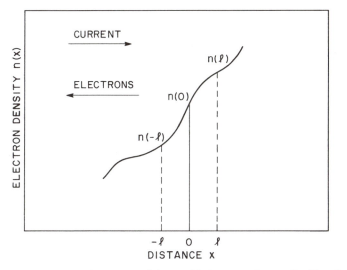

Fig. 9 Electron concentration versus distance; *l* is the mean free path. The directions of electron and current flows are indicated by arrows.

τ_c, one half of them will move across the plane $x = 0$. The average rate of electron flow per unit area F_1 of electrons crossing plane $x = 0$ from the left is then

$$F_1 = \frac{1/2 \, n(-l) \cdot l}{\tau_c} = \frac{1}{2} \, n(-l) \cdot v_{th} . \tag{23}$$

Similarly, the average rate of electron flow per unit area F_2 of electrons at $x = l$ crossing plane $x = 0$ from the right is

$$F_2 = \frac{1}{2} \, n(l) \cdot v_{th} . \tag{24}$$

The net rate of carrier flow from left to right is

$$F = F_1 - F_2 = \frac{1}{2} \, v_{th} \, [n(-l) - n(l)] . \tag{25}$$

Approximating the densities at $x = \pm l$ by the first two terms of a Taylor series expansion, we obtain

$$F = \frac{1}{2} \, v_{th} \left\{ \left[n(0) - l \, \frac{dn}{dx} \right] - \left[n(0) + l \, \frac{dn}{dx} \right] \right\}$$

$$= - v_{th} l \, \frac{dn}{dx} \equiv - D_n \, \frac{dn}{dx} \tag{26}$$

where $D_n \equiv v_{th} l$ is called the *diffusivity*. Because each electron carries a charge $- q$, the carrier flow gives rise to a current

$$J_n = -qF = qD_n \frac{dn}{dx} . \tag{27}$$

The diffusion current is proportional to the spatial derivative of the electron density. Diffusion current results from the random thermal motion of carriers in a concentration gradient. For an electron density that increases with x, the gradient is positive, and the electrons will diffuse toward the negative x-direction. The current is positive and flows in the direction opposite to that of the electrons as indicated in Fig. 9.

2.2.2 Einstein Relation

Equation 27 can be written in a more useful form using the theorem for the equipartition of energy for this one-dimensional case. We can write

$$\frac{1}{2} m_n v_{th}^2 = \frac{1}{2} kT . \tag{28}$$

Substituting Eqs. 3 and 28 into Eq. 27 and using the relationship $l = v_{th}\tau_c$ yields

$$J_n = qD_n \frac{dn}{dx} = q \left[\frac{kT}{q} \mu_n \right] \frac{dn}{dx} . \tag{29}$$

Therefore,

$$D_n = \left[\frac{kT}{q} \right] \mu_n . \tag{30}$$

Equation 30 is known as the *Einstein relation.* It relates the two important constants (diffusivity and mobility) that characterize carrier transport by diffusion and by drift in a semiconductor. The Einstein relation also applies between D_p and μ_p. Values of diffusivities for silicon and gallium arsenide are shown in Fig. 3.

2.2.3 Current Density Equations

When an electric field is present in addition to a concentration gradient, both drift current and diffusion current will flow. The total current density at any point is the sum of the drift and diffusion components:

$$J_n = q \mu_n n \mathscr{E} + qD_n \frac{dn}{dx} \tag{31}$$

where $\mathscr{E}$ is the electric field in the x-direction.

A similar expression can be obtained for the hole current:

$$J_p = q \mu_p p \mathscr{E} - qD_p \frac{dp}{dx} . \tag{32}$$

We use the negative sign in Eq. 32 because, for a positive hole gradient, the holes will diffuse in the negative x-direction. This diffusion results in a hole current that also flows in the negative x-direction.

The total conduction current density is given by the sum of Eqs. 31 and 32:

$$J_{\text{cond}} = J_n + J_p \,.$$ (33)

The three expressions (Eqs. 31–33) constitute the current density equations. These equations are very important for analyzing device operations under low electric fields. However, at sufficiently high electric fields the terms $\mu_n \mathscr{E}$ and $\mu_p \mathscr{E}$ should be replaced by the saturation velocity v_s discussed in Section 2.6.

2.3 CARRIER INJECTION

In thermal equilibrium the relationship $pn = n_i^2$ is valid. If excess carriers are introduced to a semiconductor so that $pn > n_i^2$, we have a non-equilibrium situation. The process of introducing excess carriers is called *carrier injection*. We can inject carriers by using various methods including optical excitation and forward-biasing a p–n junction (discussed in Chapter 3). In the case of optical excitation, we shine a light on a semiconductor. If the photon energy $h\nu$ of the light is greater than the bandgap energy E_g of the semiconductor, where h is the Planck constant and ν is the optical frequency, the photon is absorbed by the semiconductor and an electron–hole pair is generated. The optical excitation increases the electron and hole concentrations above their equilibrium values. These additional carriers are called *excess carriers*.

The magnitude of the excess carrier concentration relative to the majority carrier concentration determines the injection level. We shall use an example to clarify the meaning of injection level. Figure 10a shows the majority and minority carrier concentrations for an n-type silicon with $N_D = 10^{15}\,\text{cm}^{-3}$ under thermal equilibrium. The majority carrier concentration is approximately equal to the donor concentration, that is, $n_{no} = 10^{15}\,\text{cm}^{-3}$. The minority carrier concentration is given by $p_{no} = n_i^2/n_{no} = 2.1 \times 10^5\,\text{cm}^{-3}$. In this notation, the first subscript refers to the type of semiconductor and the subscript o refers to the thermal equilibrium condition. Thus, n_{no} and p_{no} denote the electron and hole concentrations, respectively, in an n-type semiconductor in equilibrium.

When we introduce (e.g., by optical excitation) excess carriers of both types into the semiconductor, the excess electron concentration Δn must equal the excess hole concentration Δp, because electrons and holes are produced in pairs. In the example shown in Fig. 10b, we have added $10^{12}/\text{cm}^2$ minority carriers (holes in an n-type semiconductor). Therefore, the hole concentration is increased by seven orders of magnitude (from $10^5/\text{cm}^3$ to $10^{12}/\text{cm}^3$). At the same time we have added $10^{12}\,\text{cm}^{-3}$ majority carriers (electrons) to the semiconductor. However, this concentration of excess electrons is negligibly small compared to the original electron concentration. The percentage change in the majority carrier concentration is only 0.1% (i.e., $10^{12}/10^{15}$). This condition,

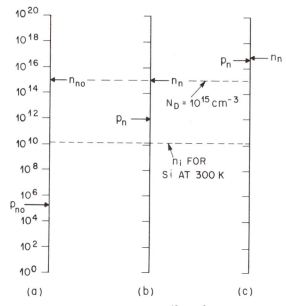

Fig. 10   Carrier concentrations in *n*-type Si with 10^{15} cm^{-3} donor concentration (*a*) at thermal equilibrium, (*b*) under low-injection condition, and (*c*) under high-injection condition.

in which the excess carrier concentration is small in comparison to the doping concentration, that is, $\Delta n = \Delta p \ll N_D$, is referred to as low-level injection.

Figure 10*c* shows an example of high-level injection in which the number of injected excess carriers is comparable to or larger than the number of carriers due to doping concentration. In this case, the injected-carrier concentrations may overwhelm the equilibrium majority carrier concentration, and p_n becomes comparable to n_n as indicated in the figure. High-level injection is sometime encountered in device operations. However, because of the complexities involved in its treatment, we shall be concerned mainly with low-level injection.

2.4   GENERATION AND RECOMBINATION PROCESSES

Whenever the thermal-equilibrium condition is disturbed (i.e., $pn \neq n_i^2$), processes exist to restore the system to equilibrium (i.e., $pn = n_i^2$). In the case of the injection of excess carriers, the mechanism that restores equilibrium is recombination of the injected minority carriers with the majority carriers. Depending on the nature of the recombination process, the released energy that results from the recombination process can be emitted as a photon or dissipated as heat to the lattice. When a photon is emitted, the process is called radiative recombination; otherwise, it is called nonradiative recombination.

Recombination phenomena can be classified as direct and indirect processes. Direct recombination, also called band-to-band recombination, usually dominates in direct-bandgap semiconductors, such as gallium arsenide; while indirect recombination via bandgap recombination centers dominates in indirect bandgap semiconductors, such as silicon.

2.4.1   Direct Recombination

Consider a direct-bandgap semiconductor in thermal equilibrium. The continuous thermal vibration of lattice atoms causes some bonds between neighboring atoms to be broken. When a bond is broken, an electron–hole pair is generated. In terms of the band diagram, the thermal energy enables a valence electron to make an upward transition to the conduction band leaving a hole in the valence band. This process is called carrier generation and is represented by the generation rate G_{th} (number of electron–hole pairs generated per cm^3 per second) in Fig. 11a. When an electron makes a transition downward from the conduction band to the valence band, an electron–hole pair is annihilated. This reverse process is called recombination; it is represented by the recombination rate R_{th} in Fig. 11a. Under thermal-equilibrium conditions, the generation rate G_{th} must equal the recombination rate R_{th}, so that the carrier concentrations remain constant and the condition $pn = n_i^2$ is maintained.

When excess carriers are introduced to a direct-bandgap semiconductor, the probability is high that electrons and holes will recombine directly, because the bottom of the conduction band and the top of the valence band are lined up and no additional crystal momentum is required for the transition across the bandgap. The rate of the direct recombination R is expected to be proportional to the number of electrons available in the conduction band and the number of holes available in the valence band; that is,

$$R = \beta n p \qquad (34)$$

where β is the proportionality constant. As discussed previously, in thermal equilibrium the recombination rate must be balanced by the generation rate. Therefore, for an n-type semiconductor, we have

$$G_{th} = R_{th} = \beta n_{no} p_{no} \qquad (35)$$

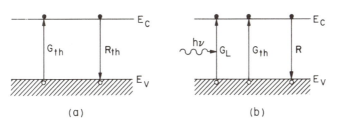

Fig. 11 Direct generation and recombination of electron-hole pairs. (*a*) at thermal equilibrium, and (*b*) under illumination.

where n_{no} and p_{no} represent electron and hole densities in an n-type semiconductor at thermal equilibrium. When we shine a light on the semiconductor to produce electron–hole pairs at a rate G_L (Fig. 11b), the carrier concentrations are above their equilibrium values. The recombination and generation rate become

$$R = \beta n_n p_n = \beta(n_{no} + \Delta n)(p_{no} + \Delta p) \tag{36}$$

$$G = G_L + G_{th} \tag{37}$$

where Δn and Δp are the excess carrier concentrations, given by

$$\Delta n = n_n - n_{no} \tag{38a}$$

$$\Delta p = p_n - p_{no} \tag{38b}$$

and $\Delta n = \Delta p$ to maintain overall charge neutrality.

The net rate of change of hole concentration is given by

$$\frac{dp_n}{dt} = G - R = G_L + G_{th} - R . \tag{39}$$

In steady state, $dp_n/dt = 0$. From Eq. 39 we have

$$G_L = R - G_{th} \equiv U \tag{40}$$

where U is the net recombination rate. Substituting Eqs. 35 and 36 into Eq. 40 yields

$$U \simeq \beta(n_{no} + p_{no} + \Delta p)\Delta p . \tag{41}$$

For low-level injection Δp, $p_{no} \ll n_{no}$, Eq. 41 is simplified to

$$U \simeq \beta n_{no} \, \Delta p = \frac{p_n - p_{no}}{\dfrac{1}{\beta n_{no}}} . \tag{42}$$

Therefore, the net recombination rate is proportional to the excess minority carrier concentration. Obviously, $U = 0$ in thermal equilibrium. The proportionality constant $1/\beta n_{no}$ is called the *lifetime* τ_p of the excess minority carriers, or

$$U = \frac{p_n - p_{no}}{\tau_p} \tag{43}$$

where

$$\tau_p \equiv \frac{1}{\beta n_{no}} . \tag{44}$$

The physical meaning of lifetime can best be illustrated by the transient response of a device after the sudden removal of the light source. Consider an n-type sample, as shown in Fig. 12a, that is illuminated with light and in which the electron–hole pairs are generated uniformly throughout the sample with a

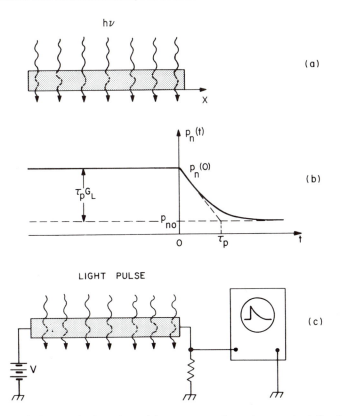

Fig. 12   Decay of photoexcited carriers. (a) n-type sample under constant illumination. (b) Decay of minority carriers (holes) with time. (c) Schematic setup to measure minority carrier lifetime.

generation rate G_L. The time-dependent expression is given by Eq. 39. In steady state, from Eqs. 40 and 43

$$G_L = U = \frac{p_n - p_{no}}{\tau_p} \tag{45}$$

or

$$p_n = p_{no} + \tau_p G_L . \tag{45a}$$

If at an arbitrary time, say, $t = 0$, the light is suddenly turned off, the boundary conditions are $p_n(t = 0) = p_{no} + \tau_p G_L$ as given by Eq. 45a, and $p_n(t \rightarrow \infty) = p_{no}$. The time-dependent expression of Eq. 39 becomes

$$\frac{dp_n}{dt} = G_{th} - R = -U = -\frac{p_n - p_{no}}{\tau_p} \tag{46}$$

and the solution is

$$p_n(t) = p_{no} + \tau_p G_L e^{-t/\tau_p} . \tag{47}$$

Figure 12b shows the variation of p_n with time. The minority carriers recombine with majority carriers and decay exponentially with a time constant τ_p which corresponds to the lifetime defined in Eq. 44.

The above case illustrates the main idea of measuring the carrier lifetime using photoconductivity method. Figure 12c shows a schematic setup. The excess carriers, generated uniformly throughout the sample by the light pulse, cause a momentary increase in the conductivity. The increase in conductivity manifests itself by a drop in voltage across the sample when a constant current is passed through it. The decay of the conductivity can be observed on an oscilloscope and is a measure of the lifetime of the excess minority carriers.

2.4.2 Indirect Recombination*

For indirect-bandgap semiconductors, such as silicon, a direct recombination process is very unlikely, because the electrons at the bottom of the conduction band have nonzero crystal momentum with respect to the holes at the top of the valence band (refer to Fig. 12 in Chapter 1). A direct transition that conserves both energy and momentum is not possible without a simultaneous lattice interaction. Therefore the dominant recombination process in such semiconductors is indirect transition via localized energy states in the forbidden energy gap.[6] These states act as stepping stones between the conduction band and the valence band. Because the transition probability depends on the energy differences between the step and the conduction and valence band edges, these intermediate states can substantially enhance the recombination process.

Figure 13 shows various transitions that occur in the recombination process through intermediate-level states (also called recombination centers).[†] We illustrate the charging state of the center before and after each of the four basic transitions takes place. The arrows in the figure designate the transition of the electron in a particular process. The illustration is for the case of a recombination center with a single energy level that is neutral when not occupied by an electron or negative when it is occupied.

Process (a) is for electron capture—an electron in the conduction band is captured by the center. Because only one electron can occupy a given center, a center that is occupied by an electron cannot capture another one. Therefore, the rate of electron capture is proportional to the concentration of centers that are not occupied by electrons (i.e., those centers that remain neutral). If the concentration of centers in the semiconductor is N_t, the concentration of unoccupied centers is given by $N_t(1 - F)$, where F is the Fermi distribution

---

[*] This section and other sections marked with an asterisk contain some graduate–level mathematics or physical concepts and may be omitted for undergraduate students. However, the concepts developed in these sections are used in later sections.

[†] Because the states behave symmetrically as sites either for generation or recombination of free carriers, they are termed generation–recombination centers. For brevity, however, they are usually called recombination centers.

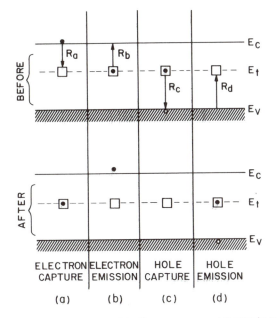

Fig. 13 Indirect generation-recombination processes at thermal equilibrium.[6]

function for the probability that a center is occupied by an electron. In equilibrium,

$$F = \frac{1}{1 + e^{(E_t - E_F)/kT}} \tag{48}$$

where E_t is the energy level of the center and E_F is the Fermi level.

Therefore, the rate of capture of electrons, process (a), is given by

$$R_a \sim nN_t(1 - F) . \tag{49}$$

We shall designate the proportionality constant by the product $v_{th}\sigma_n$, so that

$$R_a = v_{th}\sigma_n nN_t(1 - F) \tag{50}$$

where v_{th} is the thermal velocity of the carriers given in Eq. 1 and σ_n is the capture cross section. The quantity σ_n describes the effectiveness of the center to capture an electron and is a measure of how close the electron has to come to the center to be captured. We expect that the capture cross section would be of the order of the atomic dimensions, that is, of the order of 10^{-15} cm^2. The product $v_{th}\sigma_n$ may be visualized as the volume swept out per unit time by an electron with cross section σ_n. If the center lies within this volume, the electron will be captured by it.

The rate of emission of electrons from the center, process (b) in Fig. 13, is the inverse of the electron capture process. The rate is proportional to the concentration of centers occupied by electrons, that is, $N_t F$. We have

$$R_b = e_n N_t F .\qquad(51)$$

The proportionality constant e_n is called the emission probability. At thermal equilibrium the rates of capture and emission of electrons must be equal ($R_a = R_b$). Thus, the emission probability can be expressed in terms of the quantities already defined in Eq. 50:

$$e_n = \frac{v_{th}\sigma_n n (1 - F)}{F} .\qquad(52)$$

Since the electron concentration in thermal equilibrium is given by

$$n = n_i e^{(E_F - E_i)/kT}\qquad(53)$$

and

$$\frac{(1 - F)}{F} = e^{(E_t - E_F)/kT}\qquad(54)$$

we obtain

$$e_n = v_{th}\sigma_n n_i e^{(E_t - E_i)/kT} .\qquad(55)$$

Note that if the center is close to the conduction band edge (i.e., $E_t - E_i$ is larger), the electron emission from the center becomes more probable, since e_n increases exponentially with $E_t - E_i$.

The transitions between the recombination center and the valence band are analogous to those described above. Process (c) in Fig. 13 is for hole capture. The rate is proportional to the centers occupied by electrons, $N_t F$, and the hole concentration. The proportionality constant is given by the product of the thermal velocity v_{th} and the capture cross section σ_p of holes. Thus,

$$R_c = v_{th}\sigma_p p N_t F .\qquad(56)$$

The fourth process, Fig. 13d, is for hole emission, which describes the excitation of an electron from the valence band to the unoccupied center. By arguments similar to those for electron emission, the rate is

$$R_d = e_p N_t (1 - F) .\qquad(57)$$

The emission probability e_p of a hole may be expressed in terms of v_{th} and σ_p by considering the thermal-equilibrium condition for which $R_c = R_d$:

$$e_p = v_{th}\sigma_p n_i e^{(E_i - E_t)/kT} .\qquad(58)$$

Again, note that the emission probability increases exponentially as the center level E_t approaches the valence band edge. Because we can measure experimentally the capture cross sections σ_n and σ_p, the concentration of the recombination center N_t, and the center's energy level E_t, we can use Eqs. 50, 51, 56, and 57 to determine the kinetics of indirect recombination under nonequilibrium conditions.

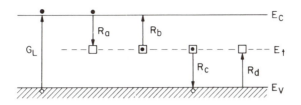

Fig. 14 Generation-recombination processes under illumination.

Figure 14 shows a nonequilibrium case in which an n-type semiconductor is illuminated uniformly to give a generation rate G_L. Thus in addition to the processes shown in Fig. 13, electron–hole pairs are generated as a result of light. In steady state the electrons entering and leaving the conduction band must be equal. This is called the *principle of detailed balance,* and it yields

$$\frac{dn_n}{dt} = G_L - (R_a - R_b) = 0 . \tag{59}$$

Similarly, in steady state the detailed balance of holes in the valence band leads to

$$\frac{dp_n}{dt} = G_L - (R_c - R_d) = 0 . \tag{60}$$

Under equilibrium conditions, that is, $G_L = 0$, $R_a = R_b$ and $R_c = R_d$. However, under steady-state nonequilibrium conditions $R_a \neq R_b$ and $R_c \neq R_d$. From Eqs. 59 and 60 we obtain

$$G_L = R_a - R_b = R_c - R_d . \tag{61}$$

Inserting the expressions for R_a through R_d into Eq. 61 gives

$$v_{th} \sigma_n N_t [n_n (1 - F) - n_i e^{(E_t - E_i)/kT} F]$$
$$= v_{th} \sigma_p N_t [p_n F - n_i e^{(E_i - E_t)/kT} (1 - F)] \tag{62}$$

We can eliminate F from Eq. 62 and solve for the net recombination rate U:

$$U \equiv R_a - R_b = \frac{v_{th} \sigma_n \sigma_p N_t (p_n n_n - n_i^2)}{\sigma_p [p_n + n_i e^{(E_i - E_t)/kT}] + \sigma_n [n_n + n_i e^{(E_t - E_i)/kT}]} \tag{63}$$

To understand the main features of the above relationship, we shall first consider a limiting case, that is, under low-injection condition in an n-type semiconductor so that $n_n \gg p_n$. Furthermore, we will assume that the centers are near the midgap so that $n_n \gg n_i e^{(E_t - E_i)/kT}$. Thus U can be approximated by

$$U \simeq v_{th} \sigma_p N_t (p_n - p_{no}) . \tag{64}$$

This expression has the same form as that of Eq. 43, and the corresponding lifetime for holes in an n-type semiconductor is

$$\tau_p \equiv \frac{1}{v_{th}\sigma_p N_t} . \tag{65}$$

We can obtain a similar expression for electrons in a p-type semiconductor. Typical lifetime is about 0.3 μs for τ_p in n-type silicon at room temperature and about 1.0 μs for τ_n in p-type silicon. The lifetime is independent of the majority carrier concentration. Because there is an abundance of electrons in an n-type semiconductor, as soon as a hole is captured by a center, an electron will immediately be captured by the same center to complete the recombination process. Thus, the rate-limiting step in the recombination process is the capture of the minority carrier.

We can simplify the general expression for the dependence of U on E_t by assuming equal electron and hole capture cross sections, that is, $\sigma_n = \sigma_p = \sigma_o$. Equation 63 then becomes

$$U = v_{th}\sigma_o N_t \; \frac{(p_n n_n - n_i)^2}{p_n + n_n + 2n_i \cosh\left[\dfrac{E_t - E_i}{kT}\right]} . \tag{66}$$

Under a low-injection condition, the recombination rate can be written as

$$U \simeq v_{th}\sigma_o N_t \; \frac{p_n - p_{no}}{1 + \left[\dfrac{2n_i}{n_{no} + p_{no}}\right]\cosh\left[\dfrac{E_t - E_i}{kT}\right]} = \frac{p_n - p_{no}}{\tau_r} \tag{67}$$

where τ_r is the recombination lifetime given by

$$\tau_r \equiv \frac{1 + \left[\dfrac{2n_i}{n_{no} + p_{no}}\right]\cosh\left[\dfrac{E_t - E_i}{kT}\right]}{v_{th}\sigma_o N_t} . \tag{68}$$

In Fig. 15 (solid curve) the recombination lifetime normalized to its minimum value is plotted against $(E_t - E_i)/kT$ for recombination in n-type silicon with $n_{no} = 10^{15}$ cm^{-3}. Note that the recombination lifetime has a broad minimum value. For $n_{no} = 10^{15}$ cm^{-3}, the lifetime τ_r remains essentially a constant ($\simeq 1/v_{th}\sigma_o N_t$) although the energy level of the center E_t varies over a fairly broad energy range ($\pm 10kT$) from the midgap. For higher majority carrier concentrations the energy range becomes even broader.

So far we have considered the injection of excess carriers into a semiconductor such that $pn > n_i^2$. Recombination is the process to restore the system to equilibrium (i.e., $pn = n_i^2$). If the pn product is less than n_i^2, carriers are extracted from the semiconductor; *carrier extraction* can occur, for example, in a p–n junction under reverse bias conditions (to be considered in Chapter 3). To restore the system to equilibrium, carriers must be generated

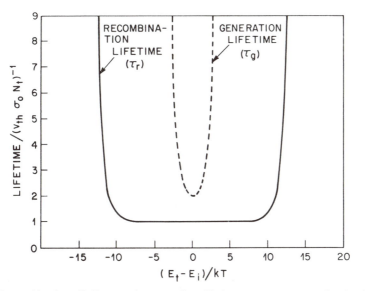

Fig. 15 Recombination lifetime and generation lifetime versus energy level of the recombination center.

by the generation–recombination centers. The generation rate can be obtained from Eq. 66 by setting $p_n < n_i$ and $n_n < n_i$:

$$G = -U \equiv \frac{(v_{th}\sigma_o N_t)n_i}{2 \cosh\left[\dfrac{E_t - E_i}{kT}\right]} = \frac{n_i}{\tau_g}. \tag{69}$$

The generation lifetime is

$$\tau_g = \frac{2 \cosh\left[\dfrac{E_t - E_i}{kT}\right]}{v_{th}\sigma_o N_t}. \tag{70}$$

Figure 15 (dotted curve) shows the normalized generation lifetime versus $(E_t - E_i)/kT$. Note that the generation lifetime depends strongly on the energy level of the center, and τ_g has a minimum value for centers having energies near the middle of the gap. The reason for this is that the energy level plays a greater role in equalizing the rate of transfer between the center and the conduction and valence bands when carrier densities are small.

From Fig. 15 we see that the generation lifetime can be substantially larger than the recombination lifetime when $E_t \neq E_i$. From Eqs. 68 and 70,

$$\frac{\tau_g}{\tau_r} \simeq 2 \cosh\left[\frac{E_t - E_i}{kT}\right]. \tag{71}$$

For example, if $E_t - E_i = 4kT$, then $\tau_g = 50\tau_r$. Equation 71 has important device design implications.[7] If we choose a generation–recombination center with the appropriate energy level, τ_g can be made much higher than τ_r. Hence, a device property that depends on τ_r, for example, the turn-off time of a diode, can be varied independent of device property that depends on τ_g, for example, diode leakage current (see Chapter 3). This is the result of the exponential dependence of τ_g on E_t and the relative independence of τ_r on E_t.

The lifetimes as given in Eqs. 68 and 70 are inversely proportional to N_t, the concentration of the recombination centers per unit volume. For device operations that require long recombination lifetimes, the concentration of the recombination centers must be minimized. On the other hand, for high-speed switching operations, short recombination lifetimes are required. In this case, the semiconductor is heavily doped with recombination centers.

Many impurities have energy levels close to the middle of the bandgap. These impurities are efficient recombination centers. A typical example is gold in silicon.[8] The gold has an acceptor state at $E_t - E_i = 0.02$ eV or $(E_t - E_i)/kT = 0.77$ at room temperature. The minority carrier lifetime decreases linearly with gold concentration as shown in Fig. 16. By increasing the gold concentration from 10^{14} cm^{-3} to 10^{18} cm^{-3}, we can reduce the minority carrier lifetime from 1 μs to 0.1 ns. Another method of changing the minority carrier lifetime is by high-energy irradiation, which causes displace-

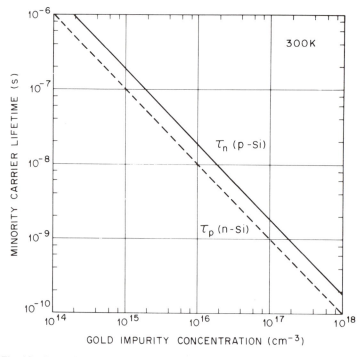

Fig. 16   Recombination lifetime versus gold impurity concentration in silicon.[8]

ment of host atoms and damage to lattices. These radiation effects, in turn, introduce energy levels in the bandgap. For example, electron irradiation in silicon results in an acceptor level at 0.4 eV above the valence band and a donor level at 0.36 eV below the conduction band; neutron irradiation creates an acceptor level very near E_i.

2.4.3   Surface Recombination*

Figure 17 shows schematically the bonds at a semiconductor surface.[9] Because of the abrupt discontinuity of the lattice structure at the surface, a large number of localized energy states or generation–recombination centers may be introduced at the surface region. These energy states may greatly enhance the recombination rate at the surface region. An understanding of the surface recombination process is important because it has a strong effect on the characteristics of many semiconductor devices.

The kinetics of surface recombination are similar to those considered before for bulk centers. The total number of carriers recombining at the surface per unit area and unit time can be expressed in a form analogous to Eq. 63:

$$U_s = \frac{v_{th}\,\sigma_n\,\sigma_p\,N_{st}(p_s n_s - n_i^2)}{\sigma_p[p_s + n_i e^{(E_i - E_t)/kT}] + \sigma_n[n_s + n_i e^{(E_t - E_i)/kT}]} \tag{72}$$

where n_s and p_s denote the electron and hole concentrations at the surface and N_{st} is the recombination center density per unit area in the surface region. For a low–injection condition, and for the limiting case where n_s is essentially equal to the bulk majority carrier concentration so that $n_s \gg p_s$ and

*See footnote to Section 2.4.2, p. 48.

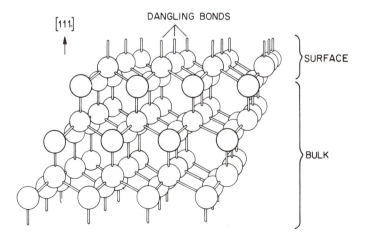

Fig. 17 Schematic diagram of bonds at a clean semiconductor surface.[9] The bonds are anisotropic and differ from those in the bulk.

$n_s \gg n_i e^{(E_t - E_i)/kT}$, Eq. 72 simplifies to

$$U_s \simeq v_{th} \sigma_p N_{st} (p_s - p_{no}) . \tag{73}$$

Since the product $v_{th} \sigma_p N_{st}$ has its dimension in centimeters per second, it is called the *low-injection surface recombination velocity S_{lr}*:

$$S_{lr} = v_{th} \sigma_p N_{st} . \tag{74}$$

The concepts of bulk generation and recombination discussed in Section 2.4.2 apply equally well to surface generation and recombination. If we assume equal capture cross sections for electrons and holes at the surface ($\sigma_n = \sigma_p = \sigma_s$), the surface recombination velocity S_r and the surface generation velocity S_g can be deduced from Eq. 72:

$$S_r = \frac{v_{th} \sigma_s N_{st}}{1 + \left[\dfrac{2n_i}{n_{no} + p_s} \right] \cosh \left[\dfrac{E_t - E_i}{kT} \right]} \tag{75}$$

$$S_g = \frac{v_{th} \sigma_s N_{st}}{2 \cosh \left[\dfrac{E_t - E_i}{kT} \right]} . \tag{76}$$

These expressions have forms that are inversely proportional to those for bulk recombination lifetime and generation lifetime, respectively. Therefore, the surface recombination velocity is expected to be higher than the surface generation velocity, that is, $S_r > S_g$. Recent advances in silicon process control have resulted[7] in an S_r value as low as 80 cm/s and an even lower S_g value of 0.1 cm/s.

2.5 CONTINUITY EQUATION

In the previous sections we have considered individual effects such as drift due to an electric field, diffusion due to a concentration gradient, and recombination of carriers through intermediate-level recombination centers. We shall now consider the overall effect when drift, diffusion, and recombination occur simultaneously in a semiconductor material. The governing equation is called the continuity equation.

To derive the one-dimensional continuity equation for electrons, consider an infinitesimal slice with a thickness dx located at x shown in Fig. 18. The number of electrons in the slice may increase due to the net current flow into the slice and the net carrier generation in the slice. The overall rate of electron increase is the algebraic sum of four components: the number of electrons flowing into the slice at x, minus the number of electrons flowing out at $x + dx$, plus the rate at which electrons are generated, minus the rate at which they are recombined with holes in the slice.

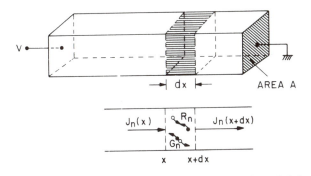

Fig. 18 Current flow and generation–recombination processes in an infinitesimal slice of thickness dx.

The first two components are found by dividing the currents at each side of the slice by the charge of an electron. The generation and recombination rates are designated by G_n and R_n, respectively. The overall rate of change in the number of electrons in the slice is then

$$\frac{\partial n}{\partial t} A dx = \left[\frac{J_n(x) A}{-q} - \frac{J_n(x + dx)A}{-q} \right] + (G_n - R_n)Adx \quad (77)$$

where A is the cross-sectional area and Adx is the volume of the slice. Expanding the expression for the current at $x + dx$ in a Taylor series yields

$$J_n(x + dx) = J_n(x) + \frac{\partial J_n}{\partial x} dx + \dots . \quad (78)$$

We thus obtain the basic *continuity equation* for electrons:

$$\frac{\partial n}{\partial t} = \frac{1}{q} \frac{\partial J_n}{\partial x} + (G_n - R_n) . \quad (79)$$

A similar continuity equation can be derived for holes, except that the sign of the first term on the right-hand side of Eq. 79 is changed because of the positive charge associated with a hole:

$$\frac{\partial p}{\partial t} = -\frac{1}{q} \frac{\partial J_p}{\partial x} + (G_p - R_p) . \quad (80)$$

We can substitute the current expressions from Eqs. 31 and 32 and the recombination expressions from Eqs. 43 and 64 into Eqs. 79 and 80. For the one-dimensional case under low-injection condition, the continuity equations for minority carriers (i.e., n_p in a p-type semiconductor or p_n in an n-type semiconductor) are

$$\frac{\partial n_p}{\partial t} = n_p \mu_n \frac{\partial \mathscr{E}}{\partial x} + \mu_n \mathscr{E} \frac{\partial n_p}{\partial x} + D_n \frac{\partial^2 n_p}{\partial x^2} + G_n - \frac{n_p - n_{po}}{\tau_n} \quad (81)$$

$$\frac{\partial p_n}{\partial t} = -p_n \mu_p \frac{\partial \mathscr{E}}{\partial x} - \mu_p \mathscr{E} \frac{dp_n}{\partial x} + D_p \frac{\partial^2 p_n}{\partial x^2} + G_p - \frac{p_n - p_{no}}{\tau_p} . \quad (82)$$

In addition to the continuity equations, Poisson's equation

$$\frac{d\mathscr{E}}{dx} = \frac{\rho_s}{\epsilon_s} \qquad (83)$$

must be satisfied, where ϵ_s is the semiconductor dielectric permittivity and ρ_s is the space charge density given by the algebraic sum of the charge carrier densities and the ionized impurity concentrations, $q(p - n + N_D^+ - N_D^-)$.

In principle, Eqs. 81 through 83 together with appropriate boundary conditions have a unique solution. Because of the algebraic complexity of this set of equations, in most cases the equations are simplified with physical approximations before a solution is attempted. We shall solve the continuity equations for three important cases.

2.5.1 Steady-State Injection From One Side

Figure 19a shows an n-type semiconductor where excess carriers are injected from one side as a result of illumination. At steady state there is a concentration gradient near the surface. From Eq. 82 the differential equation

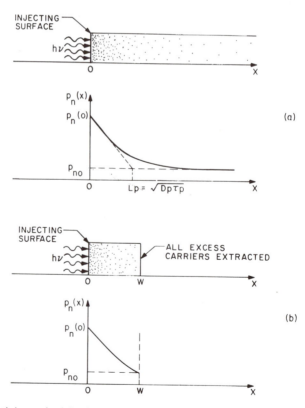

Fig. 19 Steady-state carrier injection from one side. (a) Semi-infinite sample. (b) Sample with length W.

for the minority carriers inside the semiconductor is

$$\frac{\partial p_n}{\partial t} = 0 = D_p \frac{\partial^2 p_n}{\partial x^2} - \frac{p_n - p_{no}}{\tau_p}. \tag{84}$$

The boundary conditions are $p_n(x = 0) = p_n(0) = $ constant value and $p_n(x \to \infty) = p_{no}$. The solution of $p_n(x)$ is

$$p_n(x) = p_{no} + [p_n(0) - p_{no}]e^{-x/L_p}. \tag{85}$$

The length L_p is equal to $\sqrt{D_p \tau_p}$ and is called the *diffusion length*. Figure 19a shows the variation of the minority carrier density, which decays with a characteristic length given by L_p.

If we change the second boundary condition as shown in Fig. 19b so that all excess carriers at $x = W$ are extracted, that is, $p_n(W) = p_{no}$, then we obtain a new solution for Eq. 84:

$$p_n(x) = p_{no} + [p_n(0) - p_{no}] \left[\frac{\sinh\left(\dfrac{W - x}{L_p}\right)}{\sinh(W/L_p)} \right]. \tag{86}$$

The current density at $x = W$ is given by the diffusion current expression, Eq. 32 with $\mathscr{E} = 0$:

$$J_p = -qD_p \left. \frac{\partial p_n}{\partial x} \right|_W = q[p_n(0) - p_{no}] \frac{D_p}{L_p} \frac{1}{\sinh(W/L_p)}. \tag{87}$$

2.5.2 Minority Carriers at the Surface

When surface recombination is introduced at one end of a semiconductor sample under uniform illumination (Fig. 20), the hole current density flowing into the surface from the bulk of the semiconductor is given by qU_s, where U_s

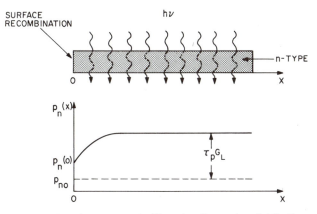

Fig. 20 Surface recombination at $x = 0$. The minority carrier distribution near the surface is affected by the surface recombination velocity.[10]

is given by Eq. 73. The surface recombination leads to a lower carrier concentration at the surface. This gradient of hole concentration yields a diffusion current density that is equal to the surface recombination current. Therefore, the boundary condition at $x = 0$ is

$$qD_p \left. \frac{dp_n}{dx} \right|_{x=0} = qU_s = qS_{lr}[p_n(0) - p_{no}] . \tag{88}$$

The boundary condition at $x = \infty$ is given by Eq. 45a. At steady state the differential equation is

$$\frac{\partial p_n}{\partial t} = 0 = D_p \frac{\partial^2 p_n}{\partial x^2} + G_L - \frac{p_n - p_{no}}{\tau_p} . \tag{89}$$

The solution of the equation, subject to the boundary conditions above, is[10]

$$p_n(x) = p_{no} + \tau_p G_L \left[1 - \frac{\tau_p S_{lr} e^{-x/L_p}}{L_p + \tau_p S_{lr}} \right] . \tag{90}$$

A plot of this equation for a finite S_{lr} is shown in Fig. 20. When $S_{lr} \to 0$, then $p_n(x) \to p_{no} + \tau_p G_L$, as obtained previously (Eq. 45a). When $S_{lr} \to \infty$, then

$$p_n(x) = p_{no} + \tau_p G_L(1 - e^{-x/L_p}) . \tag{91}$$

From Eq. 91 we can see that at the surface the minority carrier density approaches its thermal equilibrium value p_{no}.

2.5.3 The Haynes–Shockley Experiment

One of the classic experiments in semiconductor physics is the demonstration of drift and diffusion of minority carriers, first made by J. R. Haynes and W. Shockley.[11] The basic setup of the Haynes–Shockley experiment is shown in Fig. 21a. Localized light pulses generate excess carriers in a semiconductor bar. After a pulse, the transport equation is given by Eq. 82 by setting $G_L = 0$ and $\partial \mathscr{E}/\partial x = 0$ (i.e., the applied field is constant across the semiconductor bar):

$$\frac{\partial p_n}{\partial t} = -\mu_p \mathscr{E} \frac{\partial p_n}{\partial x} + D_p \frac{\partial^2 p_n}{\partial x^2} - \frac{p_n - p_{no}}{\tau_p} . \tag{92}$$

If no field is applied along the sample, the solution is given by

$$p_n(x, t) = \frac{N}{\sqrt{4\pi D_p t}} \exp \left[-\frac{x^2}{4D_p t} - \frac{t}{\tau_p} \right] + p_{no} . \tag{93}$$

where N is the number of electrons or holes generated per unit area. Figure 21b shows this solution as the carriers diffuse away from the point of injection and recombine.

If an electric field is applied along the sample, the solution is in the form of Eq. 93, except that x is replaced by $x - \mu_p \mathscr{E} t$ (Fig. 21c). Thus, all the excess

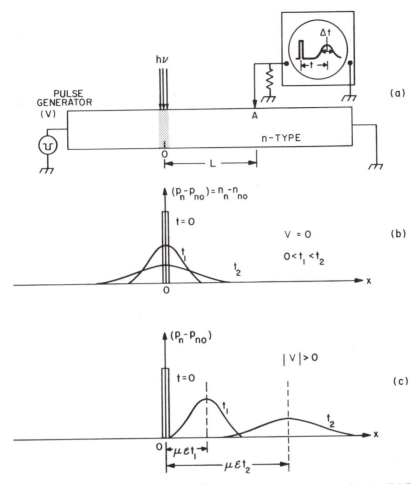

Fig. 21 The Haynes–Shockley experiment.[11] (a) Experimental setup. (b) Carrier distributions without an applied field. (c) Carrier distributions with an applied field.

carriers move toward the negative end of the sample with the drift velocity $\mu_p\mathscr{E}$. At the same time, the carriers diffuse outward and recombine as in the field-free case.

2.6 HIGH-FIELD EFFECTS

At low electric fields, the drift velocity is linearly proportional to the applied field. We assume that the time interval between collision, τ_c, is independent of the applied field. This is a reasonable assumption as long as the drift velocity is small compared to the thermal velocity of carriers, which is about 10^7 cm/s for silicon at room temperature.

As the drift velocity approaches the thermal velocity, its field dependence on the electric field will begin to depart from the linear relationship given in Section 2.1. Figure 22 shows the measured drift velocities of electrons and holes in silicon as a function of the electron field. It is apparent that initially the field dependence of the drift velocity is linear, corresponding to a constant mobility. As the electric field is further increased, the drift velocity increases less rapidly. At sufficiently large fields, the drift velocity approaches a saturation velocity. The experimental results can be approximated by the empirical expression[12]

$$v_n, \ v_p \ = \ \frac{v_s}{[1 \ + \ (\mathscr{E}_0/\mathscr{E})^\gamma]^{1/\gamma}} \tag{94}$$

where v_s is the saturation velocity (10^7 cm/s for Si at 300 K); $\mathscr{E}_0$ is a constant, equal to 7×10^3 V/cm for electrons and 2×10^4 V/cm for holes in high-purity silicon materials; and γ is 2 for electrons and 1 for holes.

The high-field transport in n-type gallium arsenide is quite different from that of silicon.[13] Figure 23 shows the measured drift velocity versus field for n-type and p-type gallium arsenide. The results for silicon are also shown in this log–log plot for comparison. Note that for n-type GaAs, the drift velocity reaches a maximum, then decreases as the field further increases. This phenomenon is due to the energy band structure of gallium arsenide that allows the transfer of conduction electrons from a high mobility energy minimum (called a valley) to low mobility, higher energy satellite valleys, that is, electron transfer from the central valley to the satellite valleys along the [111] direction shown previously in Fig. 12 of Chapter 1.

To understand this phenomenon, consider the simple two valley model of n-type gallium arsenide shown in Fig. 24. The energy separation between the two valleys is $\Delta E = 0.31$ eV. The lower valley's electron effective mass is

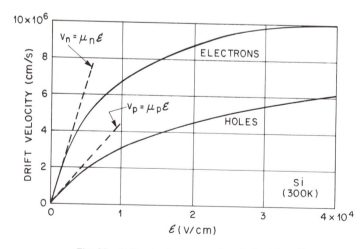

Fig. 22 Drift velocity versus electric field in Si.[12]

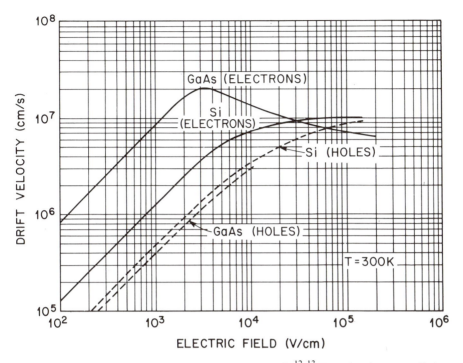

Fig. 23   Drift velocity versus electric field in GaAs and Si.[12, 13] Note that for *n*-type GaAs, there is a region of negative differential mobility.

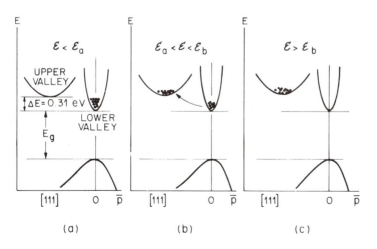

Fig. 24   Electron distributions under various conditions of electric fields for a two-valley semiconductor.

denoted by m_1, the electron mobility by μ_1, and the electron density by n_1. The upper-valley quantities are denoted by m_2, μ_2, and n_2, respectively; and the total electron concentration is given by $n = n_1 + n_2$. The steady-state conductivity of the n-type GaAs can be written as

$$\sigma = q(\mu_1 n_1 + \mu_2 n_2) = qn\bar{\mu} \tag{95}$$

where the average mobility is

$$\bar{\mu} \equiv (\mu_1 n_1 + \mu_2 n_2)/(n_1 + n_2). \tag{96}$$

The drift velocity is then

$$v_n = \bar{\mu}\,\mathscr{E}. \tag{97}$$

For simplicity we shall make the following assignments for the electron concentrations in the various ranges of electric-field values illustrated in Fig. 24. In Fig. 24a, the field is low and all electrons remain in the lower valley. In Fig. 24b, the field is higher and some electrons gain sufficient energies from the field to move to the higher valley. In Fig. 24c, the field is high enough to transfer all electrons to the higher valley. Thus, we have

$$n_1 \simeq n \quad \text{and} \quad n_2 \simeq 0 \quad \text{for } 0 < \mathscr{E} < \mathscr{E}_a$$
$$n_1 + n_2 = n \quad\quad\quad\quad \text{for } \mathscr{E}_a < \mathscr{E} < \mathscr{E}_b \tag{98}$$
$$n_1 \simeq 0 \quad \text{and} \quad n_2 \simeq n \quad \text{for } \mathscr{E} > \mathscr{E}_b.$$

Using these relations, the effective drift velocity takes on the asymptotic values

$$v_n \simeq \mu_1 \mathscr{E} \text{ for } 0 < \mathscr{E} < \mathscr{E}_a$$
$$v_n \simeq \mu_2 \mathscr{E} \text{ for } \mathscr{E} > \mathscr{E}_b. \tag{99}$$

If $\mu_1 \mathscr{E}_a$ is larger than $\mu_2 \mathscr{E}_b$, there exists a region in which the drift velocity decreases with an increasing field between $\mathscr{E}_a$ and $\mathscr{E}_b$, as shown in Fig. 25. Because of the characteristics of the drift velocity in n-type gallium arsenide, this material is used in microwave transferred-electron devices discussed in Chapter 6.

When the electric field in a semiconductor is increased above a certain value, the carriers gain enough kinetic energy to generate electron–hole pairs by an *avalanche process* that is shown schematically in Fig. 26. Consider an electron in the conduction band (designated by 1). If the electric field is high enough, this electron can gain kinetic energy before it collides with the lattice. On impact with the lattice, the electron imparts most of its kinetic energy to break a bond, that is, to ionize a valence electron from the valence band to the conduction band and thereby generate an electron–hole pair (designated by 2 and 2′). Similarly, the generated pair now begins to accelerate in the field and collides with the lattice as indicated in the figure. In turn, they will generate other electron–hole pairs (e.g., 3 and 3′, 4 and 4′), and so on. This process is called the avalanche process; it is also referred to as the impact ionization process.

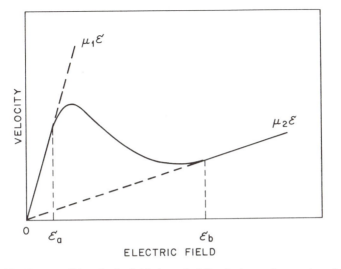

Fig. 25   One possible velocity-field characteristic of a two-valley semiconductor.

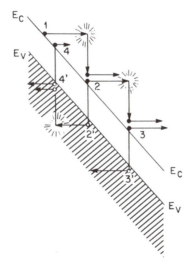

Fig. 26   Energy band diagram for avalanche process.

To gain some ideas about the ionization energy involved, let us consider the process leading to 2–2′ shown in Fig. 26. Just prior to the collision, the fast–moving electron (No. 1) has a kinetic energy $\frac{1}{2}m_1v_s^2$ and a momentum m_1v_s, where m_1 is the effective mass and v_s is the saturation velocity. After collision, there are three carriers: the original electron plus an electron–hole pair (No. 2 and No. 2′). If we assume that the three carriers have the same effective mass, the same kinetic energy, and the same momentum, the total kinetic energy is

$\frac{3}{2}m_1v_f^2$, and the total momentum is $3m_1v_f$, where v_f is the velocity after collision. To conserve both energy and momentum before and after the collision, we require that

$$\frac{1}{2}\ m_1v_s^2\ =\ E_g\ +\ \frac{3}{2}\ m_1v_f^2 \qquad (100)$$

and

$$m_1v_s\ =\ 3m_1v_f \qquad (101)$$

where in Eq. 100 the energy E_g is the bandgap corresponding to the minimum energy required to generate an electron–hole pair. Substituting Eq. 101 into Eq. 100 yields the required kinetic energy for the ionization process:

$$E_o\ =\ \frac{1}{2}\ m_1v_s^2\ =\ 1.5E_g\ . \qquad (102)$$

It is obvious that E_o must be larger than the bandgap for the ionization process to occur. The actual energy required depends on the band structure. For silicon, the value for E_o is 3.6 eV ($3.2E_g$) for electrons and 5.0 eV ($4.4E_g$) for holes.

The number of electron–hole pairs generated by an electron per unit distance traveled is called the *ionization rate* for the electron, α_n. Similarly, α_p is the ionization rate for the holes. The measured ionization rates for silicon and gallium arsenide are shown in Fig. 27. We note that both α_n and α_p are strongly dependent on the electric field. For a substantially large ionization rate (say, $\gtrsim 10^4\ \text{cm}^{-1}$), the corresponding electric field is $\gtrsim 3 \times 10^5$ V/cm for silicon and $\gtrsim 4 \times 10^5$ V/cm for gallium arsenide. The electron–hole pair generation rate G_A from the avalanche process is given by

$$G_A\ =\ \frac{1}{q}\ (\alpha_n\ |J_n|\ +\ \alpha_p\ |J_p|\ ) \qquad (103)$$

where J_n and J_p are the electron and hole current densities, respectively. This expression can be used in the continuity equation for devices operated under an avalanche condition.

REFERENCES

1   D. Halliday and R. Resnick, *Fundamental of Physics*, 2nd ed., Wiley, New York, 1981.

2   R. A. Smith, *Semiconductors*, 2nd ed., Cambridge, London, 1978.

3   J. L. Moll, *Physics of Semiconductors*, McGraw-Hill, New York, 1964.

4   W. F. Beadle, J. C. C. Tsai, and R. D. Plummer, Eds., *Quick Reference Manual for Semiconductor Engineers*, Wiley, New York, 1985.

5   H. C. Casey and M. B. Panish, *Heterostructure Lasers*, Academic, New York, 1978.

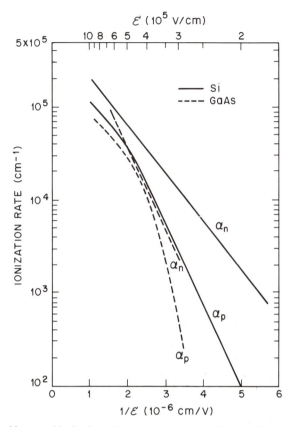

Fig. 27 Measured ionization rates versus reciprocal field for Si and GaAs.[13]

6 R. N. Hall, "Electron–Hole Recombination in Germanium," *Phys. Rev.*, **87**, 387 (1952); W. Shockley and W. T. Read, "Statistics of Recombination of Holes and Electrons," *Phys. Rev.*, **87**, 835 (1952).

7 D. K. Schroder, "The Concept of Generation and Recombination Lifetimes in Semiconductors," *IEEE Trans. Electron Devices*, **ED-29**, 1336 (1982).

8 H. F. Wolf, *Semiconductors*, Wiley, New York, 1971.

9 M. Prutton, *Surface Physics*, 2nd ed., Oxford, Clarendon, 1983.

10 A. S. Grove, *Physics and Technology of Semiconductor Devices*, Wiley, New York, 1967.

11 J. R. Haynes and W. Shockley, "The Mobility and Life of Injected Holes and Electrons in Germanium," *Phys. Rev.*, **81**, 835 (1951).

12 D. M. Caughey and R. E. Thomas, "Carrier Mobilities in Silicon Empirically Related to Doping and Field," *Proc. IEEE*, **55**, 2192 (1967).

13 S. M. Sze, *Physics of Semiconductor Devices*, 2nd ed., Wiley, New York (1981).

PROBLEMS

1 Calculate the mean free time of an electron having a mobility of $1000 \text{ cm}^2/\text{V-s}$ at 300 K; also calculate the mean free path (i.e., the distance traveled by an electron between collisions). Assume $m_n = 0.26 m_0$ in these calculations.

2 For a semiconductor with a constant mobility ratio $b \equiv \mu_n/\mu_p > 1$ independent of impurity concentration, find the maximum resistivity ρ_m in terms of the intrinsic resistivity ρ_i and the mobility ratio.

3 Minority carriers (holes) are injected into a homogeneous n-type semiconductor sample at one point. An electric field of 50 V/cm is applied across the sample, and the field moves these minority carriers a distance of 1 cm in 100 μs. Find the drift velocity and the diffusivity of the minority carriers.

4 Find the electron and hole concentrations, mobilities, and resistivities of silicon samples at 300 K, for each of the following impurity concentrations: (a) 5×10^{15} Boron atoms/cm^3. (b) 2×10^{16} Boron atoms/cm^3 and 1.5×10^{16} arsenic atoms/cm^3. (c) 5×10^{15} Boron atoms/cm^3, 10^{17} arsenic atoms/cm^3, and 10^{17} gallium atoms/cm^3.

5 A four-point probe (with probe spacing of 0.5 mm) is used to measure the resistivity of a p-type silicon sample. Find the resistivity of the sample if its diameter is 100 mm and its thickness is 50 μm. The constant current is 1 mA, and the measured voltage between the inner two probes is 10 mV. If the sample is cut into small square chips 5 mm on each side, what will be the measured voltage for a constant current of 1 mA?

6 Find the resistivities of intrinsic Si and intrinsic GaAs at 300 K.

7 Given a silicon sample of unknown doping, Hall measurement has been made and the following information obtained: $W = 0.05$ cm, $A = 1.6 \times 10^{-3}$ cm^2 (refer to Fig. 8), $I = 2.5$ mA, and the magnetic field is 30 nT (1T = 10^{-4} Wb/cm^2). If a Hall voltage of $+10$ mV is measured, find the Hall coefficient, conductivity type, majority carrier concentration, resistivity, and mobility of the semiconductor sample.

8 An n-type silicon sample has 2×10^{16} arsenic atoms/cm^3, 2×10^{15} bulk recombination centers/cm^3, and 10^{10} surface recombination centers/cm^2. (a) Find the bulk minority carrier lifetime, the diffusion length, and the surface recombination velocity under low-injection conditions. The values of σ_p and σ_s are 5×10^{-15} and 2×10^{-16} cm^2, respectively. (b) If the sample is illuminated with uniformly absorbed light which creates 10^{17} electron–hole pairs/cm^2-s, what is the hole concentration at the surface?

9 Excess carriers are injected on one surface of a thin slice of n-type silicon with length W and extracted at the opposite surface where $p_n(W) = p_{no}$. There is no electric field in the region $0 < x < W$. Derive the expression for current densities at the two surfaces. If carrier lifetime is 50 μs and $W = 0.1$ mm, calculate the portion of injected current which reaches the opposite surface by diffusion ($D = 50$ cm^2/s).

10 In a Haynes–Shockley experiment, the maximum amplitudes of the minority carriers at $t_1 = 100$ μs and $t_2 = 200$ μs differ by a factor of 5. Calculate the minority carrier lifetime.

3

p–n **Junction**

In the preceding chapters we have considered the carrier concentrations and transport phenomena in homogeneous semiconductor materials. In this chapter we shall discuss the behavior of single-crystal semiconductor material containing both p- and n-type regions that form a $p-n$ junction.

The most important characteristic of $p-n$ junctions is that they rectify, that is, they allow current to flow easily in only one direction. Figure 1 shows the current–voltage characteristics of a typical silicon $p-n$ junction. When we apply "forward bias" (we shall define bias polarity in Section 3.1) to the junction, the current increases rapidly as the voltage increases. However, when we apply a "reverse bias," virtually no current flows initially. As the reverse bias is increased, the current remains very small until a critical voltage is reached at which point the current suddenly increases. This sudden increase in current is referred to as the junction breakdown. The applied forward voltage is usually less than 1 V, but the reverse critical voltage, or breakdown voltage, can vary from just a few volts to many thousands of volts depending on the doping concentration and other device parameters.

A $p-n$ junction serves an important role both in modern electronic applications and in understanding other semiconductor devices. It is used extensively in rectification, switching, and other operations in electronic circuits. It is the basic building block for the bipolar transistor and thyristor (Chapter 4), as well as for JFETs and MOSFETs (Chapter 5). Given proper biasing conditions or when exposed to light, $p-n$ junction also functions as either microwave (Chapter 6) or photonic device (Chapter 7).

In this chapter we use the basic equations presented in the preceding chapters to develop the ideal static and dynamic characteristics of $p-n$ junctions. We discuss departures from ideal characteristics due to generation and recombination and other effects. We next discuss stored minority carriers and their influence on device transient behavior. The chapter closes with a discussion of junction breakdown.

3.1 THERMAL EQUILIBRIUM CONDITION

In Fig. 2a we see two regions of p- and n-type semiconductor materials that are uniformly doped and physically separated before the junction is formed.

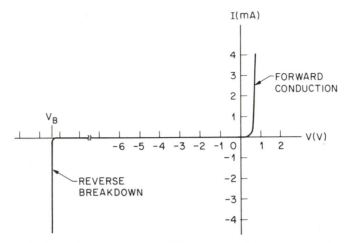

Fig. 1 Current–voltage characteristics of a typical silicon p–n junction.

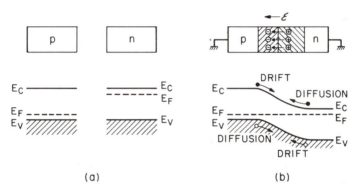

Fig. 2 (a) Uniformly doped p-type and n-type semiconductors before junction is formed. (b) The electric field in the depletion region and the energy band diagram of a p–n junction in thermal equilibrium.

Note that the Fermi level E_F is near the valence band edge in the p-type material and near the conduction band edge in the n-type material. While p-type material contains a large concentration of holes with few electrons, the opposite is true for n-type material.

A $p-n$ junction is formed when these two regions are joined. During device fabrication, this $p-n$ junction can be formed using processes such as epitaxy, diffusion, and ion implantation. These processes will be discussed in Chapters 8 and 10.

The large carrier concentration gradients at the junction cause carrier diffusion. Holes from the p-side diffuse into the n-side, and electrons from the n-side diffuse into the p-side. As holes continue to leave the p-side, some of the negative acceptor ions (N_A^-) near the junction are left uncompensated, since the acceptors are fixed in the semiconductor lattice while the holes are mobile. Similarly, some of the positive donor ions (N_D^+) near the junction are left uncompensated as the electrons leave the n-side. Consequently, a negative space charge forms near the p-side of the junction and a positive space charge forms near the n-side. This space charge region creates an electric field that is directed from the positive charge toward the negative charge, as indicated in the upper illustration of Fig. 2b.

The electric field is in the direction opposite to the diffusion current for each type of charge carrier. The lower illustration of Fig. 2b shows that the hole diffusion current flows from left to right, while the hole drift current due to the electric field flows from right to left. The electron diffusion current also flows from left to right, while the electron drift current flows in the opposite direction. Note that because of their negative charge, electrons diffuse from right to left, opposite to the direction of electron current.

At thermal equilibrium, that is, the steady-state condition at a given temperature without any external excitations, the net current flow across the junction is zero. Thus, for each type of carrier the drift current due to the electric field must exactly cancel the diffusion current due to the concentration gradient. From Eq. 32 in Chapter 2,

$$J_p = J_p(\text{drift}) + J_p(\text{diffusion})$$

$$= q\mu_p p \mathscr{E} - qD_p \frac{dp}{dx}$$

$$= q\mu_p p \left[\frac{1}{q} \frac{dE_i}{dx} \right] - kT\mu_p \frac{dp}{dx} = 0 \tag{1}$$

where we have used Eq. 8 of Chapter 2 for the electric field and the Einstein relation $D_p = kT\mu_p/q$. Substituting the expression for hole concentration

$$p = n_i e^{(E_i - E_F)/kT} \tag{2}$$

and its derivative

$$\frac{dp}{dx} = \frac{p}{kT} \left[\frac{dE_i}{dx} - \frac{dE_F}{dx} \right] \tag{3}$$

into Eq. 1 yields the net hole current density

$$J_p = \mu_p p \frac{dE_F}{dx} = 0 \tag{4}$$

or

$$\frac{dE_F}{dx} = 0 . \tag{5}$$

Similarly, we obtain for the net electron current density

$$J_n = J_n(\text{drift}) + J_n(\text{diffusion})$$

$$= q\,\mu_n n\,\mathscr{E} + qD_n\,\frac{dn}{dx}$$

$$= \mu_n n\,\frac{dE_F}{dx} = 0\,. \qquad (6)$$

Thus, for the condition of zero net electron and hole currents, the Fermi level must be constant (i.e., independent of x) throughout the sample as illustrated in the energy band diagram of Fig. 2b.

The constant Fermi level required at thermal equilibrium results in a unique space charge distribution at the junction. We repeat the one-dimensional $p-n$ junction and the corresponding equilibrium energy band diagram in Figs. 3a and 3b, respectively. The unique space charge distribution and the electrostatic potential ψ are given by Poisson's equation:

$$\frac{d^2\psi}{dx^2} \equiv -\frac{d\mathscr{E}}{dx} = -\frac{\rho_s}{\epsilon_s} = -\frac{q}{\epsilon_s}(N_D - N_A + p - n) \qquad (7)$$

here we assume that all donors and acceptors are ionized.

In regions far away from the metallurgical junction, charge neutrality is maintained and the total space charge density is zero. For these neutral regions we can simplify Eq. 7 to

$$\frac{d^2\psi}{dx^2} = 0 \qquad (8)$$

and

$$N_D - N_A + p - n = 0\,. \qquad (9)$$

For a p-type neutral region, we assume $N_D = 0$ and $p \gg n$. The electrostatic potential of the p-type neutral region with respect to the Fermi level, designated as ψ_p in Fig. 3b, can be obtained by setting $N_D = n = 0$ in Eq. 9 and by substituting the result ($p = N_A$) into Eq. 2:

$$\psi_p \equiv -\frac{1}{q}(E_i - E_F)\bigg|_{x \,\leqslant\, -x_p} = -\frac{kT}{q}\ln\frac{N_A}{n_i}\,. \qquad (10)$$

Similarly, we obtain the electrostatic potential of the n-type neutral region with respect to the Fermi level:

$$\psi_n \equiv -\frac{1}{q}(E_i - E_F)\bigg|_{x \,\geqslant\, x_n} = \frac{kT}{q}\ln\frac{N_D}{n_i}\,. \qquad (11)$$

The total electrostatic potential difference between the p-side and the n-side neutral regions at thermal equilibrium is called the *built-in potential* V_{bi}:

$$V_{bi} = \psi_n - \psi_p = \frac{kT}{q}\ln\frac{N_A N_D}{n_i^2}\,. \qquad (12)$$

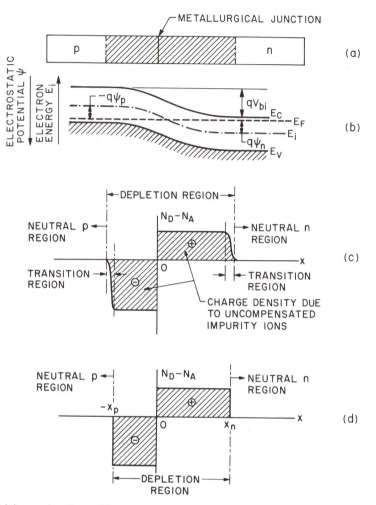

Fig. 3 (a) A *p–n* junction with abrupt doping changes at the metallurgical junction. (b) Energy band diagram of an abrupt junction at thermal equilibrium. (c) Space charge distribution. (d) Rectangular approximation of the space charge distribution.

Moving from a neutral region toward the junction, we encounter a narrow transition region, shown in Fig. 3c. Here the space charge of impurity ions is partially compensated by the mobile carriers. Beyond the transition region we enter the completely depleted region where the mobile carrier densities are zero. This is called the *depletion region* (also called the space charge region). For typical $p-n$ junctions in silicon and gallium arsenide, the width of each transition region is small compared to the width of the depletion region. Therefore, we can neglect the transition region and represent the depletion region by the rectangular distribution shown in Fig. 3d, where x_p and x_n denote the depletion layer widths of the *p*- and *n*-sides. For the completely depleted region with $p = n = 0$, Eq. 7 becomes

$$\frac{d^2\psi}{dx^2} = \frac{q}{\epsilon_s} (N_A - N_D).$$ (13)

The magnitudes of $|\psi_p|$ and ψ_n as calculated from Eqs. 10 and 11 are plotted in Fig. 4 as a function of the doping concentration of silicon and gallium arsenide. For a given doping concentration, the electrostatic potential of gallium arsenide is higher because of its smaller intrinsic concentration n_i.

Problem

Calculate the built-in potential for a silicon $p-n$ junction with $N_A = 10^{18}$ cm^{-3} and $N_D = 10^{15}$ cm^{-3} at 300 K.

Solution

From Eq. 12 we obtain

$$V_{bi} = (0.0259) \ln \frac{10^{18} \times 10^{15}}{(1.45 \times 10^{10})^2} = 0.755 \text{ v}$$

Also from Fig. 4,

$$V_{bi} = \psi_n + |\psi_p| = 0.30 \text{ v} + 0.46 \text{ v} = 0.76 \text{ v}.$$

3.2 DEPLETION REGION

To solve Poisson's equation, Eq. 13, we must know the impurity distribution. In this section we consider two important cases—the abrupt junction and the linearly graded junction. Figure 5a shows an *abrupt junction*, that is, a p–

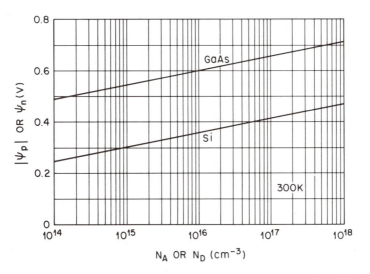

Fig. 4 Built-in potentials on the p-side and n-side of abrupt junctions in Si and GaAs as a function of impurity concentration.

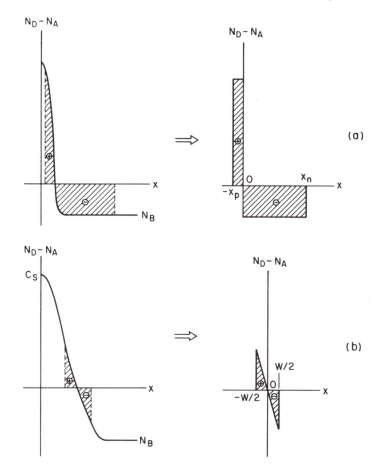

Fig. 5 Approximate doping profiles. (a) Abrupt junction. (b) Linearly graded junction.

n junction formed by shallow diffusion or low-energy ion implantation. The impurity distribution of the junction can be approximated by an abrupt transition of doping concentration between the n- and p-type regions. Figure 5b shows a linearly graded junction. For either deep diffusions or high-energy ion implantations, the impurity profiles may be approximated by linearly graded junctions, that is, the impurity distribution varies linearly across the junction. We shall consider the depletion regions of both types of junction.

3.2.1 Abrupt Junction

The space charge distribution of an abrupt junction is shown in Fig. 6a. In the depletion region, free carriers are totally depleted so that Poisson's equation, Eq. 13, simplifies to

$$\frac{d^2\psi}{dx^2} = + \frac{qN_A}{\epsilon_s} \qquad \text{for} \quad -x_p \leqslant x < 0 \qquad (14a)$$

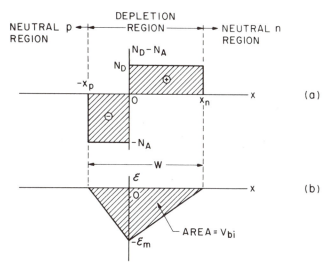

Fig. 6 (a) Space charge distribution in the depletion region at thermal equilibrium. (b) Electric–field distribution. The shaded area corresponds to the built-in potential.

$$\frac{d^2\psi}{dx^2} = -\frac{qN_D}{\epsilon_s} \qquad \text{for} \quad 0 < x \leqslant x_n .$$ (14b)

The overall space charge neutrality of the semiconductor requires that the total negative space charge per unit area in the p-side must precisely equal the total positive space charge per unit area in the n-side:

$$N_A x_p = N_D x_n .$$ (15)

The total depletion layer width W is given by

$$W = x_p + x_n .$$ (16)

The electric field shown in Fig. 6b is obtained by integrating Eqs. 14a and 14b which gives

$$\mathscr{E}(x) = -\frac{d\psi}{dx} = -\frac{qN_A(x + x_p)}{\epsilon_s} \qquad \text{for} \quad -x_p \leqslant x < 0 \quad (17a)$$

and

$$\mathscr{E}(x) = -\mathscr{E}_m + \frac{qN_D x}{\epsilon_s} = \frac{qN_D}{\epsilon_s}(x - x_n)$$

$$\text{for} \quad 0 < x \leqslant x_n \qquad (17b)$$

where $\mathscr{E}_m$ is the maximum field that exists at $x = 0$ and is given by

$$\mathscr{E}_m = \frac{qN_D x_n}{\epsilon_s} = \frac{qN_A x_p}{\epsilon_s} .$$ (18)

Integrating Eqs. 17a and 17b over the depletion region gives the total potential variation, namely, the built-in potential V_{bi}:

$$V_{bi} = - \int_{-x_p}^{x_n} \mathscr{E}(x)dx = - \int_{-x_p}^{0} \mathscr{E}(x)dx \Big|_{p-\text{side}} - \int_{0}^{x_n} \mathscr{E}(x)dx \Big|_{n-\text{side}}$$

$$= \frac{qN_A x_p^2}{2\epsilon_s} + \frac{qN_D x_n^2}{2\epsilon_s} = \frac{1}{2} \mathscr{E}_m W . \tag{19}$$

Therefore, the area of the field triangle in Fig. 6b corresponds to the built-in potential.

Combining Eqs. 15 to 19 gives the total depletion layer width as a function of the built-in potential,

$$W = \sqrt{\frac{2\epsilon_s}{q} \left[\frac{N_A + N_D}{N_A N_D} \right] V_{bi}} . \tag{20}$$

When the impurity concentration on one side of an abrupt junction is much higher than that of the other side, the junction is called a *one-sided abrupt junction* (Fig. 7a). Figure 7b shows the space charge distribution of a one-sided

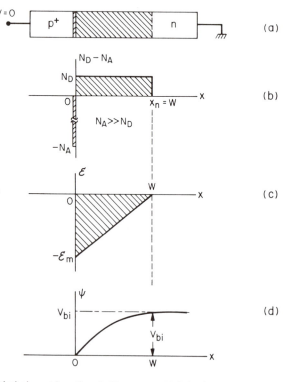

Fig. 7   (a) One-sided abrupt junction (with $N_A \gg N_D$) in thermal equilibrium. (b) Space charge distribution. (c) Electric–field distribution. (d) Potential distribution with distance, where V_{bi} is the built-in potential.

abrupt p^+-n junction, where $N_A \gg N_D$. In this case, the depletion layer width of the p-side is much smaller than that of the n-side (i.e., $x_p \ll x_n$), and the expression for W can be simplified to

$$W \simeq x_n = \sqrt{\frac{2\epsilon_s V_{bi}}{qN_D}} \; . \tag{21}$$

The expression for the electric-field distribution is the same as Eq. 17b:

$$\mathscr{E}(x) = -\mathscr{E}_m + \frac{qN_B x}{\epsilon_s} \tag{22}$$

where N_B is the lightly doped bulk concentration (i.e., N_D for a p^+-n junction). The field decreases to zero at $x = W$. Therefore,

$$\mathscr{E}_m = \frac{qN_B W}{\epsilon_s} \tag{23}$$

and

$$\mathscr{E}(x) = \frac{qN_B}{\epsilon_s} (-W + x) = -\mathscr{E}_m \left[1 - \frac{x}{W}\right] \tag{24}$$

which is shown in Fig. 7c.

Integrating Poisson's equation once more gives the potential distribution

$$\psi(x) = -\int_0^x \mathscr{E} \, dx = \mathscr{E}_m \left[x - \frac{x^2}{2W}\right] + \text{constant} \; . \tag{25}$$

With zero potential in the neutral p-region as a reference, or $\psi(0) = 0$, and employing Eq. 19,

$$\psi(x) = \frac{V_{bi} x}{W} \left[2 - \frac{x}{W}\right]. \tag{26}$$

The potential distribution is shown in Fig. 7d.

Problem

    For a silicon one-sided abrupt junction with $N_A = 10^{19}$ cm^{-3} and $N_D = 10^{16}$ cm^{-3}, calculate the depletion layer width and the maximum field at zero bias ($T = 300$ K).

Solution

    From Eqs. 12, 21, and 23, we obtain

$$V_{bi} = 0.0259 \ln \frac{10^{19} \times 10^{16}}{(1.45 \times 10^{10})^2} = 0.874 \text{ V}$$

$$W \simeq \sqrt{\frac{2\epsilon_s V_{bi}}{qN_D}} = 3.37 \times 10^{-5} \text{ cm} = 0.337 \; \mu\text{m}$$

$$\mathscr{E}_m = \frac{qN_B W}{\epsilon_s} = 5.4 \times 10^4 \text{ V/cm} \; .$$

The previous discussions are for a $p-n$ junction at thermal equilibrium without external bias. The equilibrium energy band diagram, shown again in Fig. 8a, illustrates that the total electrostatic potential across the junction is V_{bi} and that the corresponding potential energy difference from the p-side to the n-side is qV_{bi}. If we apply a positive voltage V_F to the p-side with respect to the n-side, the $p-n$ junction becomes forward-biased, as shown in Fig. 8b. The total electrostatic potential across the junction decreases by V_F, that is, it is replaced with $V_{bi} - V_F$. Thus, forward bias reduces the depletion layer width.

By contrast, as shown in Fig. 8c, if we apply positive voltage V_R to the n-side with respect to the p-side, the $p-n$ junction now becomes reverse–biased and the total electrostatic potential across the junction increases by V_R, that is, it is replaced by $V_{bi} + V_R$. Here, we find that reverse bias increases the depletion layer width. Substituting these voltage values in Eq. 21 yields the depletion layer widths as a function of the applied voltage:

$$W = \sqrt{\frac{2\epsilon_s (V_{bi} - V)}{qN_B}} \tag{27}$$

where N_B is the lightly doped bulk concentration, and V is positive for forward bias and negative for reverse bias. Note that the depletion layer width W varies as the square root of the total electrostatic potential difference across the junction.

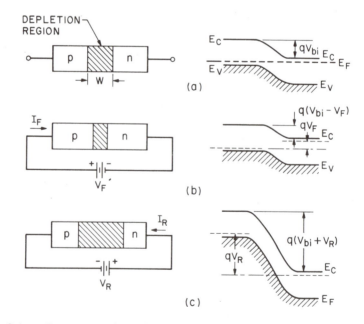

Fig. 8 Schematic representations of depletion layer width and energy band diagrams of a p–n junction under various biasing conditions. (a) Thermal-equilibrium condition. (b) Forward-bias condition. (c) Reverse-bias condition.

3.2.2 Linearly Graded Junction

We shall first consider the case of thermal equilibrium. The impurity distribution for a linearly graded junction is shown in Fig. 9a. The Poisson equation for this case is

$$\frac{d^2\psi}{dx^2} = \frac{-d\mathscr{E}}{dx} = \frac{-\rho_s}{\epsilon_s} = \frac{-q}{\epsilon_s}\,ax \qquad -\frac{W}{2} \leqslant x \leqslant \frac{W}{2} \qquad (28)$$

where a is the impurity gradient (in cm^{-4}) and W is the depletion-layer width.

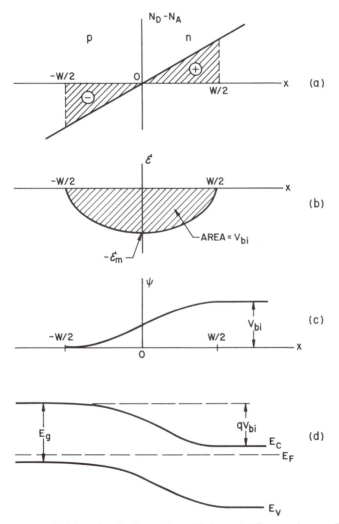

Fig. 9 Linearly graded junction in thermal equilibrium. (a) Space charge distribution. (b) Electric-field distribution. (c) Potential distribution with distance. (d) Energy band diagram.

We have assumed that the mobile carriers are negligible in the depletion region. By integrating Eq. 28 once with the boundary conditions that the electric field is zero at $\pm W/2$, we obtain the electric-field distribution shown in Fig. 9b:

$$\mathscr{E}(x) = -\frac{qa}{\epsilon_s}\left[\frac{(W/2)^2 - x^2}{2}\right]. \tag{29}$$

The maximum field at $x = 0$ is

$$\mathscr{E}_m = \frac{qaW^2}{8\epsilon_s}. \tag{29a}$$

Integrating Eq. 28 once again yields both the potential distribution and the corresponding energy band diagram shown in Figs. 9c and 9d, respectively. The built-in potential and the depletion layer width are given by

$$V_{bi} = \frac{qaW^3}{12\epsilon_s} \tag{30}$$

and

$$W = \left[\frac{12\epsilon_s V_{bi}}{qa}\right]^{1/3}. \tag{31}$$

Since the values of the impurity concentrations at the edges of the depletion region ($-W/2$ and $W/2$) are the same and both are equal to $aW/2$, the built-in potential for a linearly graded junction may be expressed in a form similar to Eq. 12:[†]

$$V_{bi} = \frac{kT}{q}\ln\left[\frac{(aW/2)(aW/2)}{n_i^2}\right] = \frac{2kT}{q}\ln\left[\frac{aW}{2n_i}\right]. \tag{32}$$

Solving the transcendental equation that results when W is eliminated from Eqs. 31 and 32 yields the built-in potential as a function of a. The results for silicon and gallium arsenide linearly graded junctions are shown in Fig. 10.

When either forward or reverse bias is applied to the linearly graded junction, the variations of the depletion layer width and the energy band diagram will be similar to those shown in Fig. 8 for abrupt junctions. However, the depletion layer width will vary as $(V_{bi} - V)^{1/3}$, where V is positive for forward bias and negative for reverse bias; for abrupt junctions, W varies as $(V_{bi} - V)^{1/2}$.

[†]Based on an accurate numerical technique, the built-in potential is given by

$$V_{bi} = \frac{2}{3}\frac{kT}{q}\ln\left(\frac{a^2\epsilon_s kT/q}{8qn_i^3}\right).$$

For a given impurity gradient, the V_{bi} is smaller than that calculated from Eq. 32 by about 0.05 to 0.1 V.

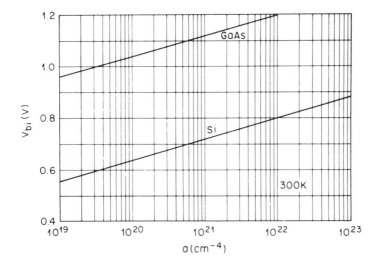

Fig. 10 Built-in potential for linearly graded junctions in Si and GaAs as a function of impurity gradient.

3.3 DEPLETION CAPACITANCE

The depletion capacitance per unit area is defined as $C_j = dQ/dV$, where dQ is the incremental change in depletion layer charge per unit area for an incremental change in the applied voltage dV.

Figure 11 illustrates the depletion capacitance of a $p-n$ junction with an arbitrary impurity distribution. The charge and electric-field distributions indicated by the solid lines correspond to a voltage V applied to the n-side. If this voltage is increased by an amount dV, the charge and field distributions will expand to those regions bounded by the dashed lines. In Fig. 11b, the incremental charge dQ corresponds to the hatched area between the two charge distribution curves on either side of the depletion region. The incremental space charges on the n- and p-sides of the depletion region are equal but with opposite charge polarity, thus maintaining overall charge neutrality. This incremental charge dQ causes an increase in the electric field by an amount $d\mathscr{E} = dQ/\epsilon_s$ (from Poisson's equation). The corresponding change in the applied voltage dV, represented by the hatched area in Fig. 11c is approximately $Wd\mathscr{E}$, which equals WdQ/ϵ_s. Therefore, the depletion capacitance per unit area is given by

$$C_j \equiv \frac{dQ}{dV} = \frac{dQ}{W\,\dfrac{dQ}{\epsilon_s}} = \frac{\epsilon_s}{W} \qquad F/cm^2 . \tag{33}$$

This equation for the depletion capacitance per unit area is the same as the

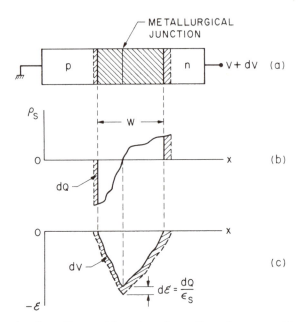

Fig. 11 (a) p–n junction with an arbitrary impurity profile under reverse bias. (b) Change in space charge distribution due to change in applied bias. (c) Corresponding change in electric–field distribution.

standard expression for a parallel-plate capacitor where the spacing between the two plates represents the depletion layer width. The equation is valid for any arbitrary impurity distribution.

In deriving Eq. 33 we have assumed that only the variation of the space charge in the depletion region contributes to the capacitance. This certainly is a good assumption for the reverse-bias condition. For forward biases, however, a large current can flow across the junction corresponding to a large number of mobile carriers present within the depletion region. The incremental change of these mobile carriers with respect to the biasing voltage contributes an additional term, called the diffusion capacitance, which will be considered in Section 3.5.

For a one-sided abrupt junction, we obtain, from Eqs. 27 and 33,

$$C_j = \frac{\epsilon_s}{W} = \sqrt{\frac{q\,\epsilon_s N_B}{2(V_{bi} - V)}} \tag{34}$$

or

$$\frac{1}{C_j^2} = \frac{2(V_{bi} - V)}{q\,\epsilon_s N_B}. \tag{35}$$

It is clear from Eq. 35 that a plot of $1/C_j^2$ versus V produces a straight line for a one-sided abrupt junction. The slope gives the impurity concentration N_B of the substrate, and the intercept (at $1/C_j^2 = 0$) gives V_{bi}.

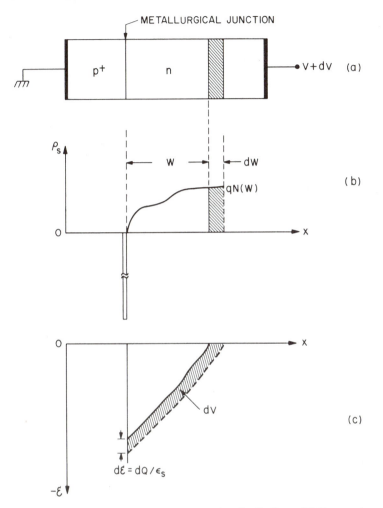

Fig. 12 (a) p^+-n junction with an arbitrary impurity distribution. (b) Change in space charge distribution in the lightly doped side due to change in applied bias. (c) Corresponding change in electric-field distribution.

The capacitance–voltage characteristics can be used to evaluate an arbitrary impurity distribution. We consider the case of a p^+-n junction with a doping profile on the n-side as shown in Fig. 12b. As before, the incremental change in depletion layer charge per unit area dQ for an incremental change in the applied voltage dV is given by $qN(W)\,dW$ (i.e., the shaded area in Fig. 12b). The corresponding change in applied voltage (shaded area in Fig. 12c) is

$$dV \simeq (d\mathscr{E})W = \left[\frac{dQ}{\epsilon_s}\right]W = \frac{qN(W)\,d(W^2)}{2\epsilon_s}. \tag{36}$$

By substituting W from Eq. 33, we obtain an expression for the impurity concentration at the edge of the depletion region:

$$N(W) = \frac{2}{q\,\epsilon_s} \left[\frac{1}{d(1/C_j^2)/dV} \right].$$

(37)

Thus, we can measure the capacitance per unit area versus reverse-bias voltage and plot $1/C_j^2$ versus V. The slope of the plot, that is, $d(1/C_j^2)/dV$, yields $N(W)$. Simultaneously, W is obtained from Eq. 33. A series of such calculations produces a complete impurity profile. This approach is referred to as the $C-V$ method for measuring impurity profiles.

For a linearly graded junction, the depletion layer capacitance is obtained from Eqs. 31 and 33:

$$C_j = \frac{\epsilon_s}{W} = \left[\frac{qa\,\epsilon_s^2}{12(V_{bi} - V)} \right]^{1/3} \quad \text{F/cm}^2.$$

(38)

For such a junction we can plot $1/C^3$ versus V and obtain the impurity gradient and V_{bi} from the slope and intercept.

Many circuit applications employ the voltage-variable properties of reverse-biased $p-n$ junctions. A $p-n$ junction designed for such a purpose is called a *varactor*, which is a shortened form of variable reactor. As previously derived, the reverse-biased depletion capacitance is given by

$$C_j \propto (V_{bi} + V_R)^{-n}$$

(39)

or

$$C_j \propto (V_R)^{-n} \quad \text{for} \quad V_R \gg V_{bi}$$

(39a)

where $n = \frac{1}{3}$ for a linearly graded junction and $n = \frac{1}{2}$ for an abrupt junction. Thus, the voltage sensitivity of C (i.e., variation of C with V_R) is greater for an abrupt junction than for a linearly graded junction. We can further increase the voltage sensitivity by using a hyperabrupt junction having an exponent n (Eq. 39) greater than $\frac{1}{2}$.

Figure 13 shows three p^+-n doping profiles with the donor distribution $N_D(x)$ given by $B(x/x_0)^m$, where B and x_0 are constants, $m = 1$ for a linearly graded junction, $m = 0$ for an abrupt junction, and $m = -\frac{3}{2}$ for a hyperabrupt junction. To obtain the capacitance–voltage relationship, we solve Poisson's equation:

$$\frac{d^2\psi}{dx^2} = -B\left[\frac{x}{x_0} \right]^m.$$

(40)

Integrating Eq. 40 twice with appropriate boundary conditions gives the dependence of the depletion layer width on the reverse bias:

$$W \propto (V_R)^{1/(m+2)}.$$

(41)

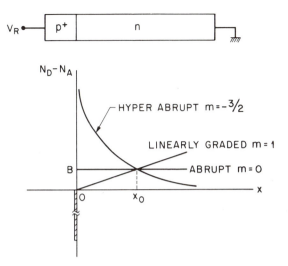

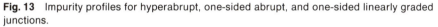

Fig. 13 Impurity profiles for hyperabrupt, one-sided abrupt, and one-sided linearly graded junctions.

Therefore,

$$C_j = \frac{\epsilon_s}{W} \propto (V_R)^{-1/(m+2)}. \qquad (42)$$

Comparing Eq. 42 with Eq. 39a yields $n = 1/(m + 2)$. For hyperabrupt junctions with $n > \frac{1}{2}$, m must be a negative number.

By choosing different values of m, we can obtain a wide variety of C_j-versus-V_R dependencies for specific applications. One interesting example, shown in Fig. 13, is the case for $m = -\frac{3}{2}$. For this case, $n = 2$. When this varactor is connected to an inductor L in a resonant circuit, the resonant frequency varies linearly with the voltage applied to the varactor:

$$\omega_r = \frac{1}{\sqrt{LC_j}} \propto \frac{1}{\sqrt{V_R^{-n}}} = V_R \qquad \text{for } n = 2. \qquad (43)$$

3.4 CURRENT—VOLTAGE CHARACTERISTICS

A voltage applied to a $p-n$ junction will disturb the precise balance between the diffusion current and drift current of electrons and holes. Under forward bias, the applied voltage reduces the electrostatic potential across the depletion region as shown in the middle of Fig. 14a. The drift current is reduced in comparison to the diffusion current. We have an enhanced hole diffusion from the p-side to the n-side and electron diffusion from the n-side to the p-side. Therefore, minority carrier injections occur, that is, electrons are injected into the p-side, while holes are injected into the n-side. Under reverse

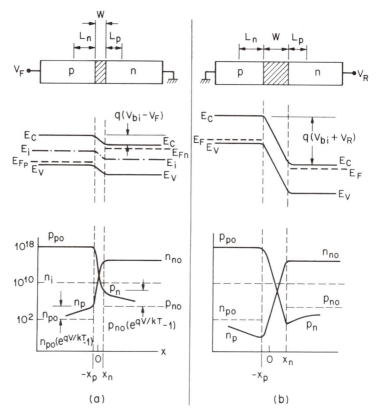

Fig. 14 Depletion region, energy band diagram, and carrier distribution. (a) Forward bias. (b) Reverse bias.

bias, the applied voltage increases the electrostatic potential across the depletion region as shown in the middle of Fig. 14b. This greatly reduces the diffusion currents, resulting in a small reverse current. In this section, we first consider the ideal current–voltage characteristics. We then discuss departures from these ideal characteristics due to generation and recombination and other effects.

3.4.1 Ideal Characteristics

We shall now derive the ideal current–voltage characteristics based on the following assumptions: (1) The abrupt depletion layer, that is, the depletion region, has abrupt boundaries, and outside the boundaries the semiconductor is assumed to be neutral. (2) The carrier densities at the boundaries are related by the electrostatic potential difference across the junction. (3) The low-injection condition, that is, the injected minority carrier densities are small compared to the majority carrier densities; in other words, the majority carrier densities are changed negligibly at the boundaries of neutral regions by the applied bias. (4) Neither generation nor recombination current exists in the depletion region, and the electron and hole currents are constant throughout the depletion region. Departures from these idealized assumptions will be considered in the next section.

At thermal equilibrium, the majority carrier density is essentially equal to the doping concentration. We shall use the subscripts n and p to denote the semiconductor type and the subscript o to specify the condition of thermal equilibrium. Hence, n_{no} and n_{po} are the equilibrium electron densities in the n- and p-sides, respectively. Similarly, p_{no} and p_{po} are the equilibrium hole densities in the n- and p-sides, respectively. The expression for the built-in potential in Eq. 12 can be rewritten as

$$V_{bi} = \frac{kT}{q} \ln \frac{p_{po} n_{no}}{n_i^2} = \frac{kT}{q} \ln \frac{n_{no}}{n_{po}} \tag{44}$$

where the mass action law $p_{po} n_{po} = n_i^2$ has been used. Rearranging Eq. 44 gives

$$n_{no} = n_{po} e^{q V_{bi}/kT} \tag{45}$$

Similarly, we have

$$p_{po} = p_{no} e^{q V_{bi}/kT} \tag{46}$$

We note from Eqs. 45 and 46 that the electron density and the hole density at the two boundaries of the depletion region are related through the electrostatic potential difference V_{bi} at thermal equilibrium. From our second assumption we expect that the same relation holds when the electrostatic potential difference is changed by an applied voltage.

When a forward bias is applied, the electrostatic potential difference is reduced to $V_{bi} - V_F$; but when a reverse bias is applied, the electrostatic potential difference is increased to $V_{bi} + V_R$. Thus, Eq. 45 is modified to

$$n_n = n_p e^{q(V_{bi} - V)/kT} \tag{47}$$

where n_n and n_p are the nonequilibrium electron densities at the boundaries of the depletion region in the n- and p-sides, respectively, with V positive for forward bias and negative for reverse bias. For the low-injection condition, the injected minority carrier density is much smaller than the majority carrier density; therefore, $n_n \simeq n_{no}$. Substituting this condition and Eq. 45 into Eq. 47 yields the electron density at the boundary of the depletion region on the p-side ($x = -x_p$):

$$n_p = n_{po} e^{q V/kT} \tag{48}$$

or

$$n_p - n_{po} = n_{po}(e^{q V/kT} - 1) . \tag{48a}$$

Similarly, we have

$$p_n = p_{no} e^{q V/kT} \tag{49}$$

or

$$p_n - p_{no} = p_{no}(e^{qV/kT} - 1) \tag{49a}$$

at $x = x_n$ for the n-type boundary. Figures 14a and 14b show band diagrams and carrier concentrations in a $p-n$ junction under forward-bias and reverse-bias conditions, respectively. Note that the minority carrier densities at the boundaries $(-x_p$ and $x_n)$ increase substantially above their equilibrium values under forward bias, while they decrease below their equilibrium values under reverse bias. Equations 48 and 49 define the minority carrier densities at the boundaries of the depletion region. These equations are the most important boundary conditions for the ideal current–voltage characteristics.

Under our idealized assumptions, no current is generated within the depletion region; all currents come from the neutral regions. In the neutral n-region, there is no electric field, thus the steady-state continuity equation reduces to

$$\frac{d^2 p_n}{dx^2} - \frac{p_n - p_{no}}{D_p \tau_p} = 0. \tag{50}$$

The solution of Eq. 50 with the boundary conditions of Eq. 49 and $p_n (x = \infty) = p_{no}$ gives

$$p_n - p_{no} = p_{no}(e^{qV/kT} - 1)e^{-(x-x_n)/L_p} \tag{51}$$

where L_p, which is equal to $\sqrt{D_p \tau_p}$, is the diffusion length of holes (minority carriers) in the n-region. At $x = x_n$,

$$J_p(x_n) = -qD_p \left. \frac{dp_n}{dx} \right|_{x_n} = \frac{qD_p p_{no}}{L_p} (e^{qV/kT} - 1). \tag{52}$$

Similarly, we obtain for the neutral p-region

$$n_p - n_{po} = n_{po}(e^{qV/kT} - 1)e^{(x+x_p)/L_n} \tag{53}$$

and

$$J_n(-x_p) = qD_n \left. \frac{dn_p}{dx} \right|_{-x_p} = \frac{qD_n n_{po}}{L_n} (e^{qV/kT} - 1) \tag{54}$$

where L_n, which is equal to $\sqrt{D_n \tau_n}$, is the diffusion length of electrons. The minority carrier densities (Eqs. 51 and 53) are shown in the middle of Fig. 15. The graphs illustrate that the injected minority carriers recombine with the majority carriers as the minority carriers move away from the boundaries. The electron and hole currents are shown at the bottom of Fig. 15. The hole and electron currents at the boundaries are given by Eqs. 52 and 54, respectively. The hole diffusion current will decay exponentially in the n-region with diffusion length L_p, and the electron diffusion current will decay exponentially in the p-region with diffusion length L_n.

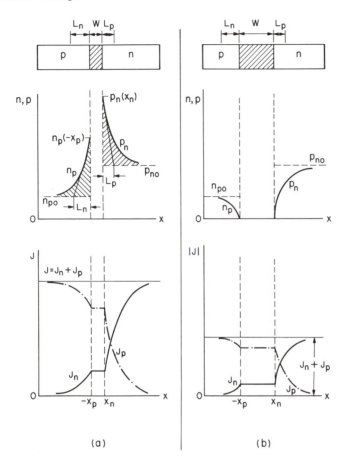

Fig. 15 Injected minority carrier distribution and electron and hole currents.[1] (a) Forward bias. (b) Reverse bias.

The total current is constant throughout the device and is the sum of Eqs. 52 and 54:

$$J = J_p(x_n) + J_n(-x_p) = J_s(e^{qV/kT} - 1) \qquad (55)$$

$$J_s \equiv \frac{qD_p p_{no}}{L_p} + \frac{qD_n n_{po}}{L_n} \qquad (55a)$$

where J_s is the saturation current density. Equation 55 is the *ideal diode equation*.[1] The ideal current–voltage characteristic is shown in Figs. 16a and 16b in the Cartesian and semilog plots, respectively. In the forward direction with positive bias on the p-side, for $V \geqslant 3kT/q$, the rate of current increase is constant as shown in Fig. 16b. At 300 K for every decade change of current, the voltage change for an ideal diode is 60 mV ($= 2.3kT/q$). In the reverse direction, the current density saturates at $-J_s$.

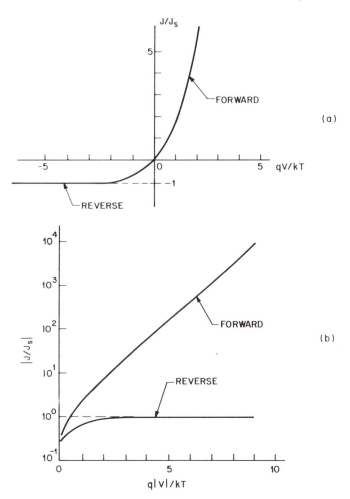

Fig. 16 Ideal current–voltage characteristics. (*a*) Cartesian plot. (*b*) Semilog plot.

3.4.2 Generation–Recombination and High-Injection Effects*

The ideal diode equation, Eq. 55, adequately describes the current–voltage characteristics of germanium $p-n$ junctions at low current densities. For silicon and gallium arsenide $p-n$ junctions, however, the ideal equation can only give qualitative agreement because of the generation or recombination of carriers in the depletion region.

Under the reverse-bias condition, carrier concentrations in the depletion region fall far below their equilibrium concentrations. The dominant generation–recombination processes discussed in Chapter 2 are those of elec-

---

* See footnote to Section 2.4.2.

tron and hole emissions through bandgap generation–recombination centers. The capture processes are not important because their rates are proportional to the concentration of free carriers, which is very small in the reverse-biased depletion region.

The two emission processes operate in the steady state by alternately emitting electrons and holes. The rate of electron–hole pair generation can be obtained from Eq. 63 of Chapter 2 with the conditions $p_n < n_i$ and $n_n < n_i$:

$$G = -U = \left[\frac{\sigma_p \sigma_n v_{th} N_t}{\sigma_n \exp\left[\dfrac{E_t - E_i}{kT}\right] + \sigma_p \exp\left[\dfrac{E_i - E_t}{kT}\right]} \right] n_i$$

$$\equiv \frac{n_i}{\tau_g} \tag{56}$$

where τ_g, the generation lifetime, is the reciprocal of the expression in the square brackets. We can arrive at an important conclusion about electron–hole generation from this expression. Let us consider a simple case where $\sigma_n = \sigma_p = \sigma_o$. For this case, Eq. 56 reduces to

$$G = \frac{\sigma_o v_{th} N_t n_i}{2 \cosh\left[\dfrac{E_t - E_i}{kT}\right]}. \tag{57}$$

The generation rate reaches a maximum value at $E_t = E_i$ and falls off exponentially as E_t moves in either direction away from the middle of the bandgap. Thus, only those centers with an energy level of E_t near the intrinsic Fermi level can contribute significantly to the generation rate.

The current due to generation in the depletion region is

$$J_{gen} = \int_0^W qG \, dx \simeq qGW = \frac{qn_i W}{\tau_g} \tag{58}$$

where W is the depletion layer width. The total reverse current for a p^+-n junction, that is, for $N_A \gg N_D$ and for $V_R > 3kT/q$, can be approximated by the sum of both the diffusion current in the neutral regions and the generation current in the depletion region:

$$J_R \simeq q \sqrt{\frac{D_p}{\tau_p}} \frac{n_i^2}{N_D} + \frac{qn_i W}{\tau_g}. \tag{59}$$

For semiconductors with large values of n_i, such as germanium, the diffusion current dominates at room temperature, and the reverse current follows the ideal diode equation. But if n_i is small, such as for silicon and gallium arsenide, the generation current in the depletion region may dominate.

At forward bias, the concentrations of both electrons and holes exceed their equilibrium values. The carriers will attempt to return to their equilibrium values by recombination. Therefore, the dominant generation–recombination processes in the depletion region are the capture processes. From Eq. 49 we obtain

$$p_n n_n \simeq p_{no} n_{no} e^{qV/kT} = n_i^2 e^{qV/kT} . \tag{60}$$

Substituting Eq. 60 in Eq. 66 of Chapter 2 and assuming $\sigma_n = \sigma_p = \sigma_o$ yields

$$U = \frac{\sigma_o v_{th} N_t n_i^2 (e^{qV/kT} - 1)}{n_n + p_n + 2n_i \cosh \dfrac{E_i - E_t}{kT}} . \tag{61}$$

The recombination rate has a broader peak than does the generation rate (refer to Fig. 15 of Chapter 2). However, in either recombination or generation, the most effective centers are those located near E_i. As practical examples, gold and copper yield effective generation-recombination centers in silicon where the values of $E_t - E_i$ are 0.02 eV for gold and -0.02 eV for copper. In gallium arsenide, chromium gives an effective center with an $E_t - E_i$ value of 0.08 eV.

Equation 61 can be simplified for the case $E_t = E_i$:

$$U = \sigma_o v_{th} N_t \frac{n_i^2 (e^{qV/kT} - 1)}{n_n + p_n + 2n_i} . \tag{62}$$

For a given forward bias, U reaches its maximum value at a location in the depletion region either where the denominator $n_n + p_n + 2n_i$ is a minimum or where the sum of the electron and hole concentrations, $n_n + p_n$, is at its minimum value. Since the product of these concentrations is a constant given by Eq. 60, the condition $d(p_n + n_n) = 0$ leads to

$$dp_n = -dn_n = \frac{p_n n_n}{p_n^2} dp_n \tag{63}$$

or

$$p_n = n_n \tag{64}$$

as the condition for this minimum. This condition exists at the location in the depletion region where E_i is halfway between E_{Fp} and E_{Fn}, as illustrated in the middle of Fig. 14a. Here, the carrier concentrations are

$$p_n = n_n = n_i e^{qV/2kT} \tag{65}$$

and therefore,

$$U_{max} = \sigma_o v_{th} N_t \frac{n_i^2 (e^{qV/kT} - 1)}{2n_i (e^{qV/2kT} + 1)} . \tag{66}$$

For $V \geqslant 3kT/q$,

$$U_{max} \simeq \frac{1}{2} \sigma_o v_{th} N_t n_i e^{qV/2kT} . \tag{67}$$

The recombination current is then

$$J_{rec} = \int_0^W qU \ dx \simeq \frac{qW}{2} \sigma_o v_{th} N_t n_i e^{qV/2kT} = \frac{qWn_i}{2\tau_r} e^{qV/2kT} \quad (68)$$

where τ_r is the effective recombination lifetime given by $1/\sigma_o v_{th} N_t$. The total forward current can be approximated by the sum of Eqs. 55 and 68, and for $p_{no} \gg n_{po}$ and $V \geqslant 3kT/q$ we have

$$J_F = q \ \sqrt{\frac{D_p}{\tau_p}} \frac{n_i^2}{N_D} e^{qV/kT} + \frac{qWn_i}{2\tau_r} e^{qV/2kT}. \quad (69)$$

In general, the experimental results can be represented empirically by

$$J_F \sim \exp\left[\frac{qV}{\eta kT}\right] \quad (70)$$

where the factor η is called the *ideality factor*. When the ideal diffusion current dominates, η equals 1; whereas when the recombination current dominates, η equals 2. When both currents are comparable, η has a value of between 1 and 2.

Figure 17 shows the measured forward characteristics of a silicon and gallium arsenide $p-n$ junction at room temperature.[2] At low current levels,

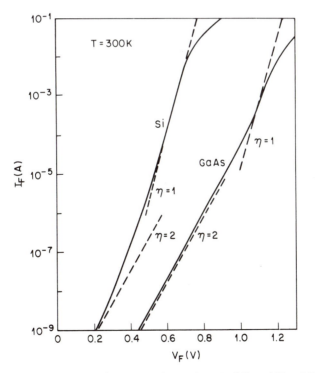

Fig. 17 Comparison of the forward current–voltage characteristics of Si and GaAs diodes[2] at 300 K. Dashed lines indicate slopes of different ideality factors η.

recombination current dominate and $\eta = 2$. At higher current levels, diffusion current dominates and η approaches 1.

At even higher current levels, we notice that the current departs from the ideal $\eta = 1$ situation and increases more gradually with forward voltage. This phenomenon is associated with two effects: series resistance and high injection. We shall first consider the series resistance effect. At both low- and medium-current levels, the IR drop across the neutral regions is usually small compared to kT/q (26 mV at 300 K), where I is the forward current and R is the series resistance. For example, for a silicon diode with $R = 1.5$ ohms, the IR drop at 1 mA is only 1.5 mV. However, at 100 mA, the IR drop becomes 0.15 V, which is six times larger than kT/q. This IR drop reduces the bias across the depletion region; therefore, the current becomes

$$I \simeq I_s \exp\left[\frac{q(V - IR)}{kT}\right] = \frac{I_s \exp(qV/kT)}{\exp\left[\frac{q(IR)}{kT}\right]} \tag{71}$$

and the ideal diffusion current is reduced by the factor $\exp[q(IR)/kT]$.

At high-current densities the injected minority carrier density is comparable to the majority concentration, that is, at the n-side of the junction $p_n(x = x_n) \simeq n_n$. This is the high-injection condition. By substituting the high injection condition in Eq. 60, we obtain $p_n(x = x_n) \simeq n_i \exp(qV/2kT)$. Using this as a boundary condition, the current becomes roughly proportional to $\exp(qV/2kT)$. Thus, the current increases at a slower rate under the high-injection condition.

3.4.3 Temperature Effect

Operating temperature has a profound effect on device performance. In both the forward-bias and reverse-bias conditions, the magnitudes of the diffusion and the recombination–generation currents depend strongly on temperature. We shall consider the forward-bias case first. The ratio of hole diffusion current to recombination current is given by

$$\frac{I_{\text{diffusion}}}{I_{\text{recombination}}} = 2\,\frac{n_i}{N_D}\,\frac{L_p}{W}\,\frac{\tau_r}{\tau_p}\,e^{qV/2kT} \sim \exp\left[-\frac{E_g - qV}{2kT}\right]. \tag{72}$$

This ratio depends on both the temperature and the semiconductor bandgap. Figure 18a shows the temperature dependence of the forward characteristics of a silicon diode. At room temperature for small forward voltages, the recombination current generally dominates, while at higher forward voltages the diffusion current usually dominates. At a given forward bias, as the temperature increases, the diffusion current will increase more rapidly than the recombination current. Therefore, the ideal diode equation will be followed over a wide range of forward biases as the temperature increases.

The temperature dependence of the saturation current density J_s (Eq. 55a) for a one-sided p^+–n junction in which diffusion current dominates is given

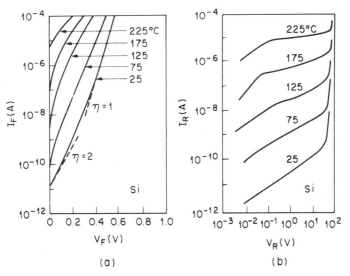

Fig. 18 Temperature dependence of the current–voltage characteristics of a Si diode.[2]
(a) Forward bias. (b) Reverse bias.

by

$$J_s \simeq \frac{q D_p p_{no}}{L_p} \sim n_i^2 \sim \exp\left(-\frac{E_g}{kT}\right). \tag{73}$$

Thus, the activation energy obtained from the slope of a plot of J_s versus $1/T$ corresponds to the energy bandgap E_g.

In the reverse-bias condition for a p^+-n junction, the ratio of the diffusion current to the generation current is

$$\frac{I_{\text{diffusion}}}{I_{\text{generation}}} = \frac{n_i L_p}{N_D W} \frac{\tau_g}{\tau_p}. \tag{74}$$

This ratio is proportional to the intrinsic carrier density n_i. As the temperature increases, the diffusion current eventually dominates. Figure 18b shows the effects of temperature on the reverse characteristics of a silicon diode. At low temperatures, the generation current dominates and the reverse current varies as $\sqrt{V_R}$ in accordance with Eq. 58 for an abrupt junction (i.e., $W \sim \sqrt{V_R}$). As the temperature increases beyond 175°C, the current demonstrates a saturation tendency for $V_R \geqslant 3kT/q$, at which point the diffusion current becomes dominant.

3.5 CHARGE STORAGE AND TRANSIENT BEHAVIOR

Under forward bias, electrons are injected from the n-region into the p-region and holes are injected from the p-region into the n-region. Once injected across the junction, the minority carriers recombine with the majority carriers and decay exponentially with distance as shown previously in Fig. 15a. These minority carrier distributions lead to current flow and to charge storage in the $p - n$ junction. We shall consider the stored charge, its effect on junction capacitance, and the transient behavior of the $p - n$ junction due to sudden changes of bias.

3.5.1 Minority Carrier Storage

The charge of injected minority carriers per unit area stored in the neutral n-region can be found by integrating the excess holes in the neutral region, shown as the shaded area in Fig. 15a, using Eq. 51:

$$Q_p = q \int_{x_n}^{\infty} (p_n - p_{no})\, dx$$

$$= q \int_{x_n}^{\infty} p_{no}(e^{qV/kT} - 1)e^{-(x - x_n)/L_p}\, dx$$

$$= qL_p p_{no}(e^{qV/kT} - 1). \tag{75}$$

A similar expression can be obtained for the stored electrons in the neutral p-region. The number of stored minority carriers depends upon both the diffusion length and the charge density at the boundary of the depletion region. We can express the stored charge in terms of the injected current. From Eqs. 52 and 75, we have

$$Q_p = \frac{L_p^2}{D_p} J_p(x_n) = \tau_p J_p(x_n). \tag{76}$$

Equation 76 states that the amount of stored charge is the product of the current and lifetime of the minority carriers. This is because the injected holes diffuse farther into the n-region before recombining if their lifetime is longer; thus, more holes are stored.

3.5.2 Diffusion Capacitance

The depletion layer capacitance considered previously accounts for most of the junction capacitance when the junction is reverse-biased. When the junction is forward-biased, there is an additional significant contribution to junction capacitance from the rearrangement of the stored charges in the neutral regions. This is called the *diffusion capacitance* denoted C_d, a term derived from the ideal-diode case in which minority carriers move across the neutral region by diffusion.

The diffusion capacitance of the stored holes in the neutral n-region is obtained by applying the definition $C_d = A dQ_p/dV$ to Eq. 75:[†]

$$C_d = \frac{Aq^2 L_p p_{no}}{kT} e^{qV/kT} \tag{77}$$

[†] A more accurate evaluation gives a diffusion capacitance half of that shown in Eq. 77 (see Ref. 8).

where A is the device cross-sectional area. We may add the contribution to C_d of the stored electrons in the neutral p-region in cases of significant storage. For a p^+-n junction, however, $n_{po} \ll p_{no}$, and the contribution to C_d of the stored electrons becomes insignificant. Under reverse bias (i.e., V is negative), Eq. 77 shows that C_d is inconsequential because of negligible minority carrier storage.

In many applications we prefer to represent a $p-n$ junction by an equivalent circuit. In addition to diffusion capacitance C_d and depletion capacitance C_j, we must include conductance to account for the current through the device. In the ideal diode the conductance can be obtained from Eq. 55:

$$ G = \frac{A\,dJ}{dV} = \frac{qA}{kT} J_s e^{qV/kT} = \frac{qA}{kT} (J + J_s) \simeq \frac{qI}{kT} \qquad (78) $$

The diode equivalent circuit is shown in Fig. 19, where C_j stands for the total depletion capacitance (i.e., the result in Eq. 33 times the device area A). For low-voltage, sinusoidal excitation of a diode that is biased quiescently (i.e., at dc) the circuit shown in Fig. 19 provides adequate accuracy. Therefore, we refer to it as the diode small-signal equivalent circuit.

3.5.3 Transient Behavior

For switching applications the forward-to-reverse-bias transition must be nearly abrupt and the transient time short. Figure 20a shows a simple circuit where a forward current I_F flows through a $p-n$ junction. At time $t = 0$, switch S is suddenly thrown to the right and an initial reverse current $I_R \simeq V/R$ flows. The transient time t_{off}, plotted in Fig. 20b, is the time required for the current to reach 10% of the initial reverse current I_R.

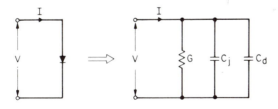

Fig. 19 Small-signal equivalent circuit of a $p-n$ junction.

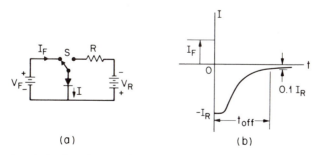

(a)   (b)

Fig. 20 Transient behavior of a $p-n$ junction. (a) Basic switching circuit. (b) Transient response of the current from forward bias to reverse bias.

The transient time may be estimated as follows. Under the forward-bias condition, the stored minority carriers in the n-region for a $p^+ - n$ junction is given by Eq. 76:

$$Q_p = \tau_p J_p = \tau_p \frac{I_F}{A} \tag{79}$$

where I_F is the total forward current and A is the device area. If the average current flowing during the turn-off period is $I_{R,\ ave}$, the turn-off time is the length of time required to remove the total stored charge Q_p:

$$t_{\text{off}} \simeq \frac{Q_p A}{I_{R,\ ave}} = \tau_p \left[\frac{I_F}{I_{R,\ ave}} \right]. \tag{80}$$

Thus, the turn-off time depends on both the ratio of forward to reverse currents and the lifetime of the minority carriers. The result of a more precise turn-off time calculation[3] accounting for the time-dependent minority carrier diffusion problem is shown in Fig. 21. For fast switching devices, we must reduce the lifetime of the minority carriers. Therefore, recombination–generation centers that have energy levels located near mid-bandgap, such as gold in silicon, are usually introduced.

3.6 JUNCTION BREAKDOWN

When a sufficiently large reverse voltage is applied to a $p - n$ junction, the junction breaks down and conducts a very large current. While the break-

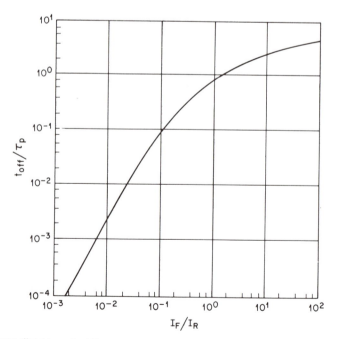

Fig. 21 Normalized transient time versus the ratio of forward current to reverse current.[3]

down process is not inherently destructive, the maximum current must be limited by an external circuit to avoid excessive junction heating. Two important breakdown mechanisms are the tunneling effect and avalanche multiplication. We shall consider the first mechanism briefly and then will discuss avalanche multiplication in detail, because avalanche breakdown imposes an upper limit on the reverse bias for most diodes. Avalanche breakdown also limits the collector voltage of a bipolar transistor (Chapter 4) and the drain voltage of a MOSFET (Chapter 5). In addition, the avalanche multiplication mechanisms can generate microwave power, as in an IMPATT diode (Chapter 6), and detect optical signals, as in an avalanche photodetector (Chapter 7).

3.6.1 Tunneling Effect

When a high electric field is applied to a $p-n$ junction in the reverse direction, a valence electron can make a transition from the valence band to the conduction band, as shown in Fig. 22a. This process, in which an electron penetrates through the energy bandgap, is called tunneling.

The tunneling process is considered in Section 6.1. Tunneling occurs only if the electric field is very high. The typical field for silicon and gallium arsenide is about 10^6 V/cm or higher. To achieve such a high field, the doping concentrations for both p- and n-regions must be quite high ($> 5 \times 10^{17}$ cm^{-3}). The breakdown mechanisms for silicon and gallium arsenide junctions with breakdown voltages of less than about $4E_g/q$, where E_g is the bandgap, are the result of the tunneling effect. For junctions with breakdown voltages in excess of $6E_g/q$, the breakdown mechanism is the result of avalanche multiplication. At voltages between 4 and $6E_g/q$, the breakdown is due to a mixture of both avalanche multiplication and tunneling.[4]

3.6.2 Avalanche Multiplication

The avalanche multiplication process is illustrated in Fig. 22b. The $p-n$ junction with moderate dopings, such as a p^+-n one-sided abrupt junction with a doping concentration of $N_D \simeq 10^{17}$ cm^{-3} or less, is under reverse bias. A thermally generated electron (designated by 1) gains kinetic energy from the

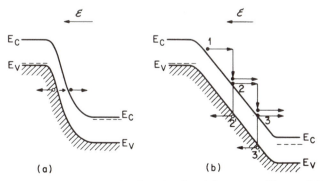

Fig. 22 Energy band diagrams under junction–breakdown conditions. (a) Tunneling effect. (b) Avalanche multiplication.

electric field. If the field is sufficiently high, the electron can gain enough kinetic energy that, upon collision with an atom, it can break the lattice bonds, creating an electron–hole pair (2 and 2'). This is called *impact ionization*. These newly created electron and hole both acquire kinetic energy from the field and create additional electron–hole pairs (e.g., 3 and 3'). These in turn continue the process, creating other electron–hole pairs. This process is therefore called *avalanche multiplication*.

To derive the breakdown condition, we assume that a current I_{no} is incident at the left-hand side of the depletion region of width W, as shown in Fig. 23. If the electric field in the depletion region is high enough to initiate the avalanche multiplication process, the electron current I_n will increase with distance through the depletion region to reach a value $M_n I_{no}$ at W, where M_n, the multiplication factor, is defined as

$$M_n \equiv \frac{I_n(W)}{I_{no}}.$$ (81)

Similarly, the hole current I_p increases from $x = W$ to $x = 0$. The total current $I = (I_p + I_n)$ is constant at steady state. The incremental electron current at x equals the number of electron–hole pairs generated per second in the distance dx:

$$d\left[\frac{I_n}{q}\right] = \left[\frac{I_n}{q}\right](\alpha_n\, dx) + \left[\frac{I_p}{q}\right](\alpha_p\, dx)$$ (82)

or

$$\frac{dI_n}{dx} + (\alpha_p - \alpha_n)I_n = \alpha_p I$$ (82a)

where α_n and α_p are the electron and hole ionization rates, respectively. If we use the simplified assumption that $\alpha_n = \alpha_p = \alpha$, the solution of Eq. 82a is

$$\frac{I_n(W) - I_n(0)}{I} = \int_0^W \alpha\, dx.$$ (83)

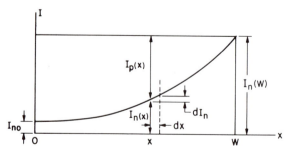

Fig. 23 Depletion region in a p–n junction with multiplication of an incident current.

From Eqs. 81 and 83, we have

$$1 - \frac{1}{M_n} = \int_0^W \alpha \, dx \; . \tag{83a}$$

The avalanche breakdown voltage is defined as the voltage where M_n approaches infinity. Hence, the breakdown condition is given by

$$\int_0^W \alpha \, dx = 1 \; . \tag{84}$$

From both the breakdown condition described above and the field dependence of the ionization rates, we may calculate the critical field (i.e., the maximum electric field at breakdown) at which the avalanche process takes place. Using measured α_n and α_p (Fig. 27 in Chapter 2) the critical field values $\mathscr{E}_c$ are calculated for silicon and gallium arsenide one-sided abrupt junctions and shown in Fig. 24 as functions of the impurity concentration of the substrate.[5] Also indicated is the critical field for the tunneling effect. It is evident that tunneling occurs only in semiconductors having high doping concentrations.

With the critical field determined, we may calculate the breakdown voltages. As discussed previously, voltages in the depletion region are determined from the solution of Poisson's equation:

$$V_B(\text{breakdown voltage}) = \frac{\mathscr{E}_c W}{2} = \frac{\epsilon_s \mathscr{E}_c^2}{2q} (N_B)^{-1} \tag{85}$$

for one-sided abrupt junctions and

$$V_B = \frac{2 \mathscr{E}_c W}{3} = \frac{4 \mathscr{E}_c^{3/2}}{3} \left[\frac{2\epsilon_s}{q} \right]^{1/2} (a)^{-1/2} \tag{86}$$

for linearly graded junctions, where N_B is the background doping of the lightly doped side, ϵ_s is the semiconductor permittivity, and a is the impurity

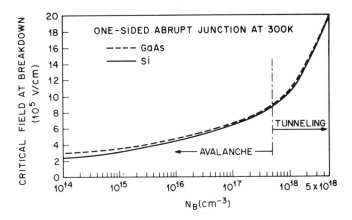

Fig. 24 Critical field at breakdown versus background doping for Si and GaAs one-sided abrupt junctions.[5]

gradient. Since the critical field is a slowly varying function of either N_B or a, the breakdown voltage, as a first order approximation, varies as N_B^{-1} for abrupt junctions and as $a^{-1/2}$ for linearly graded junctions.

Figure 25 shows the calculated breakdown voltages for silicon and gallium arsenide junctions.[5] The dash–dot line (to the right) at high dopings or high impurity gradients indicates the onset of the tunneling effect. Gallium arsenide has higher breakdown voltages than silicon for a given N_B or a, mainly because of its larger bandgap. The larger the bandgap, the larger the critical field must be for sufficient kinetic energy to be gained between collisions. As Eqs. 85 and 86 demonstrate, the larger critical field, in turn, gives rise to larger breakdown voltage.

The insert of Fig. 26 show a diffused junction with a linear gradient near the surface and a constant doping inside the semiconductor. The breakdown voltage lies between the two limiting cases of abrupt junction and linearly graded junction considered previously.[6] For large a and low N_B, the breakdown voltage of the diffused junctions is given by the abrupt junction results shown on the bottom line in Fig. 26, while for small a and high N_B, V_B is given by the linearly graded junction results indicated by the parallel lines in Fig. 26.

In Figs. 25 and 26 we assume that the semiconductor layer is thick enough to support the reverse-biased depletion layer width W_m at breakdown. If the

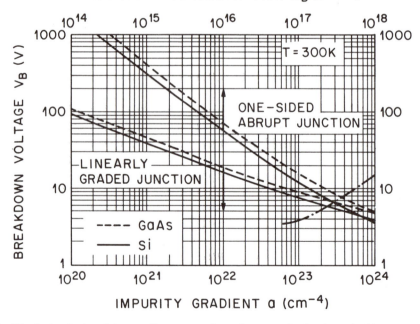

Fig. 25 Avalanche breakdown voltage versus impurity concentration for one-sided abrupt junctions and avalanche breakdown voltage versus impurity gradient for linearly graded junctions in Si and GaAs. Dash–dot line indicates the onset of the tunneling mechanism.[5]

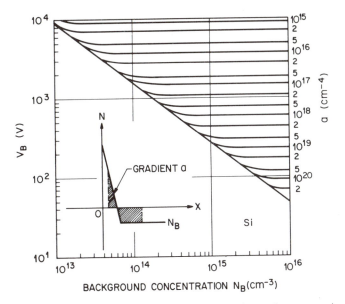

Fig. 26 Breakdown voltage for diffused junctions. Insert shows the space–charge distribution.[6]

semiconductor layer W is smaller than W_m, as shown in the insert of Fig. 27, the device will be punched through; that is, the depletion layer will reach the $n - n^+$ interface prior to breakdown. Increase the reverse bias further and the device will break down. The critical field $\mathscr{E}_c$ is essentially the same as that shown in Fig. 24. Therefore, the breakdown voltage V_B' for the punch-through diode is

$$\frac{V_B'}{V_B} = \frac{\text{shaded area in Fig. 27 insert}}{(\mathscr{E}_c W_m)/2}$$

$$= \left[\frac{W}{W_m}\right]\left[2 - \frac{W}{W_m}\right]. \tag{87}$$

Punch–through occurs when the doping concentration N_B becomes sufficiently low as in a $p^+ - \pi - n^+$ or $p^+ - \nu - n^+$ diode, where π stands for a lightly doped p-type and ν stands for a lightly doped n-type semiconductor. The breakdown voltages for such diodes calculated from Eqs. 85 and 87 are shown in Fig. 27. For a given thickness, the breakdown voltage approaches a constant value as the doping decreases.

Another important consideration of breakdown voltage is the junction curvature effect.[7] When a $p - n$ junction is formed by diffusion through a window in the insulating layer on a semiconductor, the impurities will diffuse downward and sideways (refer to Chapter 10). Hence, the junction has a plane (or flat) region with nearly cylindrical edges, as shown in Fig. 28a. If the diffusion mask contains sharp corners, the corner of the junction will acquire the

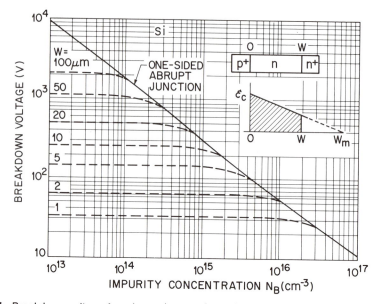

Fig. 27 Breakdown voltage for p^+–π–n^+ and p^+–ν–n^+ junctions. W is the thickness of the lightly doped p-type (π) or the lightly doped n-type (ν) region.

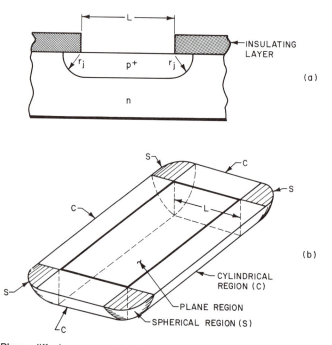

Fig. 28 (a) Planar diffusion process that forms junction curvature near the edge of the diffusion mask, where r_j is the radius of curvature. (b) Formation of cylindrical and spherical regions by diffusion through a rectangular mask.

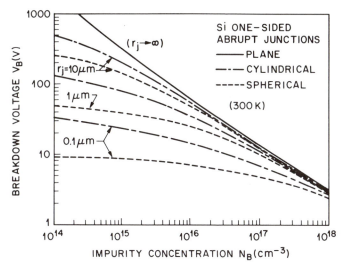

Fig. 29 Breakdown voltage versus impurity concentration for one-sided abrupt doping profile with cylindrical and spherical junction geometries,[7] where r_j is the radius of curvature as indicated in Fig. 28.

roughly spherical shape shown in Fig. 28*b*. Because the spherical or cylindrical regions of the junction have a higher field intensity, they determine the avalanche breakdown voltage. The calculated results for silicon one-sided abrupt junctions are shown in Fig. 29. The solid line represents the plane junctions considered previously. Note that as the junction radius r_j becomes smaller, the breakdown voltage decreases dramatically, especially for spherical junctions at low impurity concentrations.

REFERENCES

1 W. Shockley, *Electrons and Holes in Semiconductors*, Van Nostrand, Princeton, 1950.

2 A. S. Grove, *Physics and Technology of Semiconductor Devices*, Wiley, New York, 1967.

3 R. H. Kingston, "Switching Time in Junction Diodes and Junction Transistors," *Proc. IRE*, **42**, 829 (1954).

4 J. L. Moll, *Physics of Semiconductors*, McGraw-Hill, New York, 1964.

5 S. M. Sze and G. Gibbons, "Avalanche Breakdown Voltages of Abrupt and Linearly Graded *p–n* Junctions in Ge, Si, GaAs and GaP," *Appl. Phys. Lett.*, **8**, 111 (1966).

6 S. K. Ghandhi, *Semiconductor Power Devices*, Wiley, New York, 1977.

7 S. M. Sze and G. Gibbons, "Effect of Junction Curvature on Breakdown Voltages in Semiconductors," *Solid State Electron.*, **9**, 831 (1966).

8 S. M. Sze, *Physics of Semiconductor Devices*, 2nd ed., Wiley, New York, 1981.

PROBLEMS

1 An abrupt $p-n$ junction has a doping concentration of 10^{15}, 10^{16}, or 10^{17} cm^{-3} on the lightly doped n-side and of 10^{19} cm^{-3} on the heavily doped p-side. Obtain series of curves of $1/C^2$ versus V, where V ranges from -4 V to 0 V in steps of 0.5 V. Comment on the slopes and the interceptions at the voltage axis of these curves.

2 A diffused silicon $p-n$ junction has a linearly-graded junction on the p-side with $a = 10^{19}$ cm^{-4}, and a uniform doping of 3×10^{14} cm^{-3} on the n-side. (*a*) If the depletion layer width of the p-side is 0.8 μm at zero bias, find the total depletion layer width, built-in potential, and maximum field at zero bias. (*b*) Plot the potential distribution.

3 For an ideal silicon $p-n$ abrupt junction with $N_A = 10^{17}$ cm^{-3} and $N_D = 10^{15}$ cm^{-3}, (*a*) calculate V_{bi} at 250, 300, 350, 400, 450 and 500 K and plot V_{bi} versus T. (Hint: Use Fig. 16 of Chapter 1). (*b*) Comment on your result in terms of energy band diagram. (*c*) Find the depletion layer width and the maximum field at zero bias for $T = 300$ K.

4 Assume that the $p-n$ junction considered in Problem 3 contains 10^{15} cm^{-3} generation–recombination centers located 0.02 eV above the intrinsic Fermi level of silicon with $\sigma_n = \sigma_p = 10^{-15}$ cm^2. If $v_{th} \simeq 10^7$ cm/s, calculate the generation and recombination current at ± 0.5 V and determine the total forward and reverse current for a junction area of 10^{-4} cm^2.

5 An ideal silicon $p - n$ junction has $N_D = 10^{18}$ cm^{-3}, $N_A = 10^{16}$ cm^{-3}, $\tau_p = \tau_n = 10^{-6}$ s, and a device area of 1.2×10^{-5} cm^2. (*a*) Calculate the theoretical saturation current at 300 K. (*b*) Calculate the forward and reverse currents at ± 0.7 V.

6 (*a*) A silicon p^+-n junction has the following parameters at 300 K: $\tau_p = \tau_g = 10^{-6}$ s, $N_D = 10^{15}$ cm^{-3}. Plot diffusion current density, J gen, and total current density versus applied reverse voltage on log–log graph paper. (*b*) Repeat the above results for $N_D = 10^{17}$ cm^{-3}.

7 For an ideal abrupt silicon p^+-n junction with $N_D = 10^{16}$ cm^{-3}, find the stored minority carriers per unit area in the neutral n-region when a forward bias of 1 V is applied. The length of neutral region is 1 μm and the diffusion length of the holes is 5 μm.

8 For a silicon p^+-n one-sided abrupt junction with $N_D = 10^{15}$ cm^{-3}, find the depletion layer width at breakdown. If the n-region is reduced to 5 μm, calculate the breakdown voltage and compare your result with Fig. 27.

9 In Fig. 18b, the avalanche breakdown voltage increases with increasing temperature. Give a qualitative argument for this result.

10 If $\alpha_n = \alpha_p = 10^4(\mathscr{E}/4 \times 10^5)^6$ cm^{-1} in gallium arsenide, where $\mathscr{E}$ is in V/cm, find the breakdown voltage of (*a*) a $p-i-n$ diode with an intrinsic-layer width of 10 μm; (*b*) p^+-n junction with a doping of 2×10^{16} cm^{-3} for the lightly doped side.

4

Bipolar Devices

Bipolar devices are semiconductor devices in which both electrons and holes participate in the conduction process. This is in contrast to the "unipolar devices," to be discussed in Chapter 5, in which only one kind of carrier predominantly participates.

The bipolar transistor (contraction for *transfer resistor*), one of the most important semiconductor devices, was invented[1] by a research team at Bell Laboratories in 1947. Figure 1 shows the first transistor, a "point contact" type, which has two metal wires with sharp points making contact with a germanium substrate. The first transistor was primitive by today's standards, yet it revolutionized the electronics industry and changed our way of life.

For modern bipolar transistors, we have replaced the germanium with silicon (refer to Section 1.1) and replaced the point contacts with two closely cou-

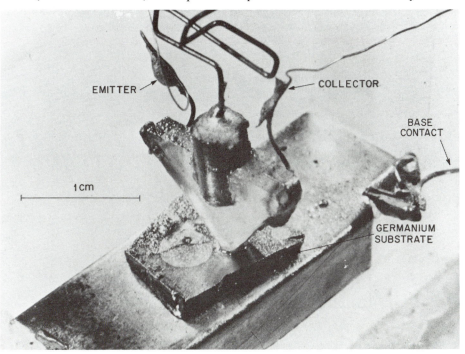

Fig. 1 The first transistor.[1]

pled $p–n$ junctions in the form of $p–n–p$ or $n–p–n$ structures. We shall consider the transistor action of the coupled junctions and derive the static characteristics from the carrier distributions. We shall also discuss the frequency response and switching behavior of the transistor and consider briefly the heterojunction bipolar transistor in which one or both of the $p–n$ junctions are formed between dissimilar semiconductors.

A related bipolar device, the thyristor, has three closely coupled $p–n$ junctions in the form of a $p–n–p–n$ structure.[2] This device exhibits bistable characteristics and can be switched between a high-impedance "off" state and a low-impedance "on" state. The name thyristor is derived from *gas thyratron,* which is a gas-filled tube with similar bistable characteristics. Because of the two stable states (on and off) and the low power dissipation in these states, thyristors are useful in applications ranging from light dimmers and speed control in home appliances to switching and power inversion in high-voltage transmission lines. We shall consider the physical operation of the thyristor and a few related switching devices.

4.1 THE TRANSISTOR ACTION

A perspective view of a silicon $p–n–p$ bipolar transistor is shown in Fig. 2a. Basically, the transistor is fabricated by first forming an n-type region in the p-type substrate; subsequently, a p^+-region is formed in the n–region. Metallic contacts are made to the p^+- and n-regions through the windows opened in

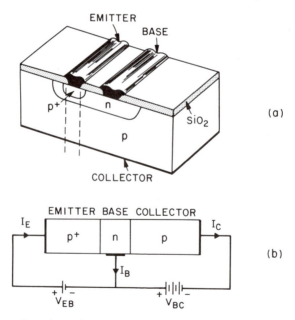

Fig. 2 (a) Perspective view of a silicon $p–n–p$ bipolar transistor. (b) Idealized one-dimensional transistor.

the oxide layer and to the *p*-region at the bottom. The details of transistor fabrication processes will be considered in later chapters.

An idealized, one-dimensional structure of the bipolar transistor, shown in Fig. 2*b*, can be considered as a section of the transistor along the dashed lines in Fig. 2*a*. The heavily doped p^+-region is called the *emitter*, the narrow central region is called the *base*, and the lightly doped *p*-region is called the *collector*. The doping concentration in each region is assumed to be uniform. The arrows in Fig. 2*b* indicate the directions of current flow under normal operating conditions (also called the *active mode*); that is, the emitter–base junction is forward-biased and the collector–base junction is reverse-biased. According to Kirchhoff's circuit laws, there are only two independent currents for this three-terminal device. If two currents are known, the third current is also known.

The complemental structure of the *p-n-p* transistor is the *n–p–n* transistor, which is obtained by interchanging *p* for *n* and *n* for *p* in Fig. 2. The current flow and voltage polarity are all reversed. However, we shall concentrate on the *p–n–p* type, because it provides a more intuitive base for understanding the carrier flow. Once we understand the *p–n–p* transistor, we need only to reverse the polarities and conduction types to describe the *n–p–n* transistor.

4.1.1 Operation in the Active Mode

Figure 3*a* shows the idealized *p–n–p* transistor in thermal equilibrium, that is, where all three leads are connected together or all are grounded. Figure 3*b* shows the impurity densities in the three doped regions, where the emitter is more heavily doped than the collector, while the base doping is less than the emitter doping, but greater than the collector doping. Figure 3*c* shows the corresponding electric-field profiles in the two depletion regions.

Figure 3*d* illustrates the energy band diagram, which is a simple extension of the thermal-equilibrium situation for the *p–n* junction as applied to a pair of closely coupled p^+–*n* and *n–p* junctions. The results obtained for the *p–n* junction in Chapter 3 are equally applicable to the emitter–base and base–collector junctions. At thermal equilibrium there is no net current flow, hence the Fermi level is a constant.

Figure 4 illustrates the corresponding situations when the transistor in Fig. 3 is biased in the active mode. Figure 4*a* is a schematic of the transistor connected as an amplifier with the *common-base configuration*, that is, the base lead is common to the input and output circuits.[3] Figures 4*b* and 4*c* show the charge densities and the electric fields under biasing conditions, respectively. Note that the depletion layer width of the emitter–base junction is narrower and that of the collector–base junction is wider, as compared to the equilibrium situation shown in Fig. 3.

Figure 4*d* shows the corresponding energy band diagram under the active mode. Since the emitter–base junction is forward-biased, holes are injected (or emitted) from the p^+ emitter into the base, and electrons are injected from the *n* base into the emitter. Under the ideal-diode condition, there is no

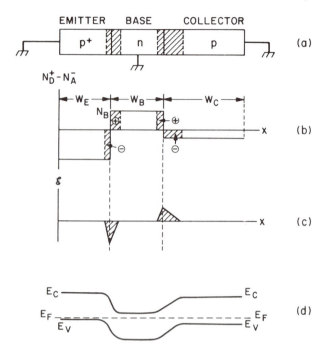

Fig. 3 (a) A p-n-p transistor with all leads grounded. (b) Doping profile of a transistor with abrupt impurity distributions. The crosshatched areas are the depletion regions. (c) Electric-field profile. (d) Energy band diagram at thermal equilibrium.

generation–recombination current in the depletion region; these two current components constitute the total emitter current. The collector–base junction is reverse-biased, and a small reverse saturation current will flow across the junction. However, if the base width is sufficiently narrow, the holes injected from the emitter can diffuse through the base to reach the base–collector depletion edge and then "float up" into the collector (recall the "bubble analogy"). This transport mechanism gives rise to the terminology of *emitter*, which emits or injects carriers, and of *collector*, which collects these carriers injected from a nearby junction. If most of the injected holes can reach the collector without recombining with electrons in the base region, then the collector hole current will be very close to the emitter hole current.

Therefore, carriers injected from a nearby emitter junction can result in a large current flow in a reverse-biased collector junction. This is the *transistor action*, and it can be realized only when the two junctions are physically close enough to interact in the manner described. If, on the other hand, the two junctions are so far apart that all the injected holes are recombined in the base before reaching the base–collector junction, then the transistor action is lost and the *p–n–p* structure becomes merely two diodes connected back to back.

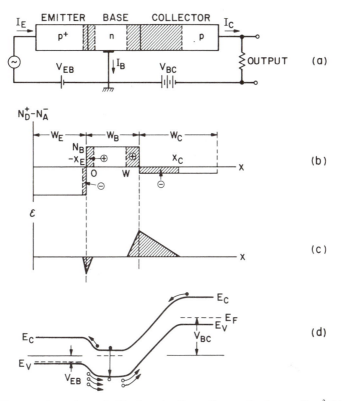

Fig. 4 (a) The transistor shown in Fig. 3 under the active mode of operation.[3] (b) Doping profiles and the depletion regions under biasing conditions. (c) Electric–field profile. (d) Energy band diagram.

4.1.2 Current Gain

Figure 5 shows the various current components in an ideal p–n–p transistor biased in the active mode. (No generation–recombination currents in the depletion regions are included; we shall consider them later.) The holes injected from the emitter constitute the current I_{Ep}, which is the largest current component in a well-designed transistor. Most of the injected holes will reach the collector junction and give rise to the current I_{Cp}. There are three base current components, which are labeled I_{En}, I_{BB}, and I_{Cn}. I_{En} corresponds to the current arising from electrons being injected from the base to the emitter. However, I_{En} is not desirable, as will be shown later; it can be minimized by using heavier emitter doping (Section 4.2) or a heterojunction (Section 4.4). I_{BB} corresponds to electrons that must be supplied by the base to replace electrons recombined with the injected holes (i.e., $I_{BB} = I_{Ep} - I_{Cp}$). I_{Cn} corresponds to thermally generated electrons that are near the collector–base junction edge and drift from the collector to the base. As indicated in the

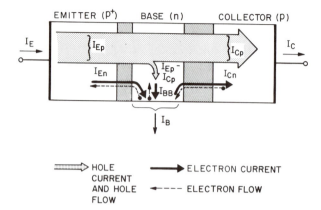

Fig. 5 Various current components in a p–n–p transistor under the active mode of operation. The electron flow is in the opposite direction to the electron current.

figure, the direction of the electron current is opposite to the direction of the electron flow.

We can now express the terminal currents in terms of the various current components described above:

$$I_E = I_{Ep} + I_{En} \tag{1}$$

$$I_C = I_{Cp} + I_{Cn} \tag{2}$$

$$I_B = I_E - I_C = I_{En} + (I_{Ep} - I_{Cp}) - I_{Cn} . \tag{3}$$

An important parameter in the characterization of bipolar transistors is the *common-base current gain* α_0. This quantity is defined as

$$\alpha_0 \equiv \frac{I_{Cp}}{I_E} . \tag{4}$$

Substituting Eq. 1 into Eq. 4 yields

$$\alpha_0 = \frac{I_{Cp}}{I_{Ep} + I_{En}} = \left[\frac{I_{Ep}}{I_{Ep} + I_{En}} \right] \left[\frac{I_{Cp}}{I_{Ep}} \right] . \tag{5}$$

The first term on the right-hand side is called the *emitter efficiency* γ, which measures the injected hole current compared to the total emitter current:

$$\gamma \equiv \frac{I_{Ep}}{I_E} = \frac{I_{Ep}}{I_{Ep} + I_{En}} . \tag{6}$$

The second term is called the *base transport factor* α_T, which is the ratio of the hole current reaching the collector to the hole current injected from the emitter:

$$\alpha_T \equiv \frac{I_{Cp}}{I_{Ep}} . \tag{7}$$

Therefore,

$$\alpha_0 = \gamma\alpha_T . \tag{8}$$

For a well-designed transistor, both γ and α_T approach unity, and α_0 is very close to 1.

We can express the collector current in terms of α_0. The collector current can be described by substituting Eqs. 6 and 7 into Eq. 2:

$$I_C = I_{Cp} + I_{Cn} = \alpha_T I_{Ep} + I_{Cn} = \gamma\alpha_T \left[\frac{I_{Ep}}{\gamma} \right] + I_{Cn} = \alpha_0 I_E + I_{Cn} . \tag{9}$$

where I_{Cn} corresponds to the collector–base current flowing with the emitter open-circuited ($I_E = 0$). We shall designate I_{Cn} as I_{CBO}, where the first two subscripts (CB) refer to the two terminals between which the current (or voltage) is measured and the third subscript (O) refers to the state of the third terminal with respect to the second. In the present case, I_{CBO} designates the leakage current between the collector and the base with the emitter–base junction open. The collector current for the common–base configuration is then given by

$$I_C = \alpha_0 I_E + I_{CBO} . \tag{10}$$

We have now developed a set of expressions for the p–n–p transistor operated in the active mode. In the next section, we shall study the static current–voltage characteristics and derive equations for the terminal currents in terms of such semiconductor parameters as doping and minority carrier lifetime.

4.2 STATIC CHARACTERISTICS OF BIPOLAR TRANSISTORS

4.2.1 Ideal Transistor Currents

To derive the current–voltage expressions for an ideal transistor, we assume the following:

(1) The device has uniform doping in each region.
(2) There is low-level injection.
(3) There are no generation–recombination currents in the depletion regions.
(4) There are no series resistances in the device.

These assumptions can be used to obtain the current–voltage relationship of the emitter–base junction and the collector–base junction in a manner similar to that for an ideal diode, as discussed in Chapter 3.

Active Mode Figure 4c shows the electric-field distributions across the junction depletion regions. The minority carrier distribution in the neutral base region can be described by the field-free steady-state continuity equation

$$D_p \left[\frac{d^2 p_n}{dx^2} \right] - \frac{p_n - p_{no}}{\tau_p} = 0 \tag{11}$$

where D_p and τ_p are the diffusion constant and the lifetime of minority carriers, respectively. The general solution of Eq. 11 is

$$p_n(x) = p_{no} + C_1 e^{x/L_p} + C_2 e^{-x/L_p} \tag{12}$$

where $L_p = \sqrt{D_p \tau_p}$ is the diffusion length of holes and C_1 and C_2 are constants to be determined by the boundary conditions for the active mode:

$$p_n(0) = p_{no} e^{qV_{EB}/kT} \tag{13a}$$

and

$$p_n(W) = 0 \tag{13b}$$

where p_{no} is the equilibrium minority carrier concentration in the base, given by $p_{no} = n_i^2/N_B$, and N_B denotes the uniform donor concentration in the base. The first boundary condition (Eq. 13a) states that under forward bias the minority carrier concentration at the edge of the emitter–base depletion region ($x = 0$) is increased above the equilibrium value by the exponential factor $e^{qV_{EB}/kT}$. The second boundary condition (Eq. 13b) states that under reverse bias the minority carrier concentration at the edge of the base–collector depletion region ($x = W$) is zero.

Equation 12, subjected to the boundary conditions expressed in Eq. 13, becomes

$$p_n(x) = p_{no}(e^{qV_{EB}/kT} - 1) \left[\frac{\sinh\left[\dfrac{W-x}{L_p}\right]}{\sinh\left[\dfrac{W}{L_p}\right]} \right] + p_{no} \left[1 - \frac{\sinh\left[\dfrac{x}{L_p}\right]}{\sinh\left[\dfrac{W}{L_p}\right]} \right]. \tag{14}$$

Figure 6 shows the calculated results of Eq. 14, where the normalized minority carrier concentration is plotted against the distance for different values of W/L_p. It is evident that for $W/L_p \gg 1$, the distribution approaches the simple exponential distribution of an isolated p–n junction. In the other extreme, where $W/L_p \ll 1$, the distribution approaches a straight line given by

$$p_n(x) = p_n(0)\left[1 - \frac{x}{W}\right] = p_{no} e^{qV_{EB}/kT}\left[1 - \frac{x}{W}\right]. \tag{15}$$

The minority carrier distributions in a typical transistor operated under active mode are shown in Fig. 7, where the central base region corresponds to that shown in Fig. 6 for $W/L_p \lesssim 0.1$. The distributions in the emitter and collector can be obtained in a manner similar to the one used to obtain the distributions for the base region. The boundary conditions are

$$n_E (x = -x_E) = n_{EO} e^{qV_{EB}/kT} \tag{16}$$

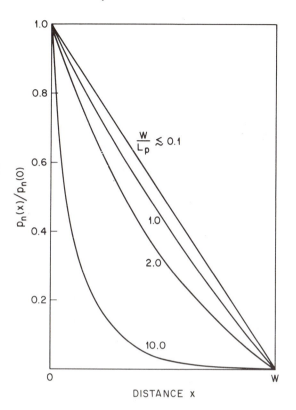

Fig. 6   Minority carrier distribution in the base region for different values of W/L_p. For $W/L_p \lesssim 0.1$, the distribution approaches a straight line.

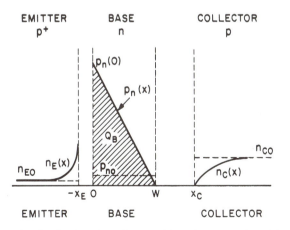

Fig. 7   Minority carrier distributions in various regions of a p–n–p transistor under active mode of operation.

and

$$n_C(x = x_C) = n_{CO}e^{-q|V_{CB}|/kT} = 0 \tag{17}$$

where n_{EO} and n_{CO} are the equilibrium electron concentrations in the emitter and collector, respectively. We assume that the emitter depth W_E and the collector depth W_C are much larger than their corresponding diffusion lengths L_E and L_C, respectively. Substituting these boundary conditions into an expression similar to Eq. 12 yields

$$n_E(x) = n_{EO} + n_{EO}(e^{qV_{EB}/kT} - 1) \exp\left[\frac{x + x_E}{L_E}\right] \quad x \leqslant -x_E \tag{18}$$

$$n_C(x) = n_{CO} - n_{CO} \exp\left[-\frac{x - x_C}{L_C}\right] \quad x \geqslant x_C \tag{19}$$

which are shown in Fig. 7. In this illustration, the stored minority charges in the base are of particular interest. Each carrier injected into the base carries a charge q. The total excess minority carrier charge in the base region Q_B is given by

$$Q_B = qA \int_0^W [p_n(x) - p_{no}] dx \tag{20}$$

where A is the device area. For $p_n(0) \gg p_{no}$, Eq. 20 can be approximated by the crosshatched area of the triangle, with 0 to W as the base width and $p_n(0)$ as the amplitude, and Q_B is given by

$$Q_B \simeq \frac{qA W p_n(0)}{2}. \tag{20a}$$

Once the minority carrier distributions are known, the various current components shown in Fig. 5 can be calculated. The hole current I_{Ep}, injected from the emitter at $x = 0$, is proportional to the gradient of the minority carrier concentration and is expressed by

$$I_{Ep} = A\left(-qD_p \frac{dp_n}{dx}\bigg|_{x=0}\right)$$
$$= qA\frac{D_p p_{no}}{L_p} \coth\left[\frac{W}{L_p}\right]\left[(e^{qV_{EB}/kT} - 1) + \frac{1}{\cosh(W/L_p)}\right] \tag{21}$$

or by using the expansion[†] of hyperbolic functions for $W/L_p \ll 1$ and replacing p_{no} by n_i^2/N_B, I_{Ep} can be given by

$$I_{Ep} \simeq \frac{qAD_p n_i^2}{N_B W}(e^{qV_{EB}/kT} - 1) + \frac{qAD_p n_i^2}{N_B W}. \tag{22}$$

[†] $\coth y \simeq 1/y$, $\sinh y \simeq y$, and $\operatorname{sech} y = 1 - y^2/2$ for $y \ll 1$.

The hole current collected by the collector at $x = W$ is

$$I_{Cp} = A \left(-qD_p \frac{dp_n}{dx} \bigg|_{x=W} \right)$$

$$= \frac{qAD_p p_{no}}{L_p} \frac{1}{\sinh\left(\dfrac{W}{L_p}\right)} \left[(e^{qV_{EB}/kT} - 1) + \cosh\left(\frac{W}{L_p}\right) \right] \qquad (23)$$

or

$$I_{Cp} \simeq \frac{qAD_p n_i^2}{N_B W} (e^{qV_{EB}/kT} - 1) + \frac{qAD_p n_i^2}{N_B W} \qquad \text{for} \quad \frac{W}{L_p} \ll 1 . \qquad (24)$$

The electron current I_{En}, which is caused by electron flow from the base to the emitter, and I_{Cn}, which is caused by electron flow from the collector to the base, are

$$I_{En} = A \left(-qD_E \frac{dn_E}{dx} \bigg|_{x=-x_E} \right) = \frac{qAD_E n_{EO}}{L_E} (e^{qV_{EB}/kT} - 1) \qquad (25)$$

$$I_{Cn} = A \left(-qD_C \frac{dn_C}{dx} \bigg|_{x=x_C} \right) = \frac{qAD_C n_{CO}}{L_C} \qquad (26)$$

where D_E and D_C are the diffusion constants in the emitter and collector, respectively.

The terminal currents can now be obtained from these equations. The emitter current is the sum of Eqs. 21 and 25:

$$I_E = a_{11}(e^{qV_{EB}/kT} - 1) + a_{12} \qquad (27)$$

where

$$a_{11} \equiv qA \left[\frac{D_p p_{no}}{L_p} \coth\left(\frac{W}{L_p}\right) + \frac{D_E n_{EO}}{L_E} \right]$$

$$\simeq qA \left(\frac{D_p n_i^2}{N_B W} + \frac{D_E n_{EO}}{L_E} \right) \qquad \text{for} \quad \frac{W}{L_p} \ll 1 \qquad (28)$$

$$a_{12} \equiv qA \frac{D_p p_{no}}{L_p} \frac{1}{\sinh\left(\dfrac{W}{L_p}\right)}$$

$$\simeq \frac{qAD_p n_i^2}{N_B W} \qquad \text{for} \quad \frac{W}{L_p} \ll 1 . \qquad (29)$$

The collector current is the sum of Eqs. 23 and 26:

$$I_C = a_{21}(e^{qV_{EB}/kT} - 1) + a_{22} \tag{30}$$

where

$$a_{21} \equiv \frac{qAD_p p_{no}}{L_p} \frac{1}{\sinh\left[\dfrac{W}{L_p}\right]}$$

$$\simeq \frac{qAD_p n_i^2}{N_B W} \quad \text{for} \quad \frac{W}{L_p} \ll 1 \tag{31}$$

$$a_{22} \equiv \frac{qAD_p p_{no}}{L_p} \coth\left[\frac{W}{L_p}\right] + \frac{qAD_C n_{CO}}{L_C}$$

$$\simeq qA\left[\frac{D_p n_i^2}{N_B W} + \frac{D_C n_{CO}}{L_C}\right] \quad \text{for} \quad \frac{W}{L_p} \ll 1 . \tag{32}$$

Note that $a_{12} = a_{21}$. The base current for the ideal transistor is obtained by subtracting Eq. 30 from Eq. 27:

$$I_B = (a_{11} - a_{21})(e^{qV_{EB}/kT} - 1) + (a_{12} - a_{22}) . \tag{33}$$

Using Eqs. 20a and 30, we can express the I_C in terms of Q_B, the total minority carrier charge stored in the base:

$$I_C \simeq \frac{qAD_p p_n(0)}{W} = \left[\frac{2D_p}{W^2}\right]Q_B . \tag{34}$$

Therefore, the collector current is directly proportional to the minority carrier charge stored in the base.

From these discussions, we see that the currents in the three terminals of a transistor are related by the minority carrier distribution in the base region. For a well-designed transistor, the expressions for the static emitter and collector current, Eqs. 27 and 30, reduce to terms proportional to the minority carrier gradient (dp/dx) at $x = 0$ and $x = W$, respectively. Thus, we can summarize the fundamental relationship of an ideal transistor as follows:

(1) The applied voltages control the boundary densities through the term $\exp(qV/kT)$.
(2) The emitter and collector currents are given by the minority carrier gradient at the junction boundaries, that is, $x = 0$ and $x = W$. These currents are proportional to the stored base charge.
(3) The base current is the difference between the emitter and collector currents.

The emitter efficiency defined in Eq. 6 can be obtained from Eqs. 22 and 25:

$$\gamma \equiv \frac{I_{Ep}}{I_{Ep} + I_{En}} \simeq \frac{D_p p_{no}/W}{\dfrac{D_p p_{no}}{W} + \dfrac{D_E n_{EO}}{L_E}} = \frac{1}{1 + \dfrac{D_E}{D_p} \cdot \dfrac{n_{EO}}{p_{no}} \cdot \dfrac{W}{L_E}} \tag{35}$$

or

$$\gamma = \frac{1}{1 + \dfrac{D_E}{D_p} \cdot \dfrac{N_B}{N_E} \cdot \dfrac{W}{L_E}} \tag{35a}$$

where N_B ($= n_i^2/p_{no}$) is the impurity doping in the base and N_E ($= n_i^2/n_{EO}$) is the impurity doping in the emitter. It is clear that to improve γ, we should decrease the ratio N_B/N_E, that is, there should be much heavier doping in the emitter than in the base. This is the reason that we use p^+-doping in the emitter.

The base transport factor defined in Eq. 7 can be obtained from Eqs. 21 and 23:

$$\alpha_T \equiv \frac{I_{Cp}}{I_{Ep}} \simeq \mathrm{sech}\left[\frac{W}{L_p}\right] \simeq 1 - \frac{W^2}{2L_p^2} \cdot \tag{36}$$

To improve α_T, we should use the smallest feasible value of W/L_p. However, L_p is more or less constant so we should reduce W.

Problem

Find the common–base current gain for an ideal p–n–p transistor with impurity dopings of 10^{19}, 10^{17}, and 5×10^{15} cm^{-3} in the emitter, base, and collector regions, respectively. The other device parameters are $D_E = 1$ cm²/s, $D_p = 10$ cm²/s, $L_E = 1.0\ \mu$m, $L_p = 1.0\ \mu$m, and $W = 0.5\ \mu$m.

Solution

From Eqs. 35a and 36,

$$\gamma = \frac{1}{1 + \dfrac{1}{10} \cdot \dfrac{10^{17}}{10^{19}} \cdot \dfrac{0.5}{1.0}} = 0.9995$$

$$\alpha_T = 1 - \frac{W^2}{2L_p^2} = 0.9987 \ .$$

Therefore,

$$\alpha_0 = \gamma \alpha_T = 0.9982 \ .$$

Modes of Operation A bipolar transistor has four modes of operation, depending on the voltage polarities on the emitter–base junction and the base–collector junction. Figure 8 shows the V_{EB} and V_{CB} voltages for the four modes of operations of a p–n–p transistor. The corresponding minority carrier distributions are also shown. So far in this chapter we have considered the

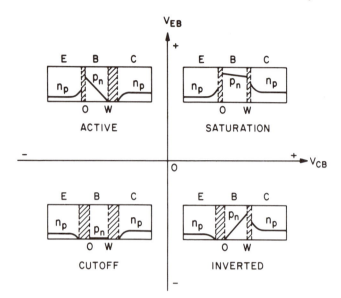

Fig. 8 Junction polarities and minority carrier distributions of a *p–n–p* transistor under various modes of operation.

active mode of transistor operation, in which the emitter–base junction is forward-biased and the collector–base junction is reverse-biased. The minority carrier distribution for the active mode is the same as that shown in Fig. 7.

In the *saturation mode*, both junctions are forward-biased, and the boundary condition at $x = W$ becomes $p_n(W) = p_{no}e^{qV_{CB}/kT}$ instead of the one given by Eq. 13*b*. The saturation mode corresponds to small biasing voltage and large output current, that is, the transistor is in a conducting state and acts as a closed (or on) switch.

In the *cutoff mode*, both junctions are reverse-biased, and the boundary conditions of Eq. 13 become $p_n(0) = p_n(W) = 0$. There is virtually no charge stored in the base region, and the collector current approaches zero. The cutoff mode corresponds to the open (or off) state of the transistor as a switch.

The fourth mode of operation is the *inverted mode*, which is sometimes called the inverted active mode. In this case, the emitter–base junction is reverse-biased and the collector–base junction is forward-biased. The inverted mode corresponds to the case where the collector acts like the emitter and the emitter acts like a collector, that is, the device is used backward. However, the current gain for the inverted mode is generally lower than that for the active mode because of poor "emitter efficiency" resulting from low collector doping with respect to the base doping (Eq. 35).

The current–voltage relationships for the various modes of operation can be obtained by following the same procedures used for the active mode, with appropriate changes of the boundary conditions as in Eq. 13. The general expressions applicable to all modes of operations are

$$I_E = a_{11}(e^{qV_{EB}/kT} - 1) - a_{12}(e^{qV_{CB}/kT} - 1) \qquad (37a)$$

and

$$I_C = a_{21}(e^{qV_{EB}/kT} - 1) - a_{22}(e^{qV_{CB}/kT} - 1) \qquad (37b)$$

where the coefficients a_{11}, a_{12}, a_{21}, and a_{22} are given by Eqs. 28, 29, 31, and 32, respectively. Note that in Eqs. 37a and 37b the biasing voltages for the junctions can be positive or negative depending on the mode of operation.

4.2.2 Modification of Static Characteristics

The ideal bipolar transistor presented in Section 4.2.1 can adequately describe the device performance of most real devices. However, at large or small voltages, the current–voltage characteristics deviate from the ideal device. We shall show the measured results of a representative transistor and consider the deviations from the ideal case.

Figure 9 shows the measured results of output current–voltage characteristics for the common-base configuration. The various modes of operation are indicated on the figure. Note that the collector current is practically equal to the emitter current (i.e., $\alpha_0 \simeq 1$) and virtually independent of V_{BC}. This is in close agreement with the ideal-transistor behavior given by Eqs. 10 and 30. The collector current remains practically constant, even down to zero volts for V_{BC}, where the holes are still extracted by the collector. This is indicated by the hole distributions shown in Fig. 10a. Since the hole gradient at $x = W$ changes only slightly from $V_{BC} > 0$ to $V_{BC} = 0$, the collector current remains essentially the same over the entire active mode of operation. To reduce the collector current to zero, we have to apply a small forward bias (~ 1 V for silicon) to the base–collector junction (in the saturation mode), as

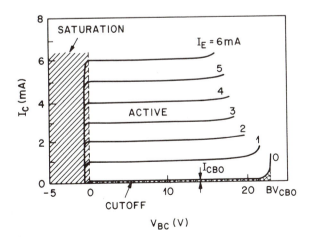

Fig. 9 Output characteristics for a p–n–p transistor in the common-base configuration.

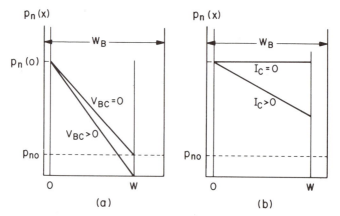

Fig. 10 Minority carrier distributions in the base region of a *p–n–p* transistor. (a) Active mode for $V_{BC} = 0$ and $V_{BC} > 0$. (b) Saturation mode with both junctions forward-biased.

shown in Fig. 10b. The forward bias will sufficiently increase the hole density at $x = W$ to make it equal to that of the emitter at $x = 0$ and thus reduce the hole gradient at $x = W$ as well as the collector current to zero.

Graded-Base Region In the ideal transistor, the impurity distribution in the base region was assumed to be uniform. However, in a real device fabricated by diffusion or by the ion implantation process of dopant into an epitaxial substrate (Fig. 11a), the impurity distribution in the base is not uniform but is strongly graded, as shown in Fig. 11b. The corresponding band diagram is shown in Fig. 11c. Because of the impurity concentration gradient, the electrons within the base tend to diffuse toward the collector. However, in thermal equilibrium there will exist a built-in electric field in the neutral base to counterbalance the diffusion current, that is, the electric field will push the electrons toward the emitter and no current will flow. The same electric field can aid the motion of injected holes. Under an active biasing condition, the injected minority carriers (holes) will now move not only by diffusion but also by drift caused by the built-in field of the base region.

The main advantage of the built-in field is to reduce the time needed for the injected holes to travel across the base region. This in turn will improve the transistor's high-frequency response, which will be considered in the next section of this chapter. An associated advantage is the improvement of the base transport factor α_T, since the holes will spend, on the average, less time in the base region and thus will be less likely to recombine with electrons there.

For the graded-base device, the total number of impurities in the neutral base region per unit area Q_G is given by

$$Q_G = \int_0^W N_B(x)\,dx \tag{38}$$

which, of course, reduces to $N_B W$ for a uniformly doped base region. If we replace the quantity $N_B W$ by Q_G in previously derived equations, for exam-

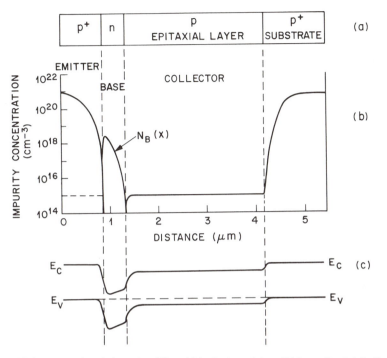

Fig. 11 (a) Cross-sectional view of a diffused bipolar transistor. (b) Impurity distribution in the transistor. (c) Corresponding band diagram in thermal equilibrium.

ple, Eqs. 28, 29, 31, 32, and 35a, results obtained for the ideal transistor can be readily adapted to diffused or ion-implanted transistors.

Base Resistance   To achieve high current gain, the base width must be very narrow. Therefore, the base resistance can be quite high. Figure 12a shows a cross section of a *p–n–p* transistor with two base contacts, one on each side of the emitter. The electrons are supplied from the base contacts and flow toward the center of the emitter causing the base–emitter voltage drop to vary with position along the base–emitter junction. As a first-order approximation with reference to Fig. 12a, the forward bias of the emitter junction above point A is

$$V_{EA} = V_{EB} - \frac{I_B}{2} (R_{AD} + R_{DB}) . \qquad (39)$$

The emitter bias voltage at point D is

$$V_{ED} = V_{EB} - \frac{I_B}{2} R_{DB} \qquad (40)$$

which can be substantially larger than V_{EA} if R_{AD} is large.

Because the forward bias is largest at the edge of the emitter (point D), the injection of holes will be greatest there. Therefore, most of the emitter current

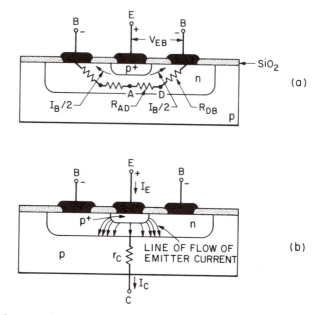

Fig. 12 (a) Cross section of a p–n–p transistor showing the base resistance. (b) Current crowding near the edge of the emitter.

will be crowded near the edge of the emitter with only a small amount of current in the central portion of the emitter (Fig. 12b). This effect is called *emitter crowding*, which causes nonuniform distribution of the emitter current. Although this crowding can have the desirable effect of reducing the base resistance, it can also give rise to undesirable high-injection effects such as reduced emitter efficiency. The most effective way to minimize the emitter crowding is to distribute the emitter current along a relatively large emitter edge, thereby reducing the current density at any one point. We thus need an emitter region with a large perimeter compared with its area. A common approach is to use interdigitated geometry as shown in Fig. 13. The emitter and base contact stripes are interlaced to provide a large perimeter for handling a large current.

Base Width Modulation   So far in this chapter we have considered transistor characteristics using the common-base configuration. However, in circuit applications the common-emitter configuration is most often used. Figure 14a shows the common-emitter configuration for a p–n–p transistor, that is, the emitter lead is common to the input and output circuits. The collector current for the common-emitter configuration can be obtained by substituting Eq. 3 into Eq. 10:

$$I_C = \alpha_0(I_B + I_C) + I_{CBO} .\tag{41}$$

Solving for I_C,

$$I_C(1 - \alpha_0) = \alpha_0 I_B + I_{CBO}\tag{42}$$

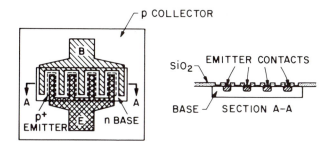

Fig. 13 Transistor with interdigitated emitter and base contacts.

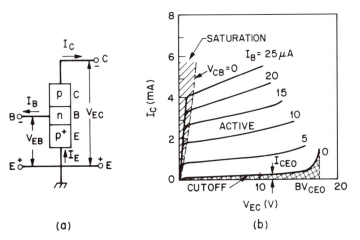

| (a) | (b) |

Fig. 14 (a) Common-emitter configuration of a p–n–p transistor. (b) Output characteristics for a p–n–p transistor in the common-emitter configuration.[4]

or

$$I_C = \frac{\alpha_0}{1 - \alpha_0} I_B + \frac{I_{CBO}}{1 - \alpha_0} . \tag{43}$$

We now designate β_0 as the *common-emitter current gain*, which is the incremental change of I_C with respect to an incremental change of I_B. From Eq. 43 we obtain

$$\beta_0 \equiv \frac{\Delta I_C}{\Delta I_B} = \frac{\alpha_0}{1 - \alpha_0} \tag{44}$$

We can also designate I_{CEO} as

$$I_{CEO} \equiv \frac{I_{CBO}}{(1 - \alpha_0)} . \tag{45}$$

This current corresponds to the collector–emitter leakage current for $I_B = 0$.

Equation 43 becomes

$$I_C = \beta_0 I_B + I_{CEO} . \tag{46}$$

Because the value of α_0 is generally close to unity, β_0 is much larger than 1. For example, if $\alpha_0 = 0.99$, β_0 is 99; and if α_0 is 0.998, β_0 is 499. Therefore, a small base current can give rise to a much larger collector current.

In an ideal transistor with the common-emitter configuration, the collector current for a given I_B is expected to be independent of V_{EC} for $V_{EC} > 0$. However, the measured results in Fig. 14b show pronounced slopes, and I_C increases with increasing V_{EC}. This deviation can be explained by base width modulation, also known as the *Early effect*.[4]

The common-emitter current gain β_0 can be obtained from Eqs. 8, 35, and 36. If the emitter efficiency γ is very close to unity, then β_0 is given by

$$\beta_0 \equiv \frac{\alpha_0}{1 - \alpha_0} = \frac{\gamma \alpha_T}{1 - \gamma \alpha_T} \simeq \frac{\alpha_T}{1 - \alpha_T} = \frac{2L_p^2}{W^2} . \tag{47}$$

Therefore, β_0 will vary as W^{-2}. As V_{EC} increases, the base width decreases from W to W', as illustrated in Fig. 15a, causing an increase in β_0. By applying Eq. 46, we can see that the collector current I_C will increase with V_{EC}.

Saturation Current and Voltage Breakdown For the common-base configuration, the saturation current I_{CBO} is measured with the emitter open-circuited. This current is considerably smaller than the ordinary reverse current of a p–n junction. This is because the emitter junction with a zero hole gradient at $x = 0$ (corresponding to zero emitter current) reduces the hole gradient at $x = W$ as shown in Fig. 15b. Therefore, the current I_{CBO} is smaller than when the emitter junction is short-circuited (corresponding to the gradient for $V_{EB} = 0$, also indicated in Fig. 15b).

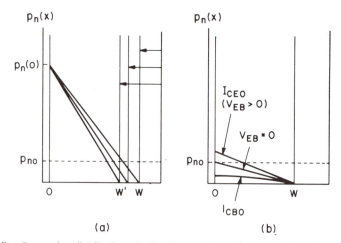

Fig. 15 Minority carrier distributions in the base region of a p–n–p transistor (a) Active mode with V_{EB} = constant, V_{EC} varying. (b) Conditions for currents I_{CBO} and I_{CEO}.

As V_{BC} increases to the value of the breakdown voltage BV_{CBO}, the collector current starts to increase rapidly (Fig. 9). Generally, this increase is caused by the avalanche breakdown of the collector–base junction. The breakdown voltage is similar to that considered in Section 3.6 for p–n junctions. For a very narrow base width or a base with relatively low doping, the breakdown may be caused by the punch-through effect; that is, the neutral base width is reduced to zero at a sufficient V_{BC} and, as a result, the collector depletion region comes in direct contact with the emitter depletion region. At this point, the collector is effectively short-circuited to the emitter, and a large current can flow.

For the common-emitter configuration, the saturation current I_{CEO} corresponds to the collector current with zero base current (the base is open-circuited). For a given V_{EC}, the emitter–base junction will be slightly forward-biased, as shown in Fig. 15b. Therefore, we expect the current I_{CEO}, which is proportional to the hole gradient at $x = W$, to be larger than I_{CBO}. Indeed, this is the case, as we see from Eq. 45, where I_{CEO} is approximately β_0 times larger than I_{CBO}.

The breakdown voltage under the open-base condition can be obtained as follows. Let the multiplication factor M at the collector junction be approximated by

$$M = \frac{1}{1 - (V/BV_{CBO})^\eta} \tag{48}$$

where BV_{CBO} is the common-base breakdown voltage (with emitter open-circuited) and η is a constant. When the base is open-circuited, we have $I_E = I_C = I$. The current I_{CBO} and $\alpha_0 I_E$ are multiplied by M when they flow across the collector junction as shown[5] in the upper left of Fig. 16:

$$M(\alpha_0 I + I_{CBO}) = I \tag{49}$$

or

$$I = \frac{MI_{CBO}}{1 - \alpha_0 M}. \tag{50}$$

Current I will be limited only by external resistances when $\alpha_0 M = 1$. By applying the condition $\alpha_0 M = 1$ and Eq. 48, the breakdown voltage BV_{CEO} for the common-emitter configuration (with base open-circuited) is given by

$$BV_{CEO} = BV_{CBO}(1 - \alpha_0)^{1/\eta} \simeq BV_{CBO}(\beta_0)^{-1/\eta}. \tag{51}$$

Because silicon has an η value between 2 and 6 and a large β_0, the common-emitter breakdown voltage BV_{CEO} is much smaller than the common-base breakdown voltage BV_{CBO}.

Generation–Recombination Current and High-Injection Effect In the ideal transistor, we neglected the generation–recombination currents in the emitter–base and collector–base depletion regions. For a real transistor, as in a p–n junction, there is a generation current in the depletion region of the

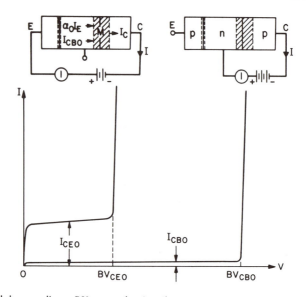

Fig. 16 Breakdown voltage BV_{CBO} and saturation current I_{CBO} for common-base configuration, and the corresponding quantities BV_{CEO} and I_{CEO} for the common-emitter configuration.[5]

reverse-biased base–collector junction. This current component is added to the leakage current. If the generation current is the dominant component in I_{CBO}, then I_{CBO} will increase as $(V_{BC})^{1/2}$ for an abrupt collector–base junction and as $(V_{BC})^{1/3}$ for a linearly graded junction. The generation current will also cause an increase of I_{CEO}, since $I_{CEO} \simeq \beta_0 I_{CBO}$.

A forward-biased emitter–base junction has a recombination current in its depletion region, and therefore a current component is added to the base current. This recombination current has a profound effect on the current gain. Figure 17a shows the collector current and the base current versus V_{EB} for a bipolar transistor operated in the active mode. At low-current levels the recombination current is the dominant current component, and the base current I_B varies as $\exp(qV_{EB}/mkT)$, with $m \simeq 2$. Note that the collector current I_C is not affected by the emitter–base recombination current, because I_C is primarily due to those holes injected into the base that diffuse to the collector.

Figure 17b shows the common-emitter current gain β_0, which is obtained from Fig. 17a by taking the ratio of ΔI_C to ΔI_B. At low collector current levels, the contribution of the recombination current in the emitter–base depletion region is larger than the diffusion current of minority carriers across the base, so that the emitter efficiency is low. By minimizing the recombination–generation centers in the device, β_0 can be improved at low-current levels. As the base diffusion current becomes dominante, β_0 increases to a high plateau.

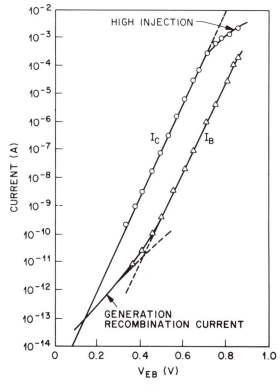

(a)

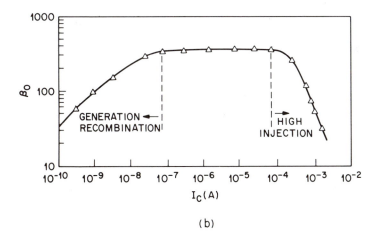

(b)

Fig. 17 (a) Collector current and base current as functions of emitter–base voltage. (b) Common–emitter current gain for the transistor shown in (a).

At higher collector current levels, β_0 starts to decrease. This is caused by the high-injection effect, where the injected minority carrier density (holes) in the base approaches the impurity concentration (N_B) and the injected carriers effectively increase the base doping, which, in turn, causes the emitter efficiency to decrease. Another factor contributing to the degradation of β_0 at high-current levels is emitter crowding, which gives rise to a nonuniform distribution of current density (or injected minority carriers) under the emitter. The current density at the emitter periphery may be much higher than the average current density. Therefore, the high-injection effect occurs at the emitter periphery, resulting in reduction of β_0.

4.3 FREQUENCY RESPONSE AND SWITCHING OF BIPOLAR TRANSISTORS

In Section 4.2 we discussed for a bipolar transistor four possible modes of operation which depend on the biasing conditions of the emitter–base and collector–base junctions. Generally, in analog or linear circuits the transistors are operated in the active mode only. However, in digital circuits all four modes of operation may be involved. In this section we shall consider a basic model for the bipolar transistor, the Ebers–Moll model, which is applicable to all modes of operation. We shall then extend the model to describe the frequency response and switching characteristics of bipolar transistors.

4.3.1 The Ebers–Moll Model

The one-dimensional p–n–p transistor (Fig. 18a) can be represented as two p–n junction diodes connected back-to-back with a common n-region (Fig. 18b). In Fig. 18c, the concept of back-to-back diodes is combined with the knowledge that most of the forward current from one diode will flow into the other diode, which is reverse-biased. In Fig. 18c, the current that crosses the emitter–base junction is shown as I_F. This current is a function of the emitter–base voltage V_{EB}. In the active mode, a large fraction of the minority carriers injected into the base from the emitter will reach the collector. This fraction is represented by the current generator $\alpha_F I_F$, where α_F is the forward common-base current gain.

In the inverted mode of operation, the collector–base junction is forward-biased and the emitter–base junction is reverse-biased. This mode is represented in Fig. 18c by the collector diode current I_R, which is a function of the collector–base voltage V_{CB}, and the current generator $\alpha_R I_R$ across the emitter–base junction. The factor α_R is called the reverse common–base current gain and is the ratio of the number of minority carriers collected at the emitter to the number of minority carriers injected into the base at the forward-biased collector–base junction.

To develop the general circuit expressions, we first write the equations for the diode currents I_F and I_R:

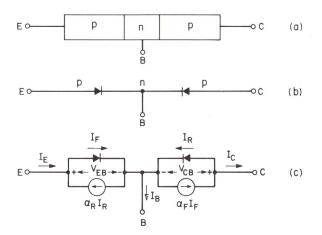

Fig. 18 (a) Cross section of a p–n–p transistor. (b) A p–n–p transistor represented as two p–n junction diodes connected back-to-back. (c) Circuit diagram of Ebers–Moll model.[6]

$$I_F = I_{FO}(e^{qV_{EB}/kT} - 1) \qquad (52)$$

$$I_R = I_{RO}(e^{qV_{CB}/kT} - 1) \qquad (53)$$

where I_{FO} and I_{RO} are the saturation currents of the normally forward- and reverse-biased diodes, respectively. The terminal currents are

$$I_E = I_F - \alpha_R I_R \qquad (54)$$

$$I_C = \alpha_F I_F - I_R \qquad (55)$$

$$I_B = I_E - I_C = (1 - \alpha_F)I_F + (1 - \alpha_R)I_R \qquad (56)$$

By combining Eqs. 52 through 55, we derive the general equations for the basic Ebers–Moll model[6]:

$$I_E = I_{FO}(e^{qV_{EB}/kT} - 1) - \alpha_R I_{RO}(e^{qV_{CB}/kT} - 1) \qquad (57)$$

$$I_C = \alpha_F I_{FO}(e^{qV_{EB}/kT} - 1) - I_{RO}(e^{qV_{CB}/kT} - 1). \qquad (58)$$

These equations show the relation between the terminal currents I_E and I_C and the terminal voltages V_{EB} and V_{CB}.

The basic Ebers–Moll model has four parameters: I_{FO}, I_{RO}, α_F and α_R. Comparing Eqs. 57 and 58 with Eq. 37 gives

$$I_{FO} = a_{11} \qquad (59a)$$

$$\alpha_R I_{RO} = a_{12} \qquad (59b)$$

$$\alpha_F I_{FO} = a_{21} \qquad (59c)$$

$$I_{RO} = a_{22} \qquad (59d)$$

where a_{11}, a_{12}, a_{21}, and a_{22} are defined previously. For the ideal transistor,

$a_{12} = a_{21}$; therefore, $\alpha_R I_{RO} = \alpha_F I_{FO}$. For a real device, the above relationship also holds because of the reciprocity characteristic of a two-port device. Therefore, only three parameters (e.g., I_{FO}, I_{RO}, and α_F) are required for the basic Ebers–Moll model.

4.3.2 Frequency Response

High-Frequency Equivalent Circuit   In the previous discussions, we were concerned with the static (or dc) characteristics of the bipolar transistor. We shall now study its ac characteristics when a small–signal voltage or current is superimposed upon the dc values. The term "small-signal" means that the peak values of the ac signal current and voltage are smaller than the dc values. Consider an amplifying circuit shown in Fig. 19a, where the transistor is connected in a common-emitter configuration. For a given dc input voltage V_{EB}, a certain dc base current I_B and dc collector current I_C flow in the transistor. These currents correspond to the operating point shown in Fig. 19b. The load line, determined by the applied voltage V_{CC} and the load resistance R_L, intercepts the V_{EC} axis at V_{CC} and has a slope of $(-1/R_L)$. When a small ac signal is superimposed on the input voltage, the base current i_B will vary as a function of time, as illustrated in Fig. 19b. This variation, in turn, brings about

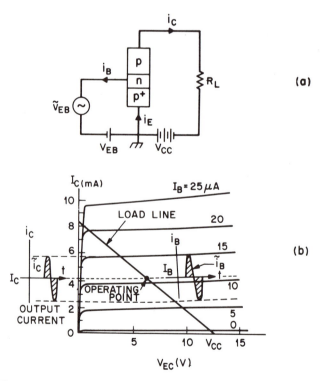

Fig. 19   (a) Bipolar transistor connected in the common-emitter configuration. (b) Small-signal operation of the transistor circuit.

a corresponding variation in the output current i_C, which however is β_0 times larger than the input current variation. Thus, the transistor amplifies the input signal.

When an incremental signal voltage $\tilde{v}_{EB}$ is applied to the emitter–base junction, as shown in Fig. 20, the injected hole density at $x = 0$ will increase; this results in an increase of the hole gradients $dp_n(x)/dx$ at $x = 0$ and $x = W$ and causes increases of both the emitter and the collector currents. We assume that the frequency of the signal voltage is low enough so the carrier distributions can follow the signal instantaneously. We define the total instantaneous value of the emitter current i_E as the sum of its dc value I_E and the ac signal $\tilde{i}_E$:

$$i_E = I_E + \tilde{i}_E . \tag{60}$$

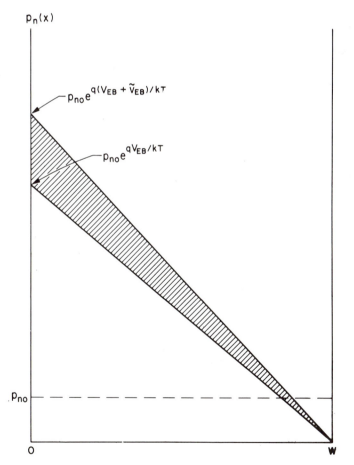

Fig. 20 Incremental change of the minority carrier distribution when an ac voltage signal is superimposed on a dc bias.

Similarly, the other voltages and currents are given by

$$v_{EB} = V_{EB} + \tilde{v}_{EB} \tag{61a}$$

$$i_B = I_B + \tilde{i}_B \tag{61b}$$

$$i_C = I_C + \tilde{i}_C \tag{61c}$$

$$v_{CB} = V_{CB} + \tilde{v}_{CB} \tag{61d}$$

The relationship between $\tilde{i}_B$ and $\tilde{i}_C$ can be developed in terms of $\tilde{v}_{EB}$ and $\tilde{v}_{EC}$ using the Ebers–Moll model. We replace the dc variables in Eqs. 57 and 58 with the total instantaneous variables of Eqs. 60 and 61. Because we use the active mode for amplifying applications, $e^{qV_{EB}/kT} \gg 1$ and $e^{qV_{CB}/kT} \ll 1$. Equations 57 and 58 can be approximated by

$$i_E = I_{FO}(e^{qv_{EB}/kT} - 1) - \alpha_R I_{RO}(e^{qv_{CB}/kT} - 1) \simeq I_{FO}e^{qv_{EB}/kT} \tag{62}$$

$$i_C = \alpha_F I_{FO}(e^{qv_{EB}/kT} - 1) - I_{RO}(e^{qv_{CB}/kT} - 1) \simeq \alpha_F I_{FO}e^{qv_{EB}/kT} \tag{63}$$

$$i_B = i_E - i_C \simeq (1 - \alpha_F)I_{FO}e^{qv_{EB}/kT} \tag{64}$$

If the signal voltage $\tilde{v}_{EB}$ is much smaller than the dc biasing voltage V_{EB}, then the currents i_C and i_B in Eqs. 63 and 64 can be expressed by the Taylor series expansion up to the linear terms of the signal voltage:

$$i_C = I_C + \left. \frac{\partial i_C}{\partial V_{EB}} \right|_{V_{EC}} \tilde{v}_{EB} = I_C + \tilde{i}_C \tag{65}$$

$$i_B = I_B + \left. \frac{\partial i_B}{\partial V_{EB}} \right|_{V_{EC}} \tilde{v}_{EB} = I_B + \tilde{i}_B . \tag{66}$$

Therefore,

$$\tilde{i}_C = \left. \frac{\partial i_C}{\partial V_{EB}} \right|_{V_{EC}} \tilde{v}_{EB} = g_m \tilde{v}_{EB} \tag{67}$$

and

$$\tilde{i}_B = \left. \frac{\partial i_B}{\partial V_{EB}} \right|_{V_{EC}} \tilde{v}_{EB} = g_{EB} \tilde{v}_{EB} \tag{68}$$

where g_m is called the *transconductance* and g_{EB} is called the *input conductance*. From Eqs. 63, 64, 67, and 68, we find

$$g_m \equiv \left. \frac{\partial i_C}{\partial V_{EB}} \right|_{V_{EC}} = \alpha_F I_{FO} \left[\frac{q}{kT} \right] e^{qV_{EB}/kT} = \alpha_F \frac{qI_E}{kT} \tag{69}$$

$$g_{EB} \equiv \left. \frac{\partial i_B}{\partial V_{EB}} \right|_{V_{EC}} = (1 - \alpha_F)I_{FO} \left[\frac{q}{kT} \right] e^{qV_{EB}/kT} = \frac{q(1 - \alpha_F)}{kT} I_E . \tag{70}$$

The equivalent circuit for this low-frequency situation is shown in Fig. 21a.

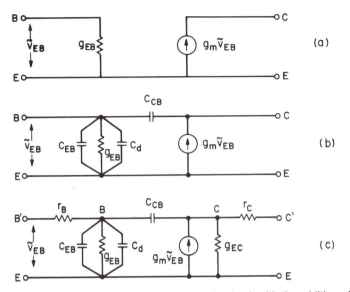

Fig. 21 (a) Basic transistor equivalent circuit. (b) Basic circuit with the addition of depletion and diffusion capacitances. (c) Basic circuit with the addition of resistance and conductance.

At higher frequencies, we can extend the equivalent circuit by adding the appropriate capacitances. Since the emitter–base junction is forward-biased, we expect to have a depletion capacitance C_{EB} and a diffusion capacitance C_d similar to that of a forward-biased diode. For the reverse-biased collector–base junction, we expect to have only a depletion capacitance C_{CB}. The high-frequency equivalent circuit with the three added capacitances is shown in Fig. 21b. To account for the base width modulation effect, there is a finite output conductance $g_{EC} \equiv \tilde{i}_C/\tilde{v}_{EC}$ when $\tilde{v}_{EB} = 0$. In addition, we have a base resistance r_B and a collector resistance r_C. Figure 21c represents the high-frequency equivalent circuit incorporating all of the above elements.

Cutoff Frequency In Fig. 21c, the transconductance g_m and the input conductance g_{EB} are dependent on the common-base current gain. At low frequencies, the current gain is a constant, independent of the operating frequency. However, the current gain will decrease after a certain critical frequency is reached. A typical plot of the current gain versus operating frequency is shown in Fig. 22. The common-base current gain α can be described as

$$\alpha = \frac{\alpha_0}{1 + j(f/f_\alpha)} \tag{71}$$

where α_0 is the low-frequency (or dc) common-base current gain and f_α is the *common-base cutoff frequency*. At $f = f_\alpha$, the magnitude of α is $0.707\alpha_0$ (3 dB down).

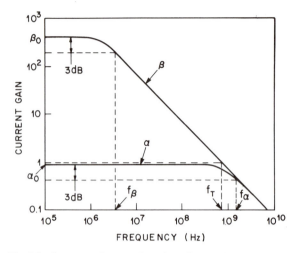

Fig. 22   Current gains as a function of operating frequency.

Also shown in Fig. 22 is the common-emitter current gain β. From Eq. 71 we have

$$\beta \equiv \frac{\alpha}{1 - \alpha} = \frac{\beta_0}{1 + j(f/f_\beta)} \tag{72}$$

where the f_β is the *common-emitter cutoff frequency* and is given by

$$f_\beta = (1 - \alpha_0)f_\alpha . \tag{73}$$

Since $\alpha_0 \approx 1$, f_β is much smaller than f_α. Another cutoff frequency is f_T when $|\beta|$ becomes unity. By setting the magnitude of the right-hand side of Eq. 72 equal to 1, we obtain

$$f_T = \sqrt{\beta_0^2 - 1} \, f_\beta \simeq \beta_0(1 - \alpha_0) f_\alpha \simeq \alpha_0 f_\alpha \tag{74}$$

Hence, f_T is very close to but is smaller than f_α.

The most important limitation on the transistor frequency response is the transit time of minority carriers across the base region. The distance traveled by a hole in a time interval dt is $dx = v(x)\, dt$, where $v(x)$ is the effective minority carrier velocity in the base. This velocity is related to the current as

$$I_p = q v(x)\, p(x)\, A \tag{75}$$

where A is the device area and $p(x)$ is the distribution of the minority carriers. The transit time τ_B required for a hole to traverse the base is given by

$$\tau_B = \int_0^W \frac{dx}{v(x)} = \int_0^W \frac{q p(x) A}{I_p}\, dx . \tag{76}$$

For a straight-line hole distribution, as given by Eq. 15, the integration of

Eq. 76 using Eq. 24 for I_p leads to

$$\tau_B = \frac{W^2}{2D_p}.$$ (77)

To improve the frequency response, the transit time of minority carriers across the base must be short. Therefore, high-frequency transistors are designed with a small base width. Because the electron diffusion constant in silicon is about three times larger than that of holes, all high-frequency silicon transistors are of the *n–p–n* type.

One way to reduce the base transit time is to use a graded base with a built-in field. For a large doping variation, as shown in Fig. 11, the base transist time can be reduced by a factor of 5.

4.3.3 Switching Transients

For digital applications, a transistor is designed to function as a switch. In these applications we use a small current (the base current), to change the collector current from a high-voltage, low-current *off* condition to a low-voltage, high-current *on* condition (or vice versa), in a very short time. A basic setup of a switching circuit is shown in Fig. 23a, where the emitter–base voltage V_{EB} is suddenly changed from a negative value to a positive value. The output

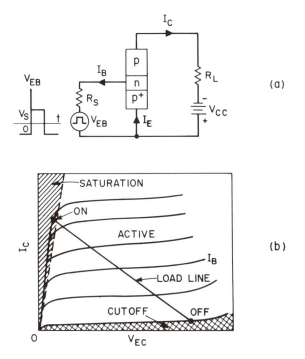

Fig. 23 (a) Schematic of a transistor switching current. (b) Switching operation from cutoff to saturation.

current of the transistor is shown in Fig. 23b. The collector current is initially very low because both the emitter–base junction and the collector–base junction are reverse-biased. The current will follow the load line through the active region and will finally reach a high current level, where both junctions become forward-biased. Thus, the transistor is virtually open-circuited between the emitter and collector terminal in the off condition (corresponding to the cutoff mode) and short-circuited in the on-condition (corresponding to the saturation mode). Therefore, a transistor operated this way can nearly duplicate the function of an ideal switch.

On- and Off-Impedance We shall now consider the switching behavior of a bipolar transistor based on the Ebers–Moll model. Referring to Eqs. 57 and 58, the coefficients α_F and α_R can be measured directly from the active mode and inverted mode of operation, respectively. The coefficients I_{FO} and I_{RO} can be obtained from two additional measurements of I_{EO} and I_{CO}, where I_{EO} is the measured reverse-saturation current of the emitter–base junction with the collector open-circuited (i.e., $e^{qV_{EB}/kT} \ll 1$ and $I_C = 0$) and I_{CO} is the measured reverse-saturation current of the collector–base junction with the emitter open-circuited (i.e., $e^{qV_{CB}/kT} \ll 1$ and $I_E = 0$).

From the quantities given above and from Eqs. 57 and 58, we find

$$I_{FO} = \frac{-I_{EO}}{1 - \alpha_F \alpha_R} \tag{78a}$$

$$I_{RO} = \frac{I_{CO}}{1 - \alpha_F \alpha_R}. \tag{78b}$$

In the active and cutoff modes, the collector–base junction is reverse-biased. Equations 57 and 58 reduce to

$$I_E = -\frac{I_{EO}}{1 - \alpha_F \alpha_R} e^{qV_{EB}/kT} + \frac{(1 - \alpha_F)I_{EO}}{1 - \alpha_F \alpha_R} \tag{79a}$$

$$I_C = -\frac{\alpha_F I_{EO}}{1 - \alpha_F \alpha_R} e^{qV_{EB}/kT} + \frac{(1 - \alpha_R)I_{CO}}{1 - \alpha_F \alpha_R}. \tag{79b}$$

In the saturation mode, it is convenient to consider the currents as independent variables. From Eq. 57 and 58 we obtain

$$V_{EB} = \frac{kT}{q} \ln \left[\frac{\alpha_R I_C - I_E}{I_{EO}} \right] \tag{80a}$$

$$V_{CB} = \frac{kT}{q} \ln \left[\frac{\alpha_F I_E - I_C}{I_{CO}} \right]. \tag{80b}$$

To characterize the switching operation, we must consider three basic quantities: the off-impedance, the on-impedance, and the switching time. The impedance at the off-condition can be obtained from Eq. 79b by setting $e^{qV_{EB}/kT} \ll 1$:

$$R\,(\text{off}) = \frac{V_C}{I_C(\text{off})} = \frac{V_C(1 - \alpha_F\alpha_R)}{I_{CO} - \alpha_F I_{EO}}. \tag{81}$$

The impedance at the on-condition can be obtained from Eq. 80:

$$R\,(\text{on}) = \frac{V_{EC}(\text{on})}{I_C} = \frac{V_{EB} - V_{CB}}{I_C}$$

$$= \frac{kT}{qI_C ln}\left\{ \frac{I_{CO}[1 + (1 - \alpha_R)\dfrac{I_C}{I_B}]}{I_{EO}[(1 - \alpha_F)\dfrac{I_C}{I_B} - \alpha_F]} \right\}. \tag{82}$$

From Eq. 81 it appears that the off-impedance will be high for small reverse-saturation currents I_{CO} and I_{EO} of the junctions. The on-impedance, Eq. 82, is approximately inversely proportional to the collector current I_C and is very small when I_C is large. In practice, the ohmic resistances of the base and collector regions should be included in the total impedances, especially for the on-impedance.

Switching Time* We shall now consider the switching time, which is the time required for a transistor to switch from the off-condition to the on-condition, or vice versa. When a current pulse is applied to the base–emitter terminal at time $t = 0$ (Fig. 24a), the transistor is being "turned on." We shall show that the transient time is determined by the variation of the stored charge in the base. The total excess minority carrier charge stored in the base is given by

$$Q_B = qA \int_0^W [p_n(x) - p_{no}]\, dx. \tag{83}$$

The time variation of the stored charge can be obtained from the continuity equation given in Eq. 80 of Chapter 2:

$$\frac{\partial p_n}{\partial t} = -\frac{1}{q}\left[\frac{\partial J_p}{\partial x}\right] - \frac{p_n - p_{no}}{\tau_p} \tag{84}$$

or

$$-\frac{1}{q}\left[\frac{\partial J_p}{\partial x}\right] = \frac{\partial p_n}{\partial t} + \frac{p_n - p_{no}}{\tau_p}. \tag{84a}$$

By integrating Eq. 84a with respect to distance from $x = 0$ to $x = W$ and

---

* See footnote to Section 2.4.2.

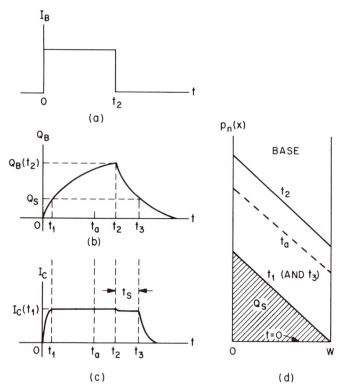

Fig. 24 (a) Input base current pulse. (b) Variations of the base-stored charge with time. (c) Variation of the collector current with time. (d) Minority carrier distributions in the base at different times.

using Eq. 83, we obtain

$$I_p(0) - I_p(W) = \frac{dQ_B}{dt} + \frac{Q_B}{\tau_p} . \tag{85}$$

This equation is referred to as the charge control equation. Since the difference between $I_p(0)$ and $I_p(W)$ is the base current, we can rewrite Eq. 85 as

$$i_B = \frac{dQ_B}{dt} + \frac{Q_B}{\tau_p} . \tag{86}$$

During the turn-on transient, for $t \geqslant 0$ the base current is a constant (Fig. 24a), and Eq. 86 becomes

$$\frac{dQ_B}{dt} = I_B - \frac{Q_B}{\tau_p} . \tag{87}$$

If $Q_B = 0$ before the base current is switched on, the solution of Eq. 87 is

$$Q_B(t) = I_B\tau_p(1 - e^{-t/\tau_p}) . \tag{88}$$

Therefore, when $t \gg \tau_p$, Q_B approaches $I_B \tau_p$. A plot of $Q_B(t)$ is shown in Fig. 24b. For $Q_B(t) < Q_S$, where Q_S is the base charge when $V_{CB} = 0$ (i.e., at the edge of saturation, as shown in Fig. 24d), the transistor is in the active mode and is moving toward saturation. The collector current $i_C(t)$ can be obtained from Eqs. 34, 77, and 88:

$$i_C = \frac{Q_B(t)}{W^2/2D_p} = \frac{Q_B(t)}{\tau_B} = \frac{I_B \tau_p}{\tau_B} (1 - e^{-t/\tau_p}). \tag{89}$$

The variation of i_C with time is plotted in Fig. 24c. At $t = t_1$, the stored base charge reaches the charge at the edge of saturation Q_S. For $Q_B > Q_S$ the device is operated in saturation mode, and both the emitter current and the collector current remain essentially constant. Figure 24d shows that for any $t > t_1$ (say, $t = t_a$), the hole distribution $p_n(x)$ will be parallel to that for $t = t_1$. Therefore, the gradients at $x = 0$ and $x = W$, as well as the currents, remain the same. We can solve for t_1 by equating $Q_B(t)$ to Q_S in Eq. 88:

$$Q_S = I_B \tau_p (1 - e^{-t_1/\tau_p}) \tag{90}$$

or

$$t_1 = \tau_p \ln \left[\frac{1}{1 - Q_S/I_B \tau_p} \right] \tag{91}$$

where Q_S and I_B can be obtained from Fig. 23a as

$$Q_S \simeq \left[\frac{V_{CC}}{R_L} \right] \tau_B \tag{92}$$

and

$$I_B \simeq \frac{V_S}{R_S}. \tag{93}$$

Therefore, to reduce the turn-on transient time t_1, a short minority lifetime τ_p, small Q_S, and large I_B should be used.

In the turn-off transient, the base current is suddenly switched to zero at $t = t_2$ (Fig. 24a). The transient equation is given by Eq. 86 with $i_B = 0$. The solution is

$$Q_B(t) = Q_B(t_2) e^{-(t - t_2)/\tau_p} \qquad \text{for } t \geq t_2 \tag{94}$$

and is plotted in Fig. 24b. Since the device is initially in the saturation mode, the collector current remains relatively unchanged until Q_B is reduced to Q_S (Fig. 24d). The time from t_2 to t_3 when $Q_B = Q_S$ is called the *storage time delay* t_S. When $Q_B = Q_S$, the device enters the active mode. The current in the time interval $t_2 < t < t_3$ is given by

$$i_C = \frac{Q_B(t - t_2)}{\tau_B} = \frac{Q_B(t_2)}{\tau_B} e^{-(t - t_2)/\tau_p} \tag{95}$$

as plotted in Fig. 24c. The storage time can be determined from Eq. 95 by equating $Q_B(t - t_2)$ to Q_S. We obtain

$$t_S \equiv t_3 - t_2 = \tau_p \ln \left[\frac{Q_B(t_2)}{Q_S} \right]. \tag{96}$$

If t_2 is much larger than τ_p, $Q_B(t_2)$ approaches $I_B \tau_p$ as given by Eq. 89, and Eq. 96 becomes

$$t_S \simeq \tau_p \ln \left[\frac{I_B \tau_p}{Q_S} \right]. \tag{97}$$

Thus, a small τ_p or I_B will reduce t_S; also, the less the device goes into saturation, the shorter the storage time will be. Once in the active mode, the collector current will decay exponentially toward zero with a time constant τ_p, as shown in Fig. 24c.

We have used the charge control equation to study the switching transients. The turn-on time depends upon how fast we can add holes (minority carriers in the p–n–p transistor) to the base region, and the turn-off time depends upon how fast we can remove the holes by recombination. One of the most important parameters for switching transistors is the minority carrier lifetime τ_p. One effective method to reduce τ_p for faster switching is to introduce efficient generation–recombination centers near the midgap.

4.4 HETEROJUNCTION BIPOLAR TRANSISTORS

A heterojunction is defined as a junction formed between two dissimilar semiconductors, such as n-type germanium on p-type gallium arsenide. Heterojunctions have many unique features that are not readily available from the conventional p–n junctions (homojunctions) discussed previously. Heterojunctions have been studied since 1951, and many important applications have been made especially in photonic devices, which will be considered in Chapter 7. In this section we first consider the basic device model and then discuss the operation of the heterojunction bipolar transistor, a potential candidate for high-speed applications.

Figure 25a shows the energy band diagram of two isolated pieces of semiconductors prior to the formation of a heterojunction. The two semiconductors are assumed to have different energy bandgaps E_g, different dielectric permittivities ϵ_s, different work functions $q \phi_s$, and different electron affinities $q \chi$. The work function is defined as the energy required to remove an electron from the Fermi level E_F to a position just outside the material (the vacuum level). The electron affinity is the energy required to remove an electron from the bottom of the conduction band E_C to the vacuum level. The difference in energy of the conduction band edges in the two semiconductors is represented by ΔE_C, and the difference in energy in the valence band edges is represented by ΔE_V. Figure 25a shows that $\Delta E_C = q(\chi_1 - \chi_2)$.

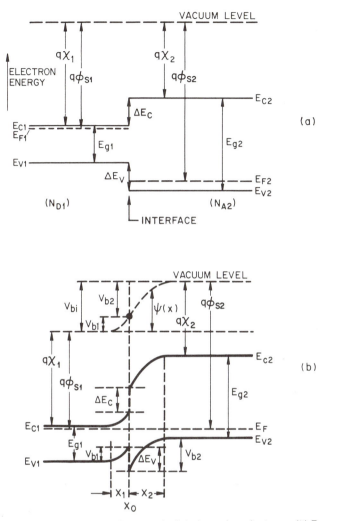

Fig. 25 (a) Energy band diagram for two isolated semiconductors. (b) Energy band diagram of an ideal n–p heterojunction at thermal equilibrium.[7]

Figure 25b shows the equilibrium band diagram of an ideal abrupt hetero-junction formed between these semiconductors.[7, 8] In this diagram it is assumed that there is a negligible number of traps or generation–recombination centers at the interface of the two dissimilar semiconductors. This assumption is valid only when heterojunctions are formed between semi-conductors with closely matched lattice constants. There are two basic requirements in the construction of the energy band diagram: (1) the Fermi level must be the same on both sides of the interface in thermal equilibrium, and (2) the vacuum level must be continuous and parallel to the band edges. Because of these requirements, the discontinuity in conduction band edges

ΔE_C and valence band edges ΔE_V will be unaffected by doping as long as the bandgap E_g and electron affinity $q\chi$ are not functions of doping (i.e., as in nondegenerate semiconductors). The total built-in potential V_{bi} is equal to the sum of the partial built-in voltage $(V_{b1} + V_{b2})$, where V_{b1} and V_{b2} are the electrostatic potentials at equilibrium in semiconductors 1 and 2, respectively.

The depletion widths and capacitance at any arbitrary biasing condition can be obtained by solving Poisson's equation for the step junction on either side of the interface. One boundary condition is the continuity of electric displacement, that is, $\epsilon_1\mathscr{E}_1 = \epsilon_2\mathscr{E}_2$, where $\mathscr{E}_1$ and $\mathscr{E}_2$ are the electric fields at the interface $(x = x_o)$ in semiconductors 1 and 2, respectively. We obtain

$$x_1 = \left[\frac{2N_{A2}\epsilon_1\epsilon_2(V_{bi} - V)}{qN_{D1}(\epsilon_1 N_{D1} + \epsilon_2 N_{A2})}\right]^{1/2} \tag{98}$$

$$x_2 = \left[\frac{2N_{D1}\epsilon_1\epsilon_2(V_{bi} - V)}{qN_{A2}(\epsilon_1 N_{D1} + \epsilon_2 N_{A2})}\right]^{1/2} \tag{99}$$

and

$$C = \left[\frac{qN_{D1}N_{A2}\epsilon_1\epsilon_2}{2(\epsilon_1 N_{D1} + \epsilon_2 N_{A2})(V_{bi} - V)}\right]^{1/2} \tag{100}$$

where V is the applied total voltage. The voltages in the semiconductors are related by

$$\frac{V_{b1} - V_1}{V_{b2} - V_2} = \frac{N_{A2}\epsilon_2}{N_{D1}\epsilon_1} \tag{101}$$

where $V = V_1 + V_2$.

The most important materials for heterojunctions are the III–V compound semiconductors such as GaAs and their solid solutions such as the ternary compound $Al_x Ga_{1-x} As$, where x can vary from 0 to 1. When $x = 0$, gallium arsenide has a bandgap of 1.42 eV and a lattice constant of 5.6533 Å at 300 K, and when $x = 1$, aluminum arsenide has a bandgap of 2.17 eV and a lattice constant of 5.6605 Å. The bandgap for the ternary $Al_x Ga_{1-x} As$ increases with x; however, the lattice constant remains essentially a constant. Even for the extreme cases where $x = 0$ and $x = 1$, the lattice mismatch is only 0.1%.

Figure 26a shows the equilibrium band diagram of a heterojunction bipolar transistor with a wide-gap emitter. This device has an n-type $Al_x Ga_{1-x} As$ emitter, a p-type GaAs base, and an n-type GaAs collector. Figure 26b shows the band diagram of the device operated in the active mode. Although the device operation is similar to that of a conventional transistor, there are many advantages to a heterojunction bipolar transistor: (1) higher emitter efficiency, because holes (minority carriers for the emitter in an n–p–n transistor) flowing from the base to the emitter are blocked by a higher barrier in the valence band; (2) decreased base resistance, because the base can be heavily doped without sacrificing emitter efficiency; (3) less emitter current crowding because

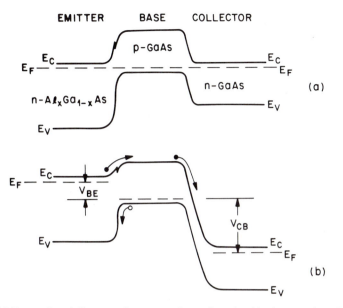

EMITTER   BASE   COLLECTOR

Fig. 26 (a) Energy band diagram of an *n–p–n* heterojunction bipolar transistor in thermal equilibrium. (*b*) Energy band diagram for the device under active mode of operation.

of a low voltage drop along the emitter–base junction; and (4) improved frequency response because of higher current gain and lower base resistance. In addition, heterojunction bipolar transistors permit use of certain materials with high-temperature capabilities. For example, a device made of materials shown in Fig. 26 can be operated at temperatures above 300°C.

Many varieties of heterojunction bipolar transistors are made possible by epitaxial techniques (refer to Chapter 8). These include transistors with a graded material composition in the base (e.g., used $Al_x Ga_{1-x} As$ for the base, with a value of x decreasing from the emitter to the collector) to provide a built-in field and to reduce base transit time; and transistors with a double heterojunction structure (e.g., wide-gap emitter and wide-gap collector) to make the emitter and collector junctions symmetrical and to improve the current gain under active and inverted modes of operation.

4.5   THE THYRISTOR

The thyristor is an important device for switching applications that require the device to change from an *off* or blocking state to an *on* or conducting state, or vice versa.[2, 9] We have considered the use of bipolar transistors in this application, in which the base current drives the transistor from cutoff to saturation for the on-state, and from saturation to cutoff for the off-state. The operations of a thyristor are intimately related to the bipolar transistor, in which both electrons and holes are involved in the transport processes. However, the switching mechanisms in a thyristor are quite different from those of a bipolar transistor. Also, because of the device construction, thyristors have a much wider range of current- and voltage-handling capabilities. Thyristors are now available with current ratings from a few milliamperes to over 5000 A and voltage ratings extending above 10,000 V. We shall first consider the operation principles of thyristors and discuss some related bidirectional and field-controlled devices.

4.5.1 Basic Characteristics

Figure 27a shows a schematic cross-sectional view of a thyristor structure that is a four–layer p–n–p–n device with three p–n junctions in series: J1, J2, and J3. The contact electrode to the outer p-layer is called the *anode* and that to the outer n-layer is called the *cathode*. This structure without any additional electrode is a two-terminal device and is called the p–n–p–n diode. If an additional electrode, called the *gate* electrode, is connected to the inner p (p2)-layer, the resulting three-terminal device is commonly called the *semiconductor-controlled rectifier* (SCR) or *thyristor*.

A typical doping profile of a thyristor is shown in Fig. 27b. An n-type, high-resistivity silicon wafer is chosen as the starting material (n-layer). To achieve doping uniformity in the starting material, a neutron transmuation process is used (refer to Section 8.2). A diffusion step is used to form the p1- and p2-layers simultaneously. Finally, an n-type layer is alloyed (or diffused) into one side of the wafer to form the n2-layer. Figure 27c shows the energy band diagram of a thyristor in thermal equilibrium. Note that at each junction there is a depletion region with a built-in potential that is determined by the impurity doping profile.

The basic current–voltage characteristic of a p–n–p–n diode is shown in Fig. 28. It exhibits five distinct regions:

0–1: The device is in the forward-blocking or off-state with very high impedance. Forward breakover (or switching) occurs where $dV/dI = 0$; and at point 1 we define a forward-breakover voltage V_{BF} and a switching current I_s.

1–2: The device is in a negative-resistance region, that is, the current increases as the voltage decreases sharply.

2–3: The device is in the forward-conducting or on-state with low impedance. At point 2, where $dV/dI = 0$, we define the holding current I_h and holding voltage V_h.

0–4: The device is in the reverse-blocking state.

4–5: The device is in the reverse-breakdown region.

Thus, a p–n–p–n diode operated in the forward region is a bistable device that can switch from a high-impedance, low-current off-state to a low-impedance, high-current on-state, or vice versa.

To understand the forward-blocking characteristics, we shall consider the device as two bipolar transistors, that is, a p–n–p transistor and an n–p–n

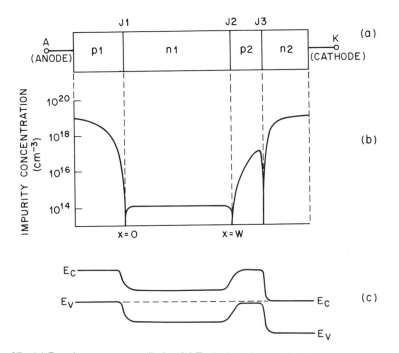

Fig. 27 (a) Four-layer p–n–p–n diode. (b) Typical doping profile of a thyristor. (c) Energy band diagram of a thyristor in thermal equilibrium.

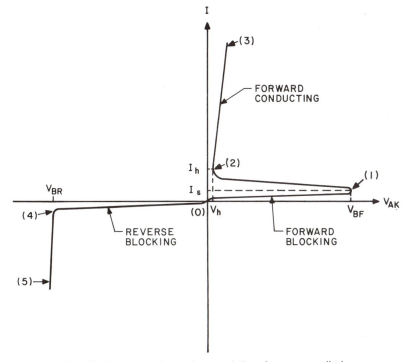

Fig. 28 Current–voltage characteristics of a p–n–p–n diode.

transistor connected with the base of one transistor attached to the collector of the other, and vice versa, as shown in Fig. 29. The relationship between emitter, collector, and base currents and the dc common-base current gain were given previously in Eqs. 3 and 10. The base current of the p–n–p transistor (transistor 1 with current gain α_1) is

$$I_{B1} = I_{E1} - I_{C1} = (1 - \alpha_1)I_{E1} - I_1$$
$$= (1 - \alpha_1)I - I_1 \tag{102}$$

where I_1 is the leakage current I_{CBO} for transistor 1. This base current is supplied by the collector of the n–p–n transistor (transistor 2 with current gain α_2). The collector current of the n–p–n transistor is

$$I_{C2} = \alpha_2 I_{E2} + I_2 = \alpha_2 I + I_2 \tag{103}$$

where I_2 is the leakage current I_{CBO} for transistor 2. By equating I_{B1} and I_{C2}, we obtain

$$(1 - \alpha_1)I - I_1 = \alpha_2 I + I_2 \tag{104}$$

or

$$I = \frac{I_1 + I_2}{1 - (\alpha_1 + \alpha_2)} . \tag{105}$$

As we have shown in Section 4.2.2, the current gains are functions of the current I and generally increase with increasing current. At low currents both α_1 and α_2 much less than 1, and the current flowing through the device is the sum of the leakage currents I_1 and I_2. As the applied voltage increases, the current I also increases, as do α_1 and α_2. This in turn causes I to increase further—a regenerative behavior. Eventually $\alpha_1 + \alpha_2$ approaches 1 and the current I increases without limit, that is, the device is at forward breakover.

The variations of the depletion layer widths of a p–n–p–n diode biased in different regions are shown in Fig. 30. At thermal equilibrium, Fig. 30a, there is no current flowing and the depletion layer widths are determined by the impurity doping profiles. In the forward-blocking state, Fig. 30b, junctions J1 and J3 are forward-biased and J2 is reverse-biased. Most of the voltage drop

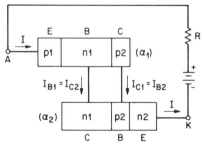

Fig. 29 Two-transistor representation of a thyristor.

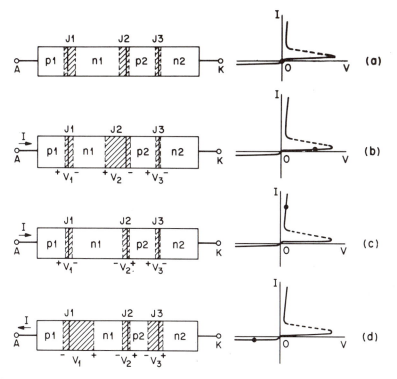

Fig. 30 Depletion layer widths and voltage drops of a thyristor operated under (a) equilibrium, (b) forward blocking, (c) forward conducting, and (d) reverse blocking.

occurs across the central junction J2. In the forward-conducting state, Fig. 30c, all three junctions are forward-biased. The two transistors ($p1-n1-p2$ and $n1-p2-n2$) are in the saturation mode of operation. Therefore, the voltage drop across the device is very low, given by ($V_1 - |V_2| + V_3$), which is approximately equal to the voltage drop across one forward-biased $p-n$ junction. In the reverse-blocking state, Fig. 30d, junction J2 is forward-biased, but both J1 and J3 are reverse-biased. For the doping profile shown in Fig. 27b, the reverse-breakdown voltage will be mainly determined by J1 because of the lower impurity concentration in the $n1$-region.

Figure 31a shows the device configuration of a thyristor that is fabricated by planar processes with a gate electrode connected to the $p2$-region. A cross section of the thyristor along the dashed lines is shown in Fig. 31b. The current–voltage characteristic of the thyristor is similar to that of the $p-n-p-n$ diode, except that the gate current I_g causes an increase of $\alpha_1 + \alpha_2$ and results in a breakover at a lower voltage. Figure 32 shows the effect of gate current on the current–voltage characteristics of a thyristor. As the gate current increases, the forward breakover voltage decreases.

A simple application of a thyristor is shown in Fig. 33a, where a variable power is delivered to a load from a constant line source. The load R_L may be

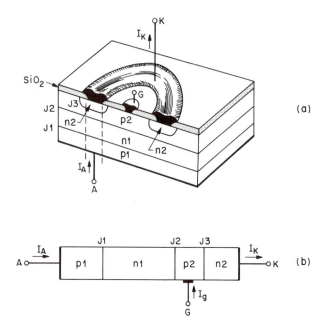

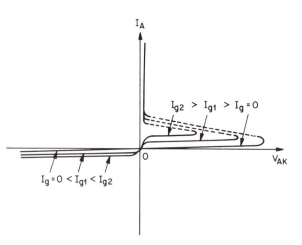

Fig. 31 (*a*) Planar three-terminal thyristor. (*b*) One-dimensional cross section of a planar thyristor.

Fig. 32 Effect of gate current on current–voltage characteristics of a thyristor.

a light bulb or a heater, such as furnace. The amount of power delivered to the load during each cycle depends on the timing of the gate-current pulses of the thyristor (Fig. 33*b*). If the current pulses are delivered to the gate near the beginning of each cycle, more power will be delivered to the load. However, if the current pulses are delayed, the thyristor will not turn on until later in the cycle, and the amount of power delivered to the load will be substantially reduced.

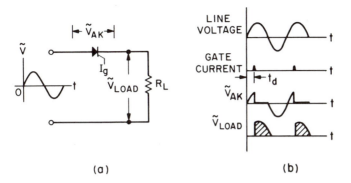

Fig. 33 (a) Schematic circuit for a thyristor application. (b) Wave forms of voltages and gate current.

4.5.2 Bidirectional and Field-Controlled Thyristors

A bidirectional thyristor is a switching device that has on- and off-states for positive and negative anode voltages and is therefore useful in ac applications. The bidirectional $p–n–p–n$ diode switch is called a *diac* (*di*ode *ac* switch). It behaves like two conventional $p–n–p–n$ diodes with the anode of the first diode connected to the cathode of the second, and vice versa, to a voltage signal of either polarity (as shown in Fig. 34a, where M1 stands for main terminal 1 and M2 for main terminal 2). When we integrate this arrangement into a single two-terminal device, we have a diac, as shown in Fig. 34b. The symmetry of this structure will result in identical performance for either polarity of applied voltage.

When a positive voltage is applied to M1 with respect to M2, junction J4 is reverse-biased so that the $n\,2'$ region does not contribute to the functioning of the device. Therefore, the $p1–n1–p2–n2$ layers constitute a $p–n–p–n$ diode that produces the forward portion of the I-V characteristic shown in Fig. 34c. If a positive voltage is applied to M2, a current will conduct in the opposite direction and J3 will be reverse-biased. Therefore, the $p\,1'–n\,1'–p\,2'–n\,2'$ layers form the reverse $p–n–p–n$ diode that produces the reverse portion of the I-V characteristics shown in Fig. 34c.

A bidirectional three-terminal thyristor is called a *triac* (*tri*ode *ac* switch).[10] The triac can switch the current in either direction by applying a low-voltage, low-current pulse of either polarity between the gate and one of the two main terminals, M1 and M2, as shown in Fig. 35. The operational principles and the I-V characteristics of a triac are similar to those of a diac. By adjusting the gate current, the breakover voltage can be varied in either polarity.

A field-controlled thyristor is a power–switching device consisting of a $p–\nu–n$ diode with multiple grids, as shown in Fig. 36a.[11] When the anode and cathode junctions are forward-biased and the grid contacts are open, electrons and holes are injected into the ν-base region, lowering its resistivity and resulting in a low-voltage drop. This is the on-state of the device. When a reverse

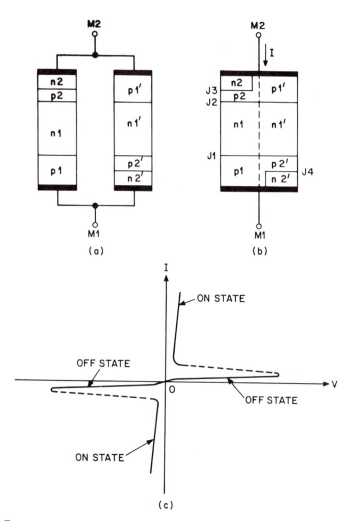

Fig. 34 (a) Two reverse–connected p–n–p–n diodes. (b) Integration of the diodes into a single two–terminal diac. (c) Current–voltage characteristics of a diac.

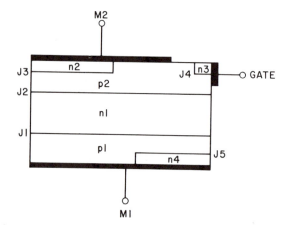

Fig. 35 Cross section of a triac, a six-layer structure having five p–n junctions.

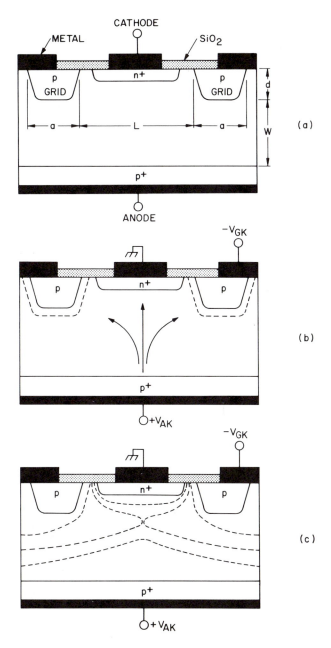

Fig. 36 (a) Cross section of a planar field-controlled thyristor. (b) Cathode current diverted to the reverse-biased grid. (c) Equipotentials in the depletion region under a forward blocking condition.[11]

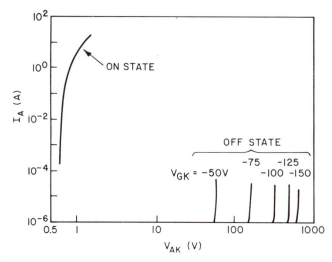

Fig. 37   On- and off-state characteristics of a field-controlled thyristor.[11]

bias is applied to the grids with respect to the cathode, the current (holes) that has been going from anode to cathode is directed to the grid, because the $p-\nu-p$ acts as a punched-through bipolar transistor, and the grid is now an efficient hole collector (Fig. 36b). If the applied grid bias is large enough, the depletion regions meet under the cathode contact and a potential barrier is established. This is illustrated in Fig. 36c, which shows the equipotential lines within the depletion region. The potential goes from a positive value associated with the anode through zero to some negative potential, then back to zero at the grounded cathode. The potential "well" thus established represents a barrier to electrons and prevents them from being injected at the cathode. Without a source of electrons, holes cannot be injected at the anode, and the device is in the forward-blocking or off-state.

The current–voltage characteristics of a field-controlled thyristor are shown in Fig. 37. For the on-state, typical forward voltage across the device is around 1 V. For the off-state, the maximum forward blocking voltage V_{AK} increases with increasing negative bias on the grid, V_{GK}. A large range of forward-blocking voltages can be obtained by adjusting the grid depth d and the grid bias V_{GK}. The field-controlled thyristor is different from the conventional thyristor because the device does not employ the regenerative cycle for switching operation. It exhibits faster turn-on and turn-off than the conventional thyristor because it can be turned on by removing the grid bias and turned off by removing the minority carriers.

REFERENCES

1   J. Bardeen and W. H. Brattain, "The Transistor, A Semiconductor Triode," *Phys. Rev.*, **74**, 230 (1948). W. Shockley, "The Theory of *p–n* Junction in Semiconductors and *p–n* Junction Transistor," *Bell Syst. Tech J.*, **28**, 435 (1949).

2   J. J. Ebers, "Four-Terminal p–n–p–n Transistor," *Proc. IEEE*, **40**, 1361 (1952).

3   S. M. Sze, *Physics of Semiconductor Devices*, 2nd ed., Wiley, New York, 1981.

4   J. M. Early, "Effects of Space-Charge Layer Widening in Junction Transistors," *Proc. IRE*, **40**, 1401 (1952).

5   W. W. Gartner, *Transistors Principles, Design and Applications*, Van Nostrand, Princeton, 1960.

6   J. J. Ebers and J. L. Moll, "Large-Signal Behavior of Junction Transistors," *Proc. IRE*, **42**, 1761 (1954).

7   R. L. Anderson, "Experiments on Ge–GaAs Heterojunction," *Solid State Electron.*, **5**, 341 (1962).

8   H. Kroemer, "Critique of Two Recent Theories of Heterojunction Lineups," *IEEE Electron Device Lett.*, **EDL-4**, 25 (1983).

9   S. K. Ghandhi, *Semiconductor Power Devices*, Wiley, New York, 1977.

10 F. E. Gentry, R. I. Scauce, and J. K. Flowers, "Bidirectional Triode p–n–p–n Switches," *Proc. IEEE*, **53**, 355 (1965).

11 D. E. Houston et al., "A Field-Terminated Diode," *IEEE Trans. Electron Devices*, **ED-23**, 905 (1976).

PROBLEMS

1   A silicon p^+–n–p transistor has impurity concentrations of 5×10^{18}, 10^{16}, and $10^{15}\,\mathrm{cm}^{-3}$ in the emitter, base, and collector, respectively. The base width W_B is 1.0 μm, and the device cross-sectional area is 3 mm². When the emitter–base junction is forward-biased to 0.5 V and the base–collector junction is reverse-biased to 5 V, calculate (*a*) the neutral base width, (*b*) the minority carrier concentration at the emitter–base junction, and (*c*) the minority carrier charge in the base region.

2   For the transistor in Problem 1, the diffusivities of minority carriers in the emitter, base, and collector are 2, 10, and 35 cm²/s, respectively; and the corresponding lifetimes are 10^{-8}, 10^{-7}, and 10^{-6} s. Find the current components I_{Ep}, I_{Cp}, I_{En}, I_{Cn}, and I_{BB} illustrated in Fig. 5. (Hint: At least six significant figures are required for the hyperbolic functions.)

3   Using the results obtained from Problems 1 and 2, (*a*) find the terminal currents I_E, I_C, and I_B of the transistor and calculate emitter efficiency, base transport factor, common-base current gain, and common-emitter current gain. (*b*) Comment on how the emitter efficiency and base transport factor can be improved. (*c*) If the transistor has a BV_{CBO} of 50 V, find the common-emitter breakdown voltage BV_{CEO} (assuming $\eta = 5$).

4   A silicon n^+–p–n transistor has abrupt dopings in both emitter and collector sides. It has impurity concentrations 10^{19}, 3×10^{16}, and $5 \times 10^{15}\,\mathrm{cm}^{-3}$ in the emitter, base, and collector, respectively. (*a*) Find the upper limit of the base–collector voltage at which the emitter bias can no longer control the collector current. Assume the base width is 0.5 μm.

4 (b) If the cutoff frequency is limited mainly by the transit time of minority carriers across the base, find the common-base and common-emitter cutoff frequencies at zero bias (the transistor has an emitter efficiency of 0.999 and a base transport factor of 0.99).

5 Plot the common-emitter current gain as a function of the base current I_B from 0 to 25 μA at a fixed V_{EC} of 5 V for the transistor shown in Fig. 14b. Explain why the current gain is not a constant.

6 A switching transistor has a base width of 0.5 μm and a diffusion constant of 10 cm^2/s. The minority carrier lifetime in the base is 10^{-7} s. The transistor is biased with a $V_{CC} = 5$ V and a load resistor of 10 kΩ. If a base current pulse of 2 μA has a duration of 1 μs, find the stored base charge and the storage time delay.

7 For an ion implanted n–p–n transistor the net impurity doping in the neutral base is given by $N(x) = N_{AO}e^{-x/l}$, where $N_{AO} = 2\times10^{18}$ cm^{-3} and $l = 0.3$ μm. (a) Find the total number of impurities in the neutral-base region per unit area for a neutral-base width of 0.8 μm. (b) Find the average impurity concentration in the neutral-base region. (c) If $L_E = 1$ μm, $N_E = 10^{19}$ cm^{-3}, $D_E = 1$ cm^2/s, the average lifetime is 10^{-6} s in the base, and the average diffusion coefficient in the base corresponds to the impurity concentration in (b), find the common-emitter current gain.

8 The current crowding in the base becomes significant when the voltage drops in the transverse base voltage is larger than kT/q. Estimate the collector current level for the transistor in Problem 7 that has an emitter area of 10^{-4} cm^2. The base resistance of the transistor can be expressed as $10^{-3} \bar{\rho}_B/W$, where W is the neutral-base width and $\bar{\rho}_B$ is the average base resistivity.

9 Derive Eqs. 98 to 101 using abrupt approximations for the impurity distributions in the heterojunction. Show that these equations will reduce to the expressions for a conventional p–n junction when both sides of the heterojunction have the same materials.

10 For the doping profile shown in Fig. 27, find the width W (> 10 μm) of the n1-region so that the thyristor has a reverse blocking voltage of 120 V. If the current gain α_2 for the n1–p2–n2 transistor is 0.4 independent of current, and α_1 of the p1–n1–p2 transistor can be expressed as $0.5 \sqrt{L_p/W} \ln(J/J_0)$, where L_p is 25 μm and J_0 is 5×10^{-6} A/cm^2, find the cross-sectional area of the thyristor that will switch at a current I_s of 1 mA.

5

Unipolar Devices

Unipolar devices are semiconductor devices in which only one type of carrier predominantly participates in the conduction process.[1] We consider five unipolar devices in this chapter: (1) the metal–semiconductor contact, (2) the junction field-effect transistor (JFET), (3) the metal–semiconductor field-effect transistor (MESFET), (4) the metal–oxide–semiconductor (MOS) diode, and (5) the metal–oxide–semiconductor field-effect transistor (MOSFET).

The first unipolar device is the metal–semiconductor contact. It is electrically similar to a one-sided abrupt *p–n* junction, yet it can be operated as a majority carrier device with inherent fast response. The metal–semiconductor contact on heavily doped semiconductors constitutes the most important form of ohmic contact.

The JFET is basically a voltage-controlled resistor. The device employs a reverse-biased *p–n* junction as a "gate" to control the resistance and thus the current flow between two ohmic contacts. The MESFET is similar to a JFET; however, it uses a metal–semiconductor rectifying contact instead of a *p–n* junction for the gate electrode. Both the JFET and the MESFET offer many attractive features for high-speed integrated circuits, because they can be made from semiconductors with high electron mobilities. Also, FETs have a negative temperature coefficient at high current levels; that is, the current decreases as temperature increases. This characteristic leads to a more uniform temperature distribution, and the device is therefore thermally stable, even when the active area is large or when many devices are connected in parallel. Further, because FETs are unipolar devices, they do not suffer from minority carrier storage effects and consequently have higher switching speeds and higher cutoff frequencies than bipolar devices.

The MOS diode is a most useful device in the study of semiconductor surfaces. Since the reliability and stability of all semiconductor devices are intimately related to their surface condition, an understanding of the surface physics, with the help of the MOS diode, is of great importance to understanding device operation. The MOS diode is also useful as a storage capacitor in integrated circuits, and it forms the basic building block for charge-coupled devices (considered in Chapter 12).

The MOSFET is basically an MOS diode with two *p–n* junctions placed immediately adjacent to the region of the MOS diode. The characteristics and operational features of the MOSFET are similar to those of the JFET and the MESFET. MOSFETs consume very low power and have high yield of working devices. Of particular importance is that a MOSFET can be readily scaled down and will take up less space than a bipolar transistor using the same design rules. The MOSFET is now the most important device for very-large-scale integrated (VLSI) circuits and is used extensively in microprocessors and semiconductor memories having thousands of individual components on a chip.

5.1 METAL–SEMICONDUCTOR CONTACTS

The first practical semiconductor device was the metal–semiconductor contact in the form of a point contact rectifier, that is, a metallic whisker pressed against a semiconductor surface. This device found many applications beginning in 1904. In 1938, Schottky suggested that the rectifying behavior could arise from a potential barrier as a result of stable space charges in the semiconductor. The model arising from this consideration is known as the *Schottky barrier*. Metal–semiconductor contacts can also be nonrectifying; that is, the contact has a negligible resistance regardless of the polarity of the applied voltage. Such a contact is called an *ohmic contact*. All semiconductor devices as well as integrated circuits need ohmic contacts to make connections to other devices in an electronic system. We shall consider the energy band diagram and the current–voltage characteristics of both the rectifying and ohmic metal–semiconductor contacts.

5.1.1 Energy Band Relation

The characteristics of point contact rectifiers were not reproducible from one device to another. They have been largely replaced by metal–semiconductor contacts fabricated by planar processes (see Chapters 8 through 12). A schematic diagram of such a device is shown in Fig. 1a. To fabricate the device, a window is opened in an oxide layer, and metal layer is deposited in a vacuum system. The metal layer covering the window is subsequently defined by a lithographic step. We shall consider a one-dimensional structure of the metal–semiconductor contact shown in Fig. 1b, which corresponds to the central section in Fig. 1a, between the dashed lines.

Figure 2a shows the energy band diagram of an isolated metal adjacent to an isolated n-type semiconductor. The metal work function is generally different from the semiconductor work function. The work function is the energy difference between the Fermi level and the vacuum level (i.e., $q \phi_m$ for the metal and $q \phi_s$ for the semiconductor). Also shown is the electron affinity $q \chi$, which is the energy difference between the conduction band edge and the vacuum level in the semiconductor. When the metal makes intimate contact with the semiconductor, the Fermi levels in the two materials must be equal at thermal equilibrium. In addition, the vacuum level must be continuous. These two requirements determine a unique energy band diagram for the ideal

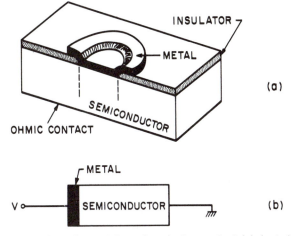

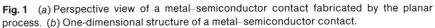

Fig. 1 (a) Perspective view of a metal–semiconductor contact fabricated by the planar process. (b) One-dimensional structure of a metal–semiconductor contact.

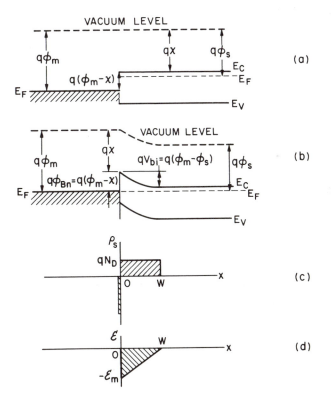

Fig. 2 (a) Energy band diagram of an isolated metal adjacent to an isolated n-type semiconductor under nonequilibrium condition. (b) Energy band diagram of a metal–semiconductor contact in thermal equilibrium. (c) Charge distribution. (d) Electric-field distribution.

metal–semiconductor contact as shown in Fig. 2b. For this ideal case the *barrier height* $q\phi_{Bn}$ is simply the difference between the metal work function and the electron affinity of the semiconductor[†]:

$$q\phi_{Bn} = q(\phi_m - \chi). \tag{1}$$

For an ideal contact between a metal and a p-type semiconductor, the barrier height $q\phi_{Bp}$ can be determined using a similar procedure:

$$q\phi_{Bp} = E_g - q(\phi_m - \chi) \tag{2}$$

where E_g is the bandgap of the semiconductor. Therefore, for a given semiconductor and for any metal, the sum of the barrier heights on n-type and p-type substrates is expected to be equal to the bandgap:

$$q(\phi_{Bn} + \phi_{Bp}) = E_g. \tag{3}$$

Figure 2c shows the charge distribution of a metal n-type semiconductor contact with a (negative) surface-charge in the metal and an equal but opposite (positive) space charge in the semiconductor. This charge distribution is identical to that of a $p^+\!-n$ junction with a corresponding identical field distribution, Fig. 2d.

Figure 3 shows the measured barrier heights for n-type silicon and n-type gallium arsenide.[2] We note that $q\phi_{Bn}$ indeed increases with increasing $q\phi_m$. However, the dependence is not as strong as predicted by Eq. 1. This is because in a practical Schottky diode, the disruption of the crystal lattice at the semiconductor surface produces a large number of surface energy states located in the forbidden bandgap. These surface states can act as donors or acceptors, which influence the final determination of the barrier height. For silicon and gallium arsenide, Eq. 1 generally underestimates the n-type barrier height and Eq. 2 overestimates the p-type barrier height. The sum of $q\phi_{Bn}$ and $q\phi_{Bp}$, however, is in agreement with Eq. 3. We can see from the previous discussion that when a metal is brought into intimate contact with a semiconductor, the conduction and valence bands of the semiconductor are brought into a definite energy relationship with the Fermi level in the metal. Once this relationship is known (i.e., the barrier heights shown in Fig. 3), it serves as a boundary condition for the solution of Poisson's equation in the semiconductor.

The energy band diagrams for metals on both n- and p-type semiconductors are shown in Fig. 4 for different biasing conditions. The built-in potential V_{bi} for the n-type semiconductor is given by

$$V_{bi} = \phi_{Bn} - V_n \tag{4}$$

where ϕ_{Bn} is the barrier height of a real metal–semiconductor contact and V_n is the potential difference between the Fermi level and E_C. Similar results can be given for p-type semiconductors. In the following discussion, however, we

---

[†] Both $q\phi_{Bn}$ (in units of eV) and ϕ_{Bn} (in units of V) are referred to as the barrier height.

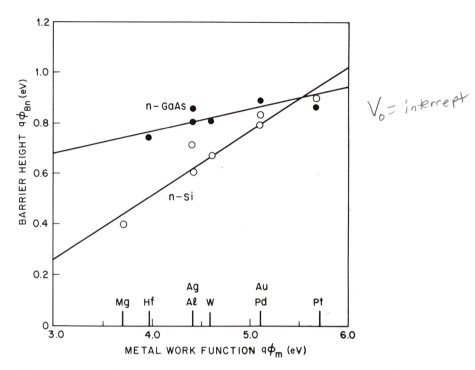

Fig. 3 Measured barrier height for metal–silicon and metal–gallium arsenide contacts.[2]

concentrate on n-type semiconductors. Of course, the results are equally applicable to the p-type with appropriate changes in symbols.

Under the abrupt-junction approximation that $\rho_s \simeq qN_D$ for $x < W$, and $\rho_s = 0$ and $d\psi/dx \simeq 0$ for $x > W$, where W is the depletion layer width, the results for the metal–semiconductor contact are similar to those of the one-sided abrupt p^+–n junction shown in Figs. 2c and 2d. We then obtain

$$W = \sqrt{\frac{2\epsilon_s}{qN_D}(V_{bi} - V)} \tag{5}$$

$$|\mathscr{E}(x)| = \frac{qN_D}{\epsilon_s}(W - x) = \mathscr{E}_m - \frac{qN_D}{\epsilon_s}x \tag{6}$$

$$\psi(x) = \frac{qN_D}{\epsilon_s}\left[Wx - \frac{1}{2}x^2\right] - \phi_{Bn} \tag{7}$$

where the applied voltage V in Eq. 5 is positive for forward bias (i.e., positive voltage on the metal with respect to the n-side) and negative for reverse bias, and $\mathscr{E}_m$ is the maximum field strength, which occurs at $x = 0$:

$$\mathscr{E}_m = \mathscr{E}(x = 0) = \sqrt{\frac{2qN_D}{\epsilon_s}(V_{bi} - V)} = \frac{2(V_{bi} - V)}{W}. \tag{8}$$

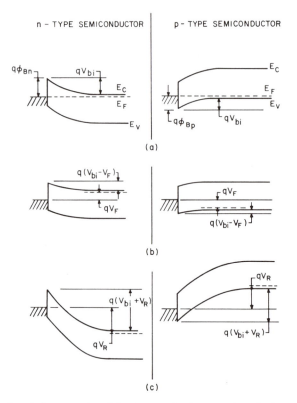

Fig. 4 Energy band diagram of metal n-type and p-type semiconductors under different biasing conditions. (*a*) Thermal equilibrium. (*b*) Forward bias. (*c*) Reverse bias.

The space charge Q_{sc} per unit area of the semiconductor and the depletion layer capacitance C per unit area are given by

$$Q_{sc} = qN_D W = \sqrt{2q \, \epsilon_s N_D (V_{bi} - V)} \qquad \text{C/cm}^2 \qquad (9)$$

$$C = \left| \frac{\partial Q_{sc}}{\partial V} \right| = \sqrt{\frac{q \, \epsilon_s N_D}{2(V_{bi} - V)}} = \frac{\epsilon_s}{W} \qquad \text{F/cm}^2 . \qquad (10)$$

Equation 10 can be written in the form

$$\frac{1}{C^2} = \frac{2(V_{bi} - V)}{q \, \epsilon_s N_D} \qquad (11a)$$

or

$$\frac{-d(1/C^2)}{dV} = \frac{2}{q \, \epsilon_s N_D} \qquad (11b)$$

$$N_D = \frac{2}{q \, \epsilon_s} \left[\frac{-1}{d(1/C^2)/dV} \right] . \qquad (11c)$$

Thus, measurements of the capacitance C per unit area as a function of voltage can provide the impurity distribution directly from Eq. 11c. If N_D is constant throughout the depletion region, we should obtain a straight line by plotting $1/C^2$ versus V. Figure 5 is a plot of measured capacitance versus voltage for tungsten—silicon and tungsten—gallium arsenide Schottky diodes.[3] From Eq. 11a, the intercept at $1/C^2 = 0$ corresponds to the built-in potential V_{bi}. Once V_{bi} is determined, the barrier height ϕ_{Bn} can be calculated from Eq. 4:

$$\phi_{Bn} = V_{bi} + V_n . \tag{12}$$

The value of V_n can be deduced from the impurity concentration.

Problem

Find the donor concentration and the barrier height of the tungsten—silicon Schottky diode shown in Fig. 5.

Solution

The plot of $1/C^2$ versus V is a straight line, which implies that the donor concentration is constant throughout the depletion region. We find

$$\frac{d(1/C^2)}{dV} = \frac{6.2 \times 10^{15} - 1.8 \times 10^{15}}{-1 - 0} = -4.4 \times 10^{15} \quad \frac{(\text{cm}^2/\text{F})^2}{\text{V}} .$$

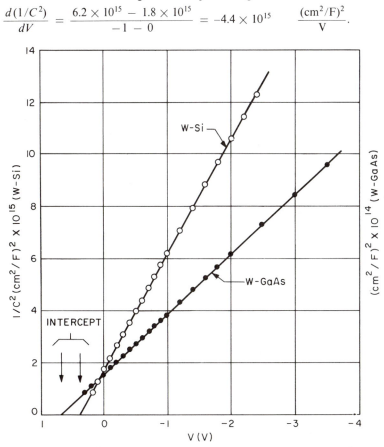

Fig. 5 $1/C^2$ versus applied voltage for W—Si and W—GaAs diode.[3]

From Eq. 11c,

$$N_D = \left[\frac{2}{1.6 \times 10^{-19} \times (11.9 \times 8.85 \times 10^{-14})} \right] \left[\frac{1}{4.4 \times 10^{15}} \right]$$

$$= 2.7 \times 10^{15} \quad cm^{-3}$$

$$V_n = \frac{kT}{q} \ln \frac{N_C}{N_D} = 0.0259 \ln \left[\frac{2.8 \times 10^{19}}{2.7 \times 10^{15}} \right] = 0.24 \, V$$

Since the intercept V_{bi} is $0.42\,V$, the barrier height is $\phi_{Bn} = V_{bi} + V_n = 0.66 \, V$

5.1.2 Current–Voltage Characteristics

The current transport in metal–semiconductor contacts is due mainly to majority carriers, in contrast to $p–n$ junctions, where current transport is due mainly to minority carriers. For Schottky diodes with moderately doped semiconductors (e.g., Si with $N_D \leqslant 10^{17} \, cm^{-3}$) operated at moderate temperature (e.g., 300 K), the dominant transport mechanism is thermionic emission of majority carriers from the semiconductor over the potential barrier into the metal.

Figure 6 illustrates the thermionic emission process.[4] At thermal equilibrium (Fig. 6a), the current density is balanced by two equal and opposite flows of carriers, thus there is zero net current. Electrons in the semiconductor tend to flow (or emit) into the metal, and there is an opposing balanced flow of electrons from the metal into the semiconductor. These current components are proportional to the density of electrons at the boundary. At the semiconductor surface the electron density n_s is

$$n_s = N_D \exp \left[\frac{-qV_{bi}}{kT} \right] = N_D \exp \left[- \frac{q(\phi_{Bn} - V_n)}{kT} \right]$$

$$= N_C \exp \left[- \frac{q\phi_{Bn}}{kT} \right] \tag{13}$$

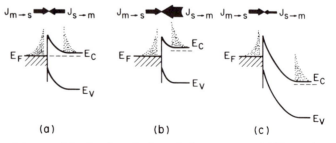

Fig. 6 Current transport by the thermionic emission process. (a) Thermal equilibrium. (b) Forward bias. (c) Reverse bias.[4]

where N_C is the density of states in the conduction band. At thermal equilibrium we have

$$| J_{m \to s} | = | J_{s \to m} | \propto n_s \tag{14}$$

or

$$| J_{m \to s} | = | J_{s \to m} | = C_1 N_C \exp \left(- \frac{q \phi_{Bn}}{kT} \right) \tag{14a}$$

where $J_{m \to s}$ is the current from the metal to the semiconductor, $J_{s \to m}$ is the current from the semiconductor to the metal, and C_1 is a proportionality constant.

When a forward bias V_F is applied to the contact (Fig. 4b), the electrostatic potential difference across the barrier is reduced, and the electron density at the surface increases to

$$n_s \simeq N_D \exp \left[- \frac{q(V_{bi} - V_F)}{kT} \right] = N_C \exp \left[- \frac{q(\phi_{Bn} - V_F)}{kT} \right]. \tag{15}$$

The current $J_{s \to m}$ that results from the electron flow out of the semiconductor is therefore altered by the same factor (Fig. 6b). The flux of electrons from the metal to the semiconductor, however, remains the same because the barrier ϕ_{Bn} remains at its equilibrium value. The net current under forward bias is then

$$\begin{aligned} J &= J_{s \to m} - J_{m \to s} \\ &= C_1 N_C \exp \left[- \frac{q(\phi_{Bn} - V_F)}{kT} \right] - C_1 N_C \exp \left[- \frac{q \phi_{Bn}}{kT} \right] \\ &= C_1 N_C e^{-q \phi_{Bn}/kT} (e^{qV_F/kT} - 1). \end{aligned} \tag{16}$$

Using the same argument for the reverse-bias condition (see Fig. 6c), the expression for the net current is identical to Eq. 16 except that V_F is replaced by $-V_R$.

The coefficient $C_1 N_C$ is found to be equal to $A^* T^2$, where A^* is called the *effective Richardson constant* (in units of $A/K^2 - cm^2$), and T is the absolute temperature. The values of A^* depend on the effective mass and are equal to 110 and 32 for n- and p-type silicon, respectively; and 8 and 74 for n- and p-type gallium arsenide, respectively.[1] The current–voltage characteristic of a metal–semiconductor contact under thermionic emission condition is therefore given by

$$J = J_s(e^{qV/kT} - 1) \tag{17}$$

and

$$J_s \equiv A^* T^2 e^{-q \phi_{Bn}/kT} \tag{18}$$

where J_s is the saturation current density and the applied voltage V is positive

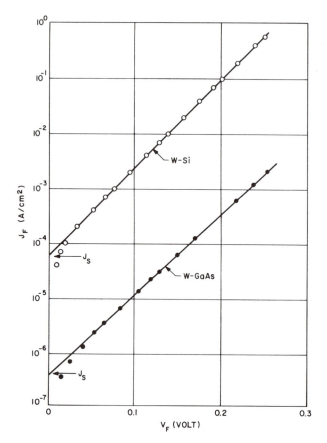

Fig. 7   Forward current density versus applied voltage of W–Si and W–GaAs diode.[3]

for forward bias and negative for reverse bias. Experimental forward *I–V* characteristics of two Schottky diodes are shown in Fig. 7. By extrapolating the forward *I–V* curve to $V = 0$, we can find J_s. From J_s and Eq. 18, we can obtain the barrier height.

In addition to the majority carrier (electron) current, in a metal *n*-type semiconductor contact, a minority carrier (hole) current exists due to hole injection from the metal to the semiconductor. The hole injection is the same as in a p^+–n junction (i.e., Eq. 52 in Chapter 3). The current density is given by

$$J_p = J_{po}(e^{qV/kT} - 1) \tag{19}$$

where

$$J_{po} \equiv \frac{qD_p n_i^2}{L_p N_D}. \tag{19a}$$

Under normal operating conditions, the minority carrier current is orders of magnitude smaller than the majority carrier current; therefore, a Schottky diode is a unipolar device.

Problem

For a tungsten–silicon Schottky diode with $N_D = 10^{16}$ cm^{-3}, find the barrier height and depletion layer width from Fig. 7. Compare the saturation current J_s with J_{po} assuming that the minority carrier lifetime in Si is 10^{-6} s.

Solution

From Fig. 7, we have $J_s = 6.5 \times 10^{-5}$ A/cm^2. The barrier height can be obtained from Eq. 18:

$$\phi_{Bn} = \frac{kT}{q} \ln \frac{A^* T^2}{J_s} = 0.0259 \ln \frac{110 \times (300)^2}{6.5 \times 10^{-5}} = 0.67 \ \text{V}$$

This result is in close agreement with the C–V measurement (see Fig. 5).

The built-in potential is given by $\phi_{Bn} - V_n$, where

$$V_n = \frac{kT}{q} \ln \frac{N_C}{N_D} = 0.17 \ \text{V} .$$

Therefore,

$$V_{bi} = 0.67 - 0.17 = 0.50 \ \text{V} .$$

The depletion layer width is given by Eq. 5 with $V = 0$:

$$W = \sqrt{\frac{2\epsilon_s V_{bi}}{qN_D}} = 2.6 \times 10^{-5} \ \text{cm} .$$

To calculate the minority carrier current density J_{po}, we need to know D_p, which is 36 cm^2/s for $N_D = 10^{16}$ cm^{-3}, and L_p, which is $\sqrt{D_p \tau_p} = 6 \times 10^{-3}$ cm. Therefore,

$$J_{po} = \frac{qD_p n_i^2}{L_p N_D} = \frac{1.6 \times 10^{-19} \times 36 \times (1.45 \times 10^{10})^2}{(6 \times 10^{-3})10^{16}} = 2 \times 10^{-11} \ \text{A/cm}^2 .$$

The ratio of the two current densities is

$$\frac{J_s}{J_{po}} = \frac{6.5 \times 10^{-5}}{2 \times 10^{-11}} = 3.2 \times 10^6 .$$

From the comparison, we see that the majority carrier current is over six orders of magnitude greater than the minority carrier current.

5.1.3 Ohmic Contact

An ohmic contact is defined as a metal–semiconductor contact that has a negligible contact resistance relative to the bulk or series resistance of the semiconductor. A satisfactory ohmic contact should not significantly degrade device performance, and it can pass the required current with a voltage drop that is small compared with the drop across the active region of the device.

A figure-of-merit for ohmic contacts is the *specific contact resistance* defined as

$$R_c \equiv \left[\frac{\partial J}{\partial V} \right]_{V=0}^{-1} \qquad \Omega - cm^2 . \tag{20}$$

For metal–semiconductor contacts with low doping concentrations, the thermionic-emission current dominates the current transport, as given by Eq. 17. Therefore,

$$R_c = \frac{k}{qA^*T} \exp \left[\frac{q\phi_{Bn}}{kT} \right] . \tag{21}$$

Equation 21 shows that low barrier height should be used to give a small R_c.

For contacts with high dopings, the barrier width becomes very narrow, and tunneling current may become dominant. The tunneling current, as indicated in the upper insert of Fig. 8, is proportional to the tunneling probability, which is given in Eq. 9 of Chapter 6:

$$I \sim \exp \left[-2W \sqrt{2m_n(q\phi_{Bn} - qV)/\hbar^2} \right] \tag{22}$$

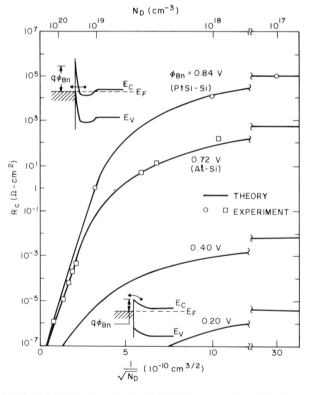

Fig. 8 Calculated and measured values of specific contact resistance. Upper insert shows the tunneling process. Lower insert shows thermionic emission over the low barrier.

where W is the depletion layer width and can be approximated as $\sqrt{(2\epsilon_s/qN_D)}\,(\phi_{Bn} - V)$. Substituting W into Eq. 22, we obtain

$$I \sim \exp\left[\frac{-C_2(\phi_{Bn} - V)}{\sqrt{N_D}}\right] \tag{23}$$

where C_2 equals $4\sqrt{m_n\epsilon_s}/\hbar$. The specific contact resistance for contacts with high dopings is thus

$$R_c \sim \exp\left[\frac{C_2\phi_{Bn}}{\sqrt{N_D}}\right]. \tag{24}$$

Equation 24 shows that in the tunneling range the specific contact resistance depends strongly on doping concentration and varies exponentially with the factor $\phi_{Bn}/\sqrt{N_D}$.

The calculated values of R_c are plotted in Fig. 8 as a function of $1/\sqrt{N_D}$. For $N_D \geqslant 10^{19}\ \text{cm}^{-3}$, R_c is dominated by the tunneling process and decreases rapidly with increased doping. On the other hand, for $N_D \leqslant 10^{17}\ \text{cm}^{-3}$, the current is due to thermionic emission, and R_c is essentially independent of doping. Also shown in Fig. 8 are experimental data for platinum silicide–silicon (PtSi–Si) and aluminum–silicon (Al–Si) diodes. They are in close agreement with the calculated values. Figure 8 shows that high doping concentration, low barrier height, or both must be used to obtain low values of R_c. These two approaches are used for all practical ohmic contacts.

5.2 THE JFET

The junction field-effect transistor (JFET) was first analyzed[5] in 1952. The JFET uses the depletion region of one or more reverse-biased p–n junctions to modulate the cross-sectional area available for current flow. The current is due to carriers of one polarity only; hence, the JFET is a unipolar device.

5.2.1 Principles of Operation

A perspective view of a JFET is shown in Fig. 9a. The JFET consists of a conductive channel with two ohmic contacts, one acting as the source and the other as the drain. When a positive voltage is applied to the drain with respect to the source, electrons flow from the source to the drain. Hence, the source acts as the origin of the carriers and the drain acts as the sink. The third electrode, the gate, forms a rectifying junction with the channel. The basic device dimensions are the channel length L, the channel width Z, and the channel depth $2a$.

To simplify the analysis, we shall consider a symmetrical structure as shown in Fig. 9b. This figure corresponds to the central section in Fig. 9a, between the dashed lines. Note that the upper and lower gates are tied together. The source is grounded, and the gate voltage V_G and drain voltage V_D are measured with respect to the source. Under normal operating conditions, the gate

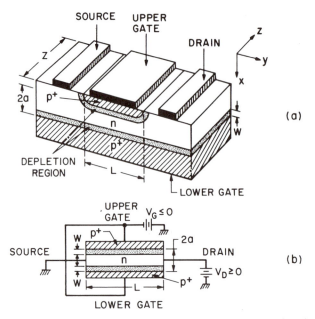

Fig. 9 (a) Perspective view of a JFET. (b) Cross section of the central region of a JFET. The source is grounded and the gate voltage and drain voltage are biased under normal operation conditions.

is zero or reverse-biased and the drain is zero or forward biased; that is, $V_G \leq 0$ and $V_D \geq 0$. Since the channel is n-type material, the device is referred to as an n-channel JFET. We will show that the n-channel JFET is preferred over the p-channel JFET because of its higher carrier mobility.

The resistance of the channel is given by

$$R = \rho \frac{L}{A} = \frac{L}{q \mu_n N_D A} = \frac{L}{2q \mu_n N_D Z (a - W)} \tag{25}$$

where N_D is the donor concentration, A is the cross-sectional area for current flow and equals $2Z(a - W)$, and W is the width of the depletion region of the upper and lower p^+–n junctions.

When no gate voltage is applied and V_D is small, as shown in Fig. 10a, a small drain current I_D flows in the channel. The magnitude of the current is given by V_D/R, where R is the channel resistance given in Eq. 25. Therefore, the current varies linearly with the drain voltage. Of course, for any given drain voltage, the voltage along the channel increases from zero at the source to V_D at the drain. Thus, the upper and lower gate junctions become increasingly reverse-biased as we proceed from the source to the drain. As V_D is increased, W increases, and the average cross-sectional area for current flow is reduced because of the increasing reverse bias of the gate junctions toward the drain. The channel resistance R also increases. As a result, the current increases at a slower rate.

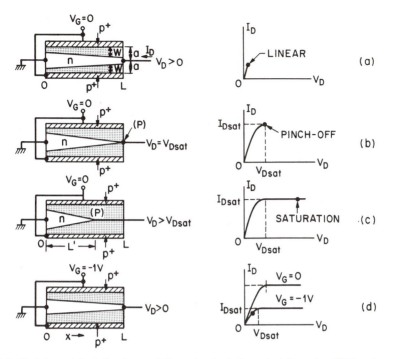

Fig. 10 Variation of depletion layer width and output characteristics of a JFET under various biasing conditions. (a) $V_G = 0$ and small V_D. (b) $V_G = 0$ and at pinch-off. (c) $V_G = 0$ and post pinch-off ($V_D > V_{D\,sat}$). (d) $V_G = -1$ V and small V_D.

As the drain voltage is further increased, the depletion layer width also increases. Eventually, the two depletion regions touch each other at the drain, as shown in Fig. 10b. This happens when $W = a$ at the drain. For an abrupt p^+–n junction, we can obtain the corresponding value of the drain voltage called the *saturation voltage* $V_{D\,sat}$:

$$V_{D\,sat} = \frac{qN_D a^2}{2\epsilon_s} - V_{bi} \qquad \text{for} \quad V_G = 0 \qquad (26)$$

where V_{bi} is the built-in potential of the gate junction. At this drain voltage, the source and the drain are *pinched off* or completely separated by a reverse-biased depletion region. The location P in Fig. 10b is called the pinch-off point. At this point, a large drain current called the *saturation current* $I_{D\,sat}$ can flow across the depletion region. This is similar to the situation caused by injecting carriers into a reverse-biased depletion region such as the collector–base depletion region of a bipolar transistor.

Beyond the pinch-off point, as V_D is increased further, the depletion region near the drain will expand and the point P will move toward the source, as indicated in Fig. 10c. However, the voltage at point P remains the same, $V_{D\,sat}$. Thus, the number of electrons per unit time arriving from the source to point

P, and hence the current flowing in the channel, remains the same, because the potential drop in the channel from source to the point P remains unaltered. Therefore, for drain voltages larger than $V_{D\,\text{sat}}$, the current remains essentially at the value $I_{D\,\text{sat}}$ and is independent of V_D.

When a gate voltage is applied to reverse-bias the gate p^+–n junction, the depletion layer width W increases. Thus, for small V_D the channel again acts as a resistor, but its resistance is larger because the cross-sectional area available for current flow is decreased. As indicated in Fig. 10d, the initial current is smaller for $V_G = -1$ V than for $V_G = 0$. When V_D is increased to a certain value, the depletion regions again touch each other. The value of such V_D is given by

$$V_{D\,\text{sat}} = \frac{qN_D a^2}{2\epsilon_s} - V_{bi} - V_G .\qquad (27)$$

For n-channel JFET, the gate voltage is negative with respect to the source, so we use the absolute value of V_G in Eq. 27 and in subsequent equations. It is clear from Eq. 27 that the application of a gate voltage V_G reduces the drain voltage required for the onset of pinch-off by an amount equal to V_G.

5.2.2 Current–Voltage Characteristics

We now consider a JFET before the onset of pinch-off as shown in Fig. 11a. The drain voltage variation along the channel is shown in Fig. 11b. The volt-

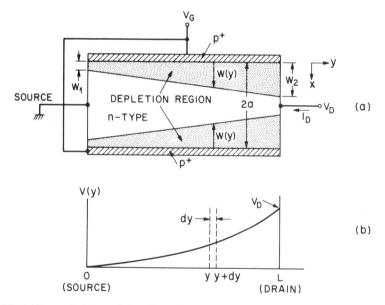

Fig. 11   (a) Expanded view of the channel region. (b) Drain voltage drop along the channel.

age drop across an elemental section dy of the channel is given by

$$dV = I_D \, dR = \frac{I_D \, dy}{2q\mu_n N_D Z [a - W(y)]} \tag{28}$$

where we have used Eq. 25 for dR and we have replaced L by dy. The depletion layer width at distance y from the source is given by

$$W(y) = \sqrt{\frac{2\epsilon_s [V(y) + V_G + V_{bi}]}{qN_D}}. \tag{29}$$

The drain current I_D is a constant, independent of y. We can rewrite Eq. 28 as

$$I_D \, dy = 2q\mu_n N_D Z [a - W(y)] \, dV. \tag{30}$$

The differentiation of the drain voltage dV is obtained from Eq. 29:

$$dV = \frac{qN_D}{\epsilon_s} W \, dW. \tag{31}$$

Substituting dV into Eq. 30 and integrating from $y = 0$ to $y = L$ yields

$$
\begin{aligned}
I_D &= \frac{1}{L} \int_{W_1}^{W_2} 2q\,\mu_n N_D Z (a - W) \frac{qN_D}{\epsilon_s} W \, dW \\
&= \frac{Z \mu_n q^2 N_D^2}{\epsilon_s L} [a(W_2^2 - W_1^2) - \frac{2}{3} (W_2^3 - W_1^3)] \\
&= I_P \left[\frac{V_D}{V_P} - \frac{2}{3} \left[\frac{V_D + V_G + V_{bi}}{V_P} \right]^{3/2} + \frac{2}{3} \left[\frac{V_G + V_{bi}}{V_P} \right]^{3/2} \right] \tag{32}
\end{aligned}
$$

where

$$I_P \equiv \frac{Z \mu_n q^2 N_D^2 a^3}{\epsilon_s L} \tag{32a}$$

and

$$V_P \equiv \frac{qN_D a^2}{2\epsilon_s}. \tag{32b}$$

The voltage V_P is called the *pinch-off voltage*, that is, the total voltage $(V_D + V_G + V_{bi})$ at which $W_2 = a$.

In Fig. 12 we show the I–V characteristics of a JFET having a pinch-off voltage of 3.2 V. The curves shown are calculated for $0 \leqslant V_D \leqslant V_{D\,sat}$ using Eq. 32. Beyond $V_{D\,sat}$ the current is taken to be constant in accordance with our previous discussion. We note that there are two different regions of the current–voltage relationship. When V_D is small, the cross-sectional area of the channel is essentially independent of V_D and the I–V characteristics are ohmic or linear. We refer to this region of operation as the linear region. In

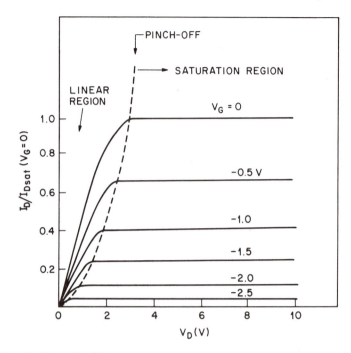

Fig. 12 Normalized ideal current–voltage characteristics with $V_P = 3.2$ V.

the other extreme, for $V_D \geqslant V_{D\,\text{sat}}$ the current saturates at $I_{D\,\text{sat}}$. We refer to this region of operation as the saturation region.

The Linear Region   In the linear region where $V_D \ll V_G + V_{bi}$, Eq. 32 can be expanded to give

$$I_D \simeq \frac{I_P}{V_P} \left[1 - \sqrt{\frac{V_G + V_{bi}}{V_P}} \right] V_D . \qquad (33)$$

The *channel conductance* g_D (also called the drain conductance) is

$$g_D \equiv \left. \frac{\partial I_D}{\partial V_D} \right|_{V_G = \text{constant}} = \frac{I_P}{V_P} \left[1 - \sqrt{\frac{V_G + V_{bi}}{V_P}} \right] . \qquad (34)$$

Figure 13 shows the normalized channel conductance in the linear region (solid curve). As the gate bias is increased, the conductance decreases until finally, at the pinchoff voltage V_P, the conductance becomes zero.

Another important parameter of a JFET is the *transconductance* g_m, which is defined by

$$g_m \equiv \left. \frac{\partial I_D}{\partial V_G} \right|_{V_D = \text{constant}} . \qquad (35)$$

The transconductance represents the change of drain current at a given drain

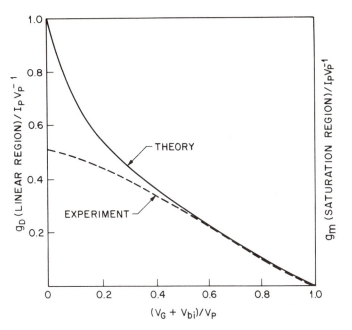

Fig. 13 Normalized drain conductance in the linear region and normalized transconductance in the saturation region versus normalized gate voltage. Solid line is for the ideal case; dotted line is for a practical device having series resistances.

voltage upon a change in gate voltage. In the linear region, g_m can be obtained from Eq. 33 as follows:

$$g_m = \frac{I_P}{2V_P^2} \sqrt{\frac{V_P}{V_G + V_{bi}}} \, V_D \,. \tag{35a}$$

The Saturation Region The drain current in the saturation region can be calculated from Eq. 32 by evaluating the current at the pinch-off point, that is, by setting $V_P = V_D + V_G + V_{bi}$:

$$I_{D\,sat} = I_P \left[\frac{1}{3} - \left[\frac{V_G + V_{bi}}{V_P} \right] + \frac{2}{3} \left[\frac{V_G + V_{bi}}{V_P} \right]^{3/2} \right]. \tag{36}$$

The corresponding saturation voltage is given by

$$V_{D\,sat} = V_P - V_G - V_{bi} \,. \tag{37}$$

The channel conductance in the saturation region is zero for the idealized situation, since $I_{D\,sat}$ in Eq. 36 is not a function of V_D. The transconductance in the saturation region can be obtained from Eqs. 35 and 36:

$$g_m = \frac{I_P}{V_P} \left[1 - \sqrt{\frac{V_G + V_{bi}}{V_P}} \right] = \frac{2Z\,\mu_n q N_D a}{L} \left[1 - \sqrt{\frac{V_G + V_{bi}}{V_P}} \right]. \tag{38}$$

Since the transconductance in the saturation region is identical to the channel conductance in the linear region (see Eq. 34), the solid curve in Fig. 13 is also the normalized transconductance curve in the saturation region.

5.2.3 Modification of Simple Theory

Small-Signal Equivalent Circuit   In Section 5.2.2 we considered the dc characteristics of a JFET, and the drain current was found to be a function of V_D and V_G, that is, $I_D = I_D(V_D, V_G)$. When ac drain and gate voltages $\tilde{v}_D$ and $\tilde{v}_G$, respectively, are superimposed on the dc voltages, the drain current is modified to $\tilde{i}_D + I_D(V_D, V_G)$, where $\tilde{i}_D$ is the ac component of the drain current. Assuming the JFET can follow the ac voltages instantaneously, we have

$$\tilde{i}_D + I_D(V_D, V_G) = I_D(V_D + \tilde{v}_D, V_G + \tilde{v}_G) \tag{39a}$$

or

$$\tilde{i}_D = I_D(V_D + \tilde{v}_D, V_G + \tilde{v}_G) - I_D(V_D, V_G). \tag{39b}$$

We can expand the first term on the right-hand side of Eq. 39b in a Taylor series and keep the dc and linear terms of the ac voltages. We obtain

$$\tilde{i}_D = \left. \frac{\partial I_D}{\partial V_D} \right|_{V_G} \tilde{v}_D + \left. \frac{\partial I_D}{\partial V_G} \right|_{V_D} \tilde{v}_G \tag{40a}$$

or

$$\tilde{i}_D = g_D \tilde{v}_D + g_m \tilde{v}_G \tag{40b}$$

where g_D and g_m are respectively the channel conductance and transconductance derived previously.

   In the previous discussion, we considered only the resistance of the channel, which can be modulated by the gate junction. In a practical device, there are series resistances near both the source and drain ends as illustrated in Fig. 14. These resistances give rise to IR drops between the source and drain contacts and the channel. The effect of these series resistances on the channel conductance and transconductance can be evaluated as follow. The center section of Fig. 14 is the "intrinsic" JFET. The voltages across the terminals of a JFET with series resistances are given by

$$V_{D'S'} = V_{DS} + I_D(R_S + R_D) \tag{41a}$$

$$V_{G'S'} = V_{GS} + I_D R_S. \tag{41b}$$

Therefore,

$$dV_{DS} = dV_{D'S'} - (R_S + R_D)dI_D \tag{42a}$$

$$dV_{GS} = dV_{G'S'} - R_S dI_D. \tag{42b}$$

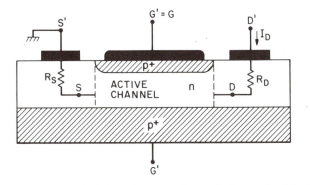

Fig. 14 JFET with source and drain resistances due to finite conductivity in the semi-conductor regions between the end of the active channel and the source and drain terminals.

In terms of the small-signal parameters, we can rewrite Eq. 42 as

$$\tilde{v}_{DS} = \tilde{v}_{D'S'} - (R_S + R_D)\tilde{i}_D \tag{43a}$$

$$\tilde{v}_{GS} = \tilde{v}_{G'S'} - R_S\tilde{i}_D \ . \tag{43b}$$

Substituting Eq. 43 into Eq. 40b yields

$$\tilde{i}_D = g_m(\tilde{v}_{G'S'} - R_S\tilde{i}_D) + g_D[\tilde{v}_{D'S'} - (R_S + R_D)\tilde{i}_D] \tag{44a}$$

or

$$\tilde{i}_D = \left[\frac{g_m}{1 + R_S g_m + (R_S + R_D)g_D}\right]\tilde{v}_{G'S'}$$

$$+ \left[\frac{g_D}{1 + R_S g_m + (R_S + R_D)g_D}\right]\tilde{v}_{D'S'}. \tag{44b}$$

Thus, the terminal transconductance and channel conductance for the JFET with series resistances are given by

$$g_m' = \frac{g_m}{1 + R_S g_m + (R_S + R_D)g_D} \tag{45a}$$

and

$$g_D' = \frac{g_D}{1 + R_S g_m + (R_S + R_D)g_D} \tag{45b}$$

which shows that the measured g_m' and g_D' will be reduced by the series resistances. A typical result for g_m' is shown in Fig. 13 (dashed curve).

A simple equivalent circuit of the JFET is shown in Fig. 15a. The output portion (right side) of the circuit corresponds to Eq. 40b with the addition of the drain resistance. The input portion (left side) is open-circuited because the

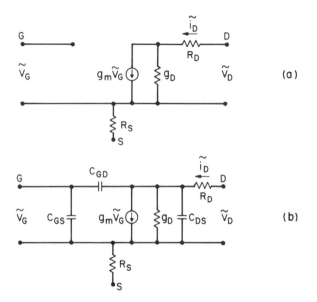

Fig. 15 (a) Low-frequency, small-signal equivalent circuit of the JFET. (b) High-frequency, small-signal equivalent circuit of the JFET.

JFET is assumed to have very high input impedance. We have also added the source resistance to the equivalent circuit. At high frequencies, we add the junction capacitances to the equivalent circuit as shown in Fig. 15b.

Cutoff Frequency   The maximum operating frequency f_T is defined as the frequency at which the JFET can no longer amplify the input signal. We will use the equivalent circuit (Fig. 15b) with the output short-circuited and no series resistances. The unity gain condition is reached when the current through the input capacitance is equal to the output drain current. The input current is

$$\tilde{i}_{in} = 2\pi f_T (C_{GS} + C_{GD})\tilde{v}_g \equiv 2\pi f_T C_G \tilde{v}_g \qquad (46)$$

where C_{GS} is the junction capacitance between gate and source, C_{GD} is the junction capacitance between gate and drain, $C_G = C_{GS} + C_{GD}$, and the output current is

$$\tilde{i}_{out} = g_m \tilde{v}_g . \qquad (47)$$

Equating Eqs. 46 to 47, we obtain the cutoff frequency

$$f_T = \frac{g_m}{2\pi C_G} \leqslant \frac{(I_P/V_P)}{2\pi ZL\,\epsilon_s/2a} = \frac{2\mu_n q N_D a^2}{\pi \epsilon_s L^2} . \qquad (48)$$

From Eq. 48 we can see that to improve high-frequency performance, we should use a JFET having high carrier mobility and short channel length. This is the reason that n-channel JFET, which has higher electron mobility, is preferred.

Problem

For an *n*-channel silicon JFET with $N_D = 10^{16}$ cm^{-3}, $N_A = 10^{19}$ cm^{-3}, $a = 1$ μm, $L = 20$ μm, $Z = 100$ μm, and $\mu_n = 1350$ cm^2/V-s, find the pinch-off voltage, the corresponding current at pinch-off with $V_G = 0$, and the cutoff frequency.

Solution

$$V_P = \frac{qN_D a^2}{2\epsilon_s} = 7.6 \text{ V} .$$

To obtain $I_{D\,\text{sat}}$ we need to know I_P and V_{bi}:

$$I_P \equiv \frac{Z \mu_n q^2 N_D^2 a^3}{\epsilon_s L} = 6.4 \times 10^{-3} \text{ A}$$

$$V_{bi} = \frac{kT}{q} \ln \frac{N_D N_A}{n_i^2} = 0.87 \text{ V} .$$

Therefore, $I_{D\,\text{sat}}$ from Eq. 36 is

$$I_{D\,\text{sat}} = I_P \left[\frac{1}{3} - \frac{V_{bi}}{V_P} + \frac{2}{3} \left[\frac{V_{bi}}{V_P} \right]^{3/2} \right] = 1.56 \text{ mA} .$$

The cutoff frequency from Eq. 48 is

$$f_T \leqslant \frac{2\mu_n q N_D a^2}{\pi \epsilon_s L^2} = 3.3 \times 10^9 \text{ Hz} = 3.3 \text{ GHz} .$$

Channel Conductance and Breakdown Voltage   Figure 16 shows the measured output characteristics of a JFET having device parameters identical to those shown in Fig. 12. The constant-current approximation for the saturation region is reasonably valid. The slight upward tilt of the current (corresponding to a nonzero channel conductance) results mainly from the reduction of the effective channel length as was indicated in Fig. 10c. The drain current in the linear region is smaller than that shown in Fig. 12 due to series resistances.

Eventually as the drain voltage increases, avalanche breakdown of the gate-to-channel diode occurs, and the drain current suddenly increases. The breakdown occurs at the drain end of the channel where the reverse voltage is highest:

$$V_B(\text{breakdown voltage}) = V_D + |V_G| . \tag{49}$$

For example, in Fig. 16 the breakdown voltage is 12 V for $V_G = 0$. At $|V_G| = 1$, the breakdown voltage is still 12 V and the drain voltage at breakdown is $(V_B - |V_G|)$, or 11 V.

5.3   THE MESFET

The metal–semiconductor field-effect transistor (MESFET) was proposed[6] in 1966. The operation of a MESFET is identical to that of a JFET. The MESFET, however, has a metal–semiconductor rectifying contact instead of a

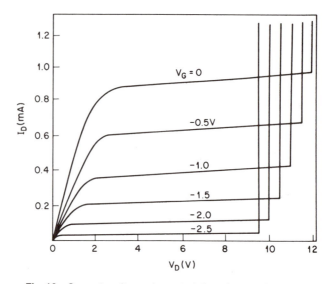

Fig. 16   Current–voltage characteristics of a practical JFET.

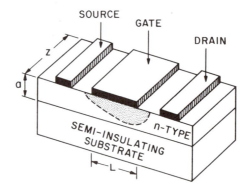

Fig. 17   Perspective view of an MESFET.

p–n junction for the gate electrode. Practical MESFETs are fabricated by using epitaxial layers on semi-insulating substrates to minimize parasitic capacitances. A perspective view of such an MESFET is shown in Fig. 17. Most MESFETs are made of *n*-type III–V compound semiconductors, such as gallium arsenide, because of their high electron mobilities which help to minimize series resistances, and their high saturation velocities which result in increased cutoff frequencies.

5.3.1   The Normally-Off MESFET

In Section 5.2 we considered only a normally-on (or depletion mode) device; that is, the device has a conductive channel at $V_G = 0$. For high-speed, low-power applications, the normally-off device is preferred. This device does not have a conductive channel at $V_G = 0$; that is, the built-in potential V_{bi} of the gate junction is sufficient to deplete the channel region. This is possible, for example, in a gallium arsenide MESFET with a very thin epitaxial layer on a semi-insulating substrate. For a normally-off MESFET, a positive bias must be applied to the gate before the channel current begins to flow. The required voltage, called the *threshold voltage* V_T, is given by

$$V_T = V_{bi} - V_P \qquad (50a)$$

or

$$V_{bi} = V_T + V_P \qquad (50b)$$

where V_P is the pinch-off voltage defined in Eq. 32b. Near the threshold, the drain current in the saturation region can be obtained by substituting V_{bi} of Eq. 50b in Eq. 36 and by using the Taylor series expansion assuming $(V_G - V_T)/V_P \ll 1$. We obtain

$$I_{D\,\text{sat}} = \frac{I_P}{2} \left\{ \frac{1}{3} - \left[1 - \left[\frac{V_G - V_T}{V_P} \right] \right] \right. + \left. \frac{2}{3} \left[1 - \left[\frac{V_G - V_T}{V_P} \right] \right]^{3/2} \right\} \simeq \frac{Z \mu_n \epsilon_s}{2aL} (V_G - V_T)^2 . \qquad (51)$$

In deriving Eq. 51 we have divided I_P by 2 to account for the upper channel in the MESFET (there is no lower channel in Fig. 17) and have used a negative sign for V_G to account for its polarity.

The basic current–voltage characteristics of normally-on and normally-off devices are similar. Figure 18 compares these two modes of operation. The main difference is the shift of threshold voltage along the V_G axis. The normally-off device (Fig. 18b) has no current conduction at $V_G = 0$, and the current varies as in Eq. 51 when $V_G > V_T$. Since the built-in potential of the gate is less than about 1 V, the forward bias on the gate is limited to about 0.5 V to avoid excessive gate current.

5.3.2 The Heterojunction MESFET

To improve MESFET device performance, various heterojunction MESFETs have been studied recently. Figure 19a shows a cross section of a double heterojunction device with III–V ternary compound $Ga_{0.47}In_{0.53}As$ as the active channel layer.[7] The semiconductor layers are grown successively on $<100>$ indium phosphide semi-insulating substrates using a molecular-beam epitaxy technique (see Chapter 8). The semiconductor layers have a good lattice match to the indium phosphide substrate, implying a low density of interface traps (refer to Section 5.4). Figure 19b shows the energy band diagram at equilibrium. The top $Al_{0.48}In_{0.52}As$ layer forms a Schottky barrier with the aluminum gate ($\phi_{Bn} = 0.8$ V), so that electrons in the channel are confined in the active $Ga_{0.47}In_{0.53}As$ layer. If the aluminum makes direct contact with the

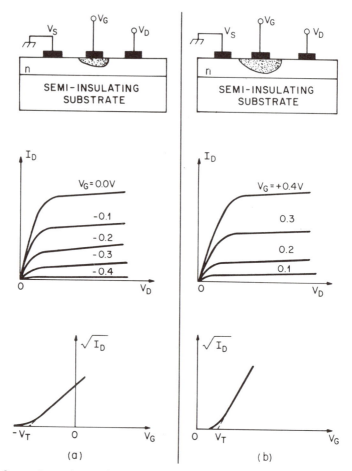

Fig. 18 Comparison of *I–V* characteristics. (*a*) Normally-on MESFET. (*b*) Normally-off MESFET.

active layer, the barrier height is too low to be considered a rectifying contact. Since the mobility and peak velocity of the active layer are higher than those for gallium arsenide, higher transconductance and higher operating speed are obtained.

Figure 20*a* shows another version of the heterojunction MESFET. A Schottky contact is made to a high bandgap material (e.g., $Al_x Ga_{1-x} As$), which is grown epitaxially on a lower bandgap material (e.g., GaAs).[8] By proper control of the bandgaps and doping concentrations of these two semiconductors, we can form an inversion layer at the interface of the two semiconductors as indicated in Fig. 20*b*. An inversion layer is a region in which the minority carrier concentration (electrons for a *p*-type semiconductor) is higher than the equilibrium majority carrier concentration (holes). We shall consider in detail the concept of inversion in Section 5.4. Because of high conductivity

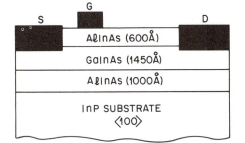

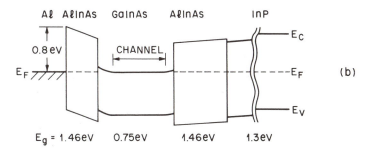

Fig. 19 (a) Cross section of a double heterojunction MESFET. (b) Energy band diagram at thermal equilibrium.[7]

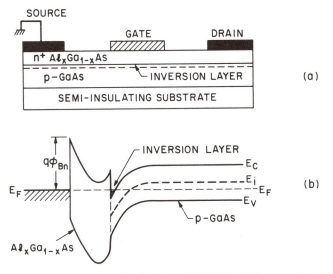

Fig. 20 (a) Cross section of a heterojunction MESFET having an inversion layer. (b) Energy band diagrams at thermal equilibrium.[8]

in this inversion layer, a large current can flow through it from source to drain. When a gate voltage is applied, the conductivity of the inversion layer will be modulated by the gate bias, which results in changes of the drain current. The current–voltage characteristics are similar to those of a normally-on (depletion mode) MESFET shown in Fig. 18a. If the lower bandgap material is lightly doped, the mobility in the inversion layer will be high. This in turn can give rise to large transconductance and high operating speed, since both parameters are directly proportional to the mobility (see Eqs. 38 and 48).

5.4 THE MOS DIODE

The metal–oxide–semiconductor, or MOS diode, is of paramount importance in semiconductor device physics.[†] This is because the device has proved to be extremely useful in the study of semiconductor surfaces, and the MOS diode forms the heart of the most important device for very-large-scale integration—the MOSFET. In this section we first consider its characteristics in the ideal case; then we extend our consideration to include the effect of metal–semiconductor work function differences, interface traps, and oxide charges.[9]

5.4.1 The Ideal MOS Diode

A perspective view of an MOS diode is shown in Fig. 21a. The cross section of the device is shown in Fig. 21b, where d is the thickness of the oxide and V is the applied voltage on the metal field plate. Throughout this section we shall use the convention that the voltage V is positive when the metal plate is positively biased with respect to the ohmic contact and V is negative when the metal plate is negatively biased with respect to the ohmic contact.

The energy band diagram of an ideal p-type semiconductor MOS diode at $V = 0$ is shown in Fig. 22. An ideal MOS diode is defined as follows. (1) At zero applied bias, the energy difference between the metal work function $q\phi_m$ and the semiconductor work function $q\phi_s$ is zero, or the work function difference $q\phi_{ms}$ for a p-type semiconductor is zero:

$$q\phi_{ms} \equiv (q\phi_m - q\phi_s) = q\phi_m - \left[q\chi + \frac{E_g}{2} + q\psi_B \right] = 0 \quad (52)$$

where $q\chi$ is the semiconductor electron affinity and $q\psi_B$ is the energy difference between the Fermi level E_F and the intrinsic Fermi level E_i. In other words, the energy band is flat (flat-band condition) when there is no applied voltage. (2) The only charges that exist in the diode under any biasing conditions are those in the semiconductor and those with equal but opposite sign on the metal surface adjacent to the oxide. (3) There is no carrier tran-

---

[†] A more general class of device is the metal–insulator–semiconductor (MIS) diode. However, because in most experimental studies the insulator has been silicon dioxide, the term MOS diode will be used interchangeably with MIS diode.

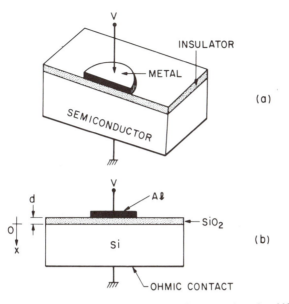

Fig. 21 (a) Perspective view of an MOS diode. (b) Cross section of an MOS diode.

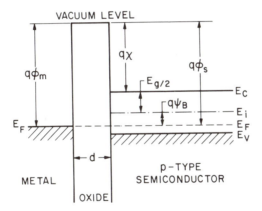

Fig. 22 Energy band diagram of an ideal MOS diode at $V = 0$.

sport through the oxide under dc-biasing conditions, or the resistivity of the oxide is infinite. The ideal MOS diode theory will serve as a foundation for understanding practical MOS devices.

When an ideal MOS diode is biased with positive or negative voltages, three cases may exist at the semiconductor surface. When a negative voltage $(V < 0)$ is applied to the metal plate, the bands near the semiconductor surface are bent upward, as shown in Fig. 23a. For an ideal MOS diode, no current flows in the device regardless of the value of the applied voltage; thus, the Fermi level in the semiconductor will remain constant. Previously, we determined that the carrier density in the semiconductor depends exponentially on the energy difference $E_i - E_F$, that is,

$$p_p = n_i e^{(E_i - E_F)/kT}. \tag{53}$$

The upward bending of the energy bands at the semiconductor surface causes an increase in the energy difference $E_i - E_F$ there, which in turn gives rise to an enhanced concentration, an accumulation of holes near the oxide–semiconductor interface. This is called the *accumulation* case. The corresponding charge distribution is shown on the right side of Fig. 23a.

When a small positive voltage ($V > 0$) is applied to an ideal MOS diode, the energy bands bend downward, and the majority carriers (holes) are depleted (Fig. 23b). This is called the *depletion* case. The space charge per unit area,

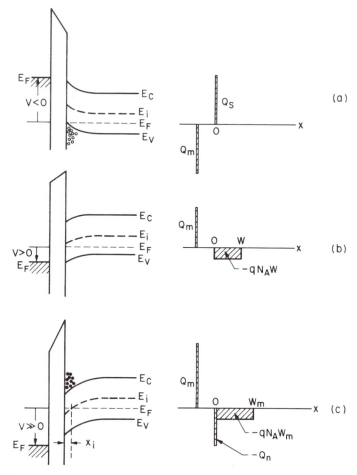

Fig. 23 Energy band diagrams and charge distributions of an ideal MOS diode. (a) Accumulation. (b) Depletion. (c) Inversion.

Q_{sc}, in the semiconductor is given by the charge within the depletion region:

$$Q_{sc} = -qN_A W \tag{54}$$

where W is the width of the surface depletion region.

When a larger positive voltage is applied, the bands bend downward even more so that the intrinsic level E_i at the surface crosses over the Fermi level as shown in Fig. 23c. The electron concentration depends exponentially on the energy difference $E_F - E_i$ and is given by

$$n_p = n_i e^{(E_F - E_i)/kT} . \tag{55}$$

In the situation shown in Fig. 23c, $(E_F - E_i) > 0$. Therefore, the electron concentration n_p at the surface is larger than n_i, and the hole concentration given by Eq. 53 becomes less than n_i. The number of electrons (minority carriers) at the surface is greater than the number of holes (majority carriers); the surface is thus inverted. This is called the *inversion* case. As the bands are bent further, eventually the conduction band edge comes close to the Fermi level. At this point the electron concentration near the surface increases very rapidly. After this point most of the additional negative charges in the semiconductor consist of the charge Q_n (Fig. 23c) due to the electrons in a very narrow n-type inversion layer $0 \leqslant x \leqslant x_i$, where x_i is the width of the inversion region. Typically, the value of x_i ranges from 10 to 100 Å and is always much smaller than the surface depletion layer width.

Once an inversion layer is formed, the surface depletion layer width reaches a maximum. This is because when the bands are bent downward far enough for strong inversion to occur, even a very small increase in band bending (corresponding to a very small increase in depletion layer width) results in a large increase in the charge Q_n in the inversion layer. Thus, under a strong inversion condition the charge per unit area in the semiconductor is given by

$$Q_s = Q_n + Q_{sc} \tag{56}$$

and

$$Q_{sc} = -qN_A W_m \tag{56a}$$

where W_m is the maximum width of the surface depletion region.

The Surface Depletion Region Figure 24 shows a more detailed band diagram at the surface of a p-type semiconductor. The electrostatic potential ψ, defined as zero in the bulk of the semiconductor. At the semiconductor surface, $\psi = \psi_s$; ψ_s is called the *surface potential*. We can express electron and hole concentrations in Eqs. 53 and 55 as a function of ψ:

$$n_p = n_i e^{q(\psi - \psi_B)/kT} \tag{57a}$$

$$p_p = n_i e^{q(\psi_B - \psi)/kT} \tag{57b}$$

where ψ is positive when the band is bent downward (as shown in Fig. 24). At

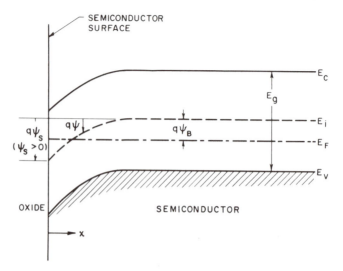

Fig. 24   Energy band diagram at the surface of a p-type semiconductor.

the surface the densities are

$$n_s = n_i e^{q(\psi_s - \psi_B)/kT} \tag{58a}$$

$$p_s = n_i e^{q(\psi_B - \psi_s)/kT} . \tag{58b}$$

From this discussion and with the help of Eq. 58, the following regions of surface potential can be distinguished:

$\psi_s < 0$	Accumulation of holes (bands bend upward)
$\psi_s = 0$	Flat-band condition
$\psi_B > \psi_s > 0$	Depletion of holes (bands bend downward)
$\psi_s = \psi_B$	Midgap with $n_s = n_p = n_i$ (intrinsic concentration)
$\psi_s > \psi_B$	Inversion (bands bend downward as shown in Fig. 24).

The potential ψ as a function of distance can be obtained by using the one-dimensional Poisson equation:

$$\frac{d^2\psi}{dx^2} = \frac{-\rho_s(x)}{\epsilon_s} \tag{59}$$

where $\rho_s(x)$ is the total space charge density. We shall use the depletion approximation that we have employed in the study of p–n junctions. When the semiconductor is depleted and the charge within the semiconductor is given by $\rho_s = -qN_A$, integration of Poisson's equation gives the electrostatic potential distribution in the surface depletion region:

$$\psi = \psi_s \left[1 - \frac{x}{W} \right]^2 . \tag{60}$$

The surface potential ψ_s is

$$\psi_s = \frac{qN_A W^2}{2\epsilon_s} . \tag{61}$$

Note that the potential distribution is identical to that for a one-sided n^+–p junction.

The surface is inverted whenever ψ_s is larger than ψ_B. However, we need a criterion for the onset of strong inversion after which the charges in the inversion layer become significant. A simple criterion is that the electron concentration at the surface is equal to the substrate impurity concentration

$$n_s = N_A . \tag{62}$$

Since $N_A = n_i e^{q\psi_B/kT}$, from Eq. 58a and 62 we obtain

$$\psi_s(\text{inv}) \simeq 2\psi_B = \frac{2kT}{q} \ln\left[\frac{N_A}{n_i}\right]. \tag{63}$$

Equation 63 states that a potential ψ_B is required to bend the bands down to the intrinsic condition at the surface ($E_i = E_F$), and bands must then be bent downward by another $q\psi_B$ at the surface to obtain the condition of strong inversion.

As we discussed previously, the surface depletion layer width reaches a maximum when the surface is strongly inverted. Accordingly, the maximum width of the surface depletion region is given by Eq. 61 in which ψ_s equals $\psi_s(\text{inv})$, or

$$W_m = \sqrt{\frac{2\epsilon_s\psi_s(\text{inv})}{qN_A}} \simeq \sqrt{\frac{2\epsilon_s(2\psi_B)}{qN_A}} = \sqrt{\frac{4\epsilon_s kT \ln(N_A/n_i)}{q^2 N_A}} \tag{64}$$

and

$$Q_{sc} = -qN_A W_m \simeq -\sqrt{2q\epsilon_s N_A(2\psi_B)} . \tag{64a}$$

The relationship between W_m and the impurity concentration is shown in Fig. 25 for silicon and gallium arsenide, where N_B is equal to N_A for p-type and N_D for n-type semiconductors.

Ideal MOS Curves Figure 26a shows the band diagram of an ideal MOS diode with the bending of the bands identical to that shown in Fig. 24. The charge distribution is shown in Fig. 26b. Clearly, in the absence of any work function differences, the applied voltage will appear partly across the oxide and partly across the semiconductor. Thus,

$$V = V_o + \psi_s \tag{65}$$

where V_o is the potential across the oxide and is given (Fig. 26c) by

$$V_o = \mathscr{E}_o d = \frac{|Q_s|d}{\epsilon_{ox}} \equiv \frac{|Q_s|}{C_o} \tag{66}$$

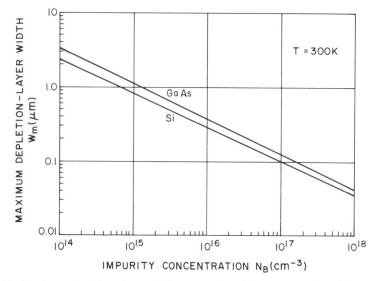

Fig. 25 Maximum depletion layer width versus impurity concentration of Si and GaAs semiconductors under strong-inversion condition.

where $\mathscr{E}_o$ is the field in the oxide, Q_s is the charge per unit area in the semiconductor, and C_o ($= \epsilon_{ox}/d$) is the oxide capacitance per unit area. The corresponding electrostatic potential distribution is shown in Fig. 26d.

The total capacitance C of the MOS diode is a series combination (Fig. 27a, insert) of the oxide capacitance C_o and the semiconductor depletion-layer capacitance C_j:

$$C = \frac{C_o C_j}{C_o + C_j} \quad \text{F/cm}^2 \tag{67}$$

where $C_j = \epsilon_s/W$, the same as for abrupt p–n junction.

From Eqs. 61, 65, 66, and 67, we can eliminate W and obtain the formula for the capacitance:

$$\frac{C}{C_o} = \frac{1}{\sqrt{1 + \frac{2\epsilon_{ox}^2 V}{qN_A \epsilon_s d^2}}} \tag{68}$$

which predicts that the capacitance will decrease with increasing gate voltage while the surface is being depleted. When the applied voltage is negative, there is no depletion region, and we have an accumulation of holes at the semiconductor surface. As a result, the total capacitance is close to the oxide capacitance ϵ_{ox}/d.

In the other extreme when strong inversion occurs, the width of the depletion region will not increase with a further increase in applied voltage. This condition takes place at a gate voltage that causes the surface potential ψ_s to

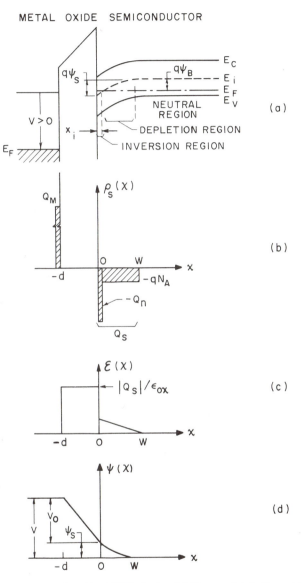

Fig. 26 (a) Band diagram of an ideal MOS diode. (b) Charge distribution under inversion condition. (c) Electric-field distribution. (d) Potential distribution.

reach ψ_s (inv) as given in Eq. 63. Substituting ψ_s (inv) into Eq. 65 and noting that the corresponding charge per unit area is $qN_A W_m$ yields the gate voltage at the onset of strong inversion. This gate voltage is called the *threshold voltage*:

$$V_T = \frac{qN_A W_m}{C_o} + \psi_s(\text{inv}) \simeq \frac{\sqrt{2\epsilon_s qN_A(2\psi_B)}}{C_o} + 2\psi_B . \qquad (69)$$

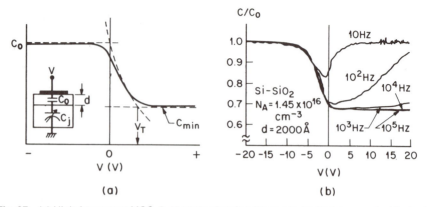

Fig. 27 (a) High-frequency MOS C–V curve showing its approximated segments (dashed lines). Insert shows the series connection of the capacitors. (b) Effect of frequency on the C–V curve.[10]

Once the strong inversion takes place, the total capacitance will remain at a minimum value C_{min} given by Eq. 67 with $C_j = \epsilon_s / W_m$:

$$C_{min} = \frac{\epsilon_{ox}}{d + (\epsilon_{ox}/\epsilon_s)W_m}.$$
(70)

A typical capacitance–voltage characteristic of an ideal MOS diode is shown in Fig. 27a based on both the depletion approximation (Eqs. 68 to 70) and exact calculations (solid curve). Note the close correlation between the depletion approximation and the exact calculations.

Although we have considered only the *p*-type substrate, all of the considerations are equally valid for an *n*-type substrate with the proper changes in signs and symbols (e.g., Q_p for Q_n). The capacitance–voltage characteristics will have identical shapes but will be mirror images of each other, and the threshold voltage is a negative quantity for an ideal MOS diode on an *n*-type substrate.

In Fig. 27a we assumed that when gate voltages change, all the incremental charge appears at the edge of the depletion region. Indeed, this happens when the measurement frequency is high. If however the measurement frequency is low enough so that generation–recombination rates in the surface depletion region are equal to or faster than the gate voltage variation, then the electron concentration (minority carriers) can follow the ac gate signal and lead to charge exchange with the inversion layer in step with the measurement signal. As a result the capacitance in strong inversion will be that of the oxide layer alone, C_o. Figure 27b shows the measured MOS C–V curves at different frequencies.[10] Note that the onset of the low-frequency curves occurs at $f \lesssim 100$ Hz.

Problem

For an ideal metal–SiO$_2$–Si diode having $N_A = 10^{16}$ cm^{-3} and $d = 250$ Å, calculate the minimum capacitance on the C–V curve of Fig. 27a. The relative dielectric constant of SiO$_2$ is 3.9.

Solution

$$C_o = \frac{\epsilon_{ox}}{d} = \frac{3.9 \times (8.85 \times 10^{-14})}{250 \times 10^{-8}} = 1.38 \times 10^{-7} \; \text{F/cm}^2$$

$$Q_{sc} = -qN_A W_m = -1.6 \times 10^{-19} \times 10^{16} \times (3 \times 10^{-5}) = -4.8 \times 10^{-8} \; \text{C/cm}^2 .$$

We used Fig. 25 for W_m;

$$\psi_s(\text{inv}) = 2\psi_B = \frac{2kT}{q} \ln \left[\frac{N_A}{n_i} \right] = 0.69 \; \text{V}$$

$$V_T = -\frac{Q_{sc}}{C_o} + 2\psi_B = 0.35 + 0.69 = 1.04 \; \text{V} .$$

The minimum capacitance $C_{\min}$ at V_T is

$$C_{\min} = \frac{\epsilon_{ox}}{d + (\epsilon_{ox}/\epsilon_s)W_m} = \frac{3.9 \times (8.85 \times 10^{-14})}{2.5 \times 10^{-6} + (3.9/11.9)3 \times 10^{-5}}$$

$$= 2.8 \times 10^{-8} \; \text{F/cm}^2 .$$

Therefore, $C_{\min}$ is about 20% of C_o.

5.4.2 The SiO$_2$–Si MOS Diode

Of all the MOS diodes, the metal–SiO$_2$–Si diode is the most extensively studied. The electrical characteristics of the SiO$_2$–Si system approach those of the ideal MOS diode. However, for commonly used metal electrodes, the work function difference $q\phi_{ms}$ is generally not zero; and there are various charges inside the oxide or at the SiO$_2$–Si interface that will, in one way or another, affect the ideal MOS characteristics.

The Work Function Difference The work function of a semiconductor $q\phi_s$, which is the energy difference between the vacuum level and the Fermi level (Fig. 22), varies with the doping concentration. For a given metal with a fixed work function $q\phi_m$ we expect that the work function difference $q\phi_{ms} \equiv q\phi_m - q\phi_s$ will vary depending on the doping of the semiconductor. One of the most common metal electrodes is aluminum, with $q\phi_m = 4.1$ eV. Another material also used extensively is the heavily doped polycrystalline silicon (also called polysilicon). The work function for n^+-polysilicon is 3.95 eV.

Figure 28 shows the work function differences for aluminum and n^+-polysilicon on silicon as the doping is varied. It is interesting to note that $q\phi_{ms}$ is always negative and is most negative for n^+-polysilicon on p-type silicon.

To construct the band diagram of an MOS diode, we start from an isolated metal and an isolated semiconductor with an oxide layer sandwiched between

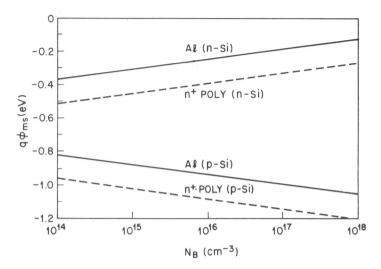

Fig. 28 Work function difference as a function of background impurity concentration for Al and n^+-polysilicon gates.

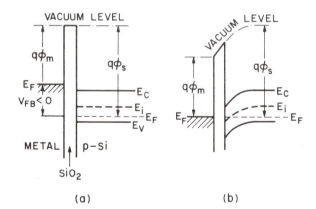

Fig. 29 (a) Energy band diagram of an isolated metal and an isolated semiconductor with an oxide layer between them. (b) Energy band diagram of an MOS diode in thermal equilibrium.

them (Fig. 29a). In this isolated situation, all bands are flat; this is the flat-band condition. As shown in the figure, $q\phi_m$ is smaller than $q\phi_s$, their difference $q\phi_{ms}$ is negative and its magnitude is equal to the voltage V_{FB}:

$$V_{FB} = \frac{q\phi_{ms}}{q} = \phi_m - \phi_s \qquad (71)$$

where V_{FB} is called the *flat-band voltage*. At thermal equilibrium, the Fermi level must be a constant and the vacuum level must be continuous. To accommodate the work function difference, the semiconductor bands bend down as shown in Fig. 29b. Thus, the metal is positively charged and the semiconductor surface is negatively charged at thermal equilibrium. It is clear that to achieve the ideal flat-band condition of Fig. 22, we have to apply a voltage equal to the work function difference $q(\phi_m - \phi_s)$, and this corresponds exactly to the situation shown in Fig. 29a, where we must apply a negative voltage V_{FB} to the metal ($V_{FB} = \phi_{ms}$).

Interface Traps and Oxide Charge In addition to the work function difference, the equilibrium MOS diode is affected by charges in the oxide and traps at the Si–SiO$_2$ interface. The basic classifications of these traps and charges are shown in Fig. 30. They are interface-trapped charge, fixed-oxide charge, oxide-trapped charge and mobile ionic charge.[11]

Interface-trapped charges Q_{it} are due to Si–SiO$_2$ interface properties and dependent on the chemical composition of this interface. The traps are located at the Si–SiO$_2$ interface with energy states in the silicon forbidden bandgap. The interface trap density (i.e., number of interface traps per unit area) is orientation dependent. In <100> orientation, the interface trap density is about an order of magnitude smaller than that in <111>. Present-day MOS diodes with thermally grown silicon dioxide on silicon have most of the interface-trapped charge neutralized by low temperature (450°C) hydrogen annealing. The value of Q_{it} for <100>-oriented silicon can be as low as 10^{10} cm^{-2}, which amounts to about one interface-trapped charge per 10^5 surface atoms. For <111>-oriented silicon, Q_{it} is about 10^{11} cm^{-2}.

The fixed-oxide charge Q_f is located within approximately 30 Å of the Si–SiO$_2$ interface. This charge is fixed and cannot be charged or discharged over a wide variation of surface potential ψ_s. Generally, Q_f is positive and depends on oxidation and annealing conditions and on silicon orientation. It has been suggested that when the oxidation is stopped, some ionic silicon is left near the

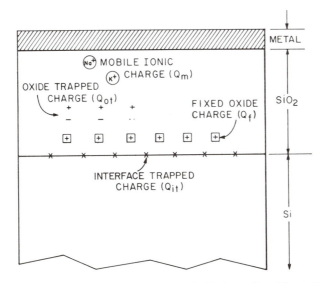

Fig. 30 Terminology for charges associated with thermally oxidized silicon.[11]

interface. These ions, along with uncompleted silicon bonds (e.g., Si–Si or Si–O bonds) at the surface, may result in the positive fixed-oxide charge Q_f. Q_f can be regarded as a charge sheet located at the Si–SiO$_2$ interface. Typical fixed-oxide charge densities for carefully treated Si–SiO$_2$ systems are about 10^{10} cm^{-2} for a <100> surface and about 5×10^{10} cm^{-2} for a <111> surface. Because of the lower values of Q_{it} and Q_f, the <100> orientation is preferred for silicon MOSFETs.

The oxide-trapped charges Q_{ot} are associated with defects in silicon dioxide. These charges can be created, for example, by X-ray radiation or high-energy electron bombardment. The traps are distributed inside the oxide layer. Most of the process-related Q_{ot} can be removed by low-temperature annealing.

The mobile ionic charges Q_m such as sodium or other alkali ions are mobile within the oxide under high-temperature and high-voltage operations. Trace contamination by alkali metal ions may cause reliability problems in semiconductor devices operated under high bias-temperature conditions. Under high bias-temperature conditions mobile ionic charges move back and forth through the oxide layer, depending on biasing conditions, and thus give rise to shifts of the C–V curve along the voltage axis. Special attention must therefore be paid to the elimination of mobile ions in device fabrication.

The foregoing charges are the effective net charges per unit area (in C/cm^2). We shall evaluate the influence of these charges on the flat-band voltage. Consider a positive sheet charge per unit area, Q_o, within the oxide as shown in Fig. 31. This positive sheet charge will induce negative charges partly in the metal and partly in the semiconductor (Fig. 31a). The resulting field distribution, obtained from integrating Poisson's equation once, is shown in the lower part of Fig. 31a where we have assumed that there is no work function difference, or $q \phi_{ms} = 0$.

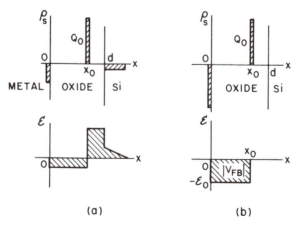

Fig. 31 Effect of a sheet charge within the oxide.[10] (a) Conditions for $V_G = 0$. (b) Flat-band condition.

To reach the flat-band condition (i.e., no charge induced in the semi-conductor), we must apply a negative voltage to the metal, as shown in Fig. 31b. As the negative voltage increases, more negative charges are put on the metal and thereby the electric-field distribution shifts downward until the electric field at the semiconductor surface is zero. Under this condition the area contained under the electric-field distribution corresponds to the flat-band voltage V_{FB}:

$$V_{FB} = -\mathscr{E}_o x_o = -\frac{Q_o}{\epsilon_{ox}} x_o = -\frac{Q_o}{C_o} \frac{x_o}{d}. \tag{72}$$

The flat-band voltage is thus dependent on both the density of the sheet charge Q_o and its location x_o within the oxide. When the sheet charge is located very close to the metal—that is, if $x_o = 0$—it will induce no charges in the silicon and therefore have no effect on the flat-band voltage. On the other hand, when Q_o is located very close to the semiconductor—$x_o = d$, (such as the fixed oxide charges Q_f)—it will exert its maximum influence and give rise to a flat-band voltage

$$V_{FB} = -\frac{Q_o}{C_o}\left[\frac{d}{d}\right] = -\frac{Q_o}{C_o}. \tag{73}$$

For the more general case of an arbitrary space charge distribution within the oxide, the flat-band voltage is given by an expression similar to Eq. 72:

$$V_{FB} = -\frac{1}{C_o}\left[\frac{1}{d}\int_0^d x\,\rho(x)\,dx\right] \tag{74}$$

where $\rho(x)$ is the volume charge density in the oxide. Once we know $\rho_{ot}(x)$, the volume charge density for the oxide-trapped charges, and $\rho_m(x)$, the volume charge density for the mobile ionic charges, we can obtain Q_{ot} and Q_m and their corresponding contribution to the flat-band voltage:

$$Q_{ot} \equiv \frac{1}{d}\int_0^d x\,\rho_{ot}(x)\,dx \tag{75a}$$

$$Q_m \equiv \frac{1}{d}\int_0^d x\,\rho_m(x)\,dx. \tag{75b}$$

If the value of the work function difference $q\,\phi_{ms}$ is not zero, and if the values of Q_f, Q_m, and Q_{ot} are significant (assuming negligible interface traps), the experimental capacitance–voltage curve will be shifted from the ideal theoretical curve by an amount

$$V_{FB} = \phi_{ms} - \frac{Q_f + Q_m + Q_{ot}}{C_o}. \tag{76}$$

The curve labeled (a) in Fig. 32 shows the ideal C–V characteristics of an MOS diode. Due to nonzero ϕ_{ms}, Q_f, Q_m, or Q_{ot}, the C–V curve will be shifted by an amount given by Eq. 76. This parallel shift of the C–V curve is

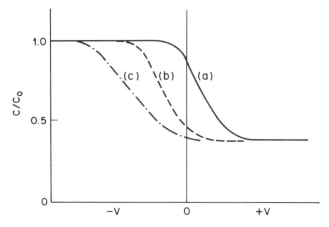

Fig. 32 Effect of fixed oxide charges and interface traps on the C–V characteristics of an MOS diode.

illustrated in curve (*b*). If, in addition, there are large amounts of interface-trapped charges, the charges in the interface traps will vary with the surface potential. The C–V curve will be displaced by an amount which itself changes with the surface potential. Therefore, curve (*c*) is distorted as well as shifted due to interface-trapped charges.

5.5 THE MOSFET: BASIC CHARACTERISTICS

The metal–oxide–semiconductor field-effect transistor (MOSFET) is the most important device for very-large-scale integrated circuits such as microprocessors and semiconductor memories. The MOSFET is also becoming an important power device. It has many acronyms, including IGFET (insulated-gate field-effect transistor), MISFET (metal–insulator–semiconductor field-effect transistor), and MOST (metal–oxide–semiconductor transistor). The MOSFET is a member of the family of field-effect transistors. The other members, JFETs and MESFETs, have already been considered in Sections 5.2 and 5.3, respectively. The current–voltage characteristics of the MOSFET are similar to those of the JFET and MESFET.

Figure 33 shows the first MOSFET, fabricated in 1960 using a thermally oxidized silicon substrate.[12] The device had a channel length of over 20 μm and a gate oxide thickness over 1000 Å. Although present-day MOSFETs have been scaled down considerably, the choice of silicon and thermally-grown silicon dioxide used in the first MOSFET remains to be the most important combination. Hence, most of the results in the section are obtained from the Si–SiO$_2$ system.

5.5.1 Linear and Saturation Regions

A perspective view for an MOSFET is shown in Fig. 34. It is a four-terminal device and consists of a *p*-type semiconductor substrate in which two

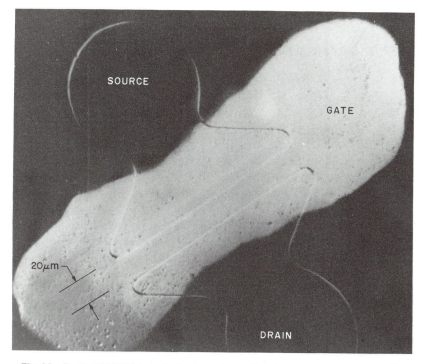

Fig. 33 First MOSFET fabricated using a thermally oxidized silicon substrate.[12]

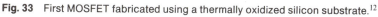

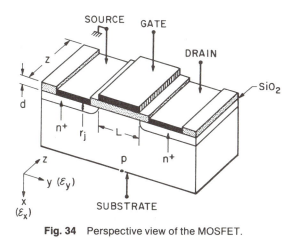

Fig. 34 Perspective view of the MOSFET.

n^+ regions, the source and drain, are formed.[†] The metal contact on the oxide is called the gate. Heavily doped polysilicon or a combination of silicide (e.g., $MoSi_2$) and polysilicon can also be used as the gate electrode. The basic dev-

---

[†] This is an *n*-channel device; one may consider a *p*-channel device by substituting *p* for *n* and reversing the polarities of all the voltages.

ice parameters are the channel length L, which is the distance between the two metallurgical n^+–p junctions; the channel width Z; the oxide thickness d; the junction depth r_j; and the substrate doping N_A. Note that the central section of the device corresponds to the MOS diode discussed in Section 5.4.

The source contact will be used as the voltage reference throughout this section. When no voltage is applied to the gate, the source-to-drain electrodes correspond to two p–n junctions connected back to back. The only current that can flow from source to drain is the reverse leakage current.[§] When we apply a sufficiently large positive bias to the gate, the central MOS structure is inverted so that a surface inversion layer (or channel) is formed between the two n^+-regions. The source and drain are then connected by a conducting-surface n-channel through which a large current can flow. The conductance of this channel can be modulated by varying the gate voltage. The back-surface contact (or substrate contact) can be at the reference voltage or is reverse-biased; the back-surface voltage will also affect the channel conductance.

We shall now present a qualitative discussion of MOSFET operation. Let us consider that a voltage is applied to the gate, causing an inversion at the semiconductor surface (Fig. 35a). If a small drain voltage is applied, electrons will flow from the source to the drain (the corresponding current will flow from drain to source) through the conducting channel. Thus, the channel acts as a resistance, and the drain current I_D is proportional to the drain voltage. This is the linear region as indicated by the constant-resistor line in the right-hand diagram of Fig. 35a.

When the drain voltage increases, eventually it reaches a point at which the width of the inversion layer x_i at $y = L$ is reduced to zero; this is called the pinch-off point (Fig. 35b). Beyond the pinch-off point, the drain current remains essentially the same, because for $V_D > V_{D\text{sat}}$, at point P the voltage $V_{D\text{sat}}$ remains the same. Thus, the number of carriers arriving at point P from the source, and hence the current flowing from the drain to the source, remains the same. The major change is a decrease of L to the value L' shown in Fig. 35c. Carrier injection from P into the drain depletion region is quite similar to that of carrier injection from an emitter–base junction to the base–collector depletion region of a bipolar transistor. Note that the resulting I–V curves are very similar to those for the JFET.

We shall now derive the basic MOSFET characteristics under the following ideal conditions: (1) The gate structure corresponds to an ideal MOS diode as defined in Section 5.4, that is, there are no interface traps, fixed-oxide charges, or work function differences. (2) Only drift current will be considered. (3) Carrier mobility in the inversion layer is constant. (4) Doping in the channel is uniform. (5) Reverse-leakage current is negligibly small. (6) The transverse field ($\mathscr{E}_x$ in the x-direction, shown in Fig. 34, which is perpendicular to the current flow) in the channel is much larger than the longitudinal field ($\mathscr{E}_y$ in the y-direction, which is in parallel to the current flow). The last condi-

---

[§] This is true for n-channel normally-off MOSFET. Other types of MOSFETs will be discussed later.

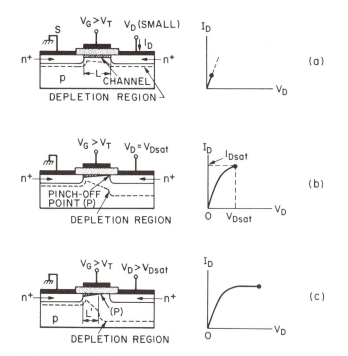

Fig. 35 Operations of the MOSFET and output I–V characteristics. (a) Low drain voltage. (b) Onset of saturation. Point P indicates the pinch-off point. (c) Beyond saturation.

tion is called the gradual-channel approximation and generally is valid for long-channel MOSFETs.

Figure 36a shows a MOSFET operated in the linear region. Under the above ideal conditions, the total charge induced in the semiconductor per unit area, Q_s, at a distance y from the source is shown in Fig. 36b, which is an enlarged central section of Fig. 36a. Q_s is given from Eqs. 65 and 66 by

$$Q_s(y) = -[V_G - \psi_s(y)]C_o \tag{77}$$

where $\psi_s(y)$ is the surface potential at y and $C_o = \epsilon_{ox}/d$ is the gate capacitance per unit area. The charge in the inversion layer is given by Eqs. 56 and 77:

$$Q_n(y) = Q_s(y) - Q_{sc}(y)$$
$$= -[V_G - \psi_s(y)]C_o - Q_{sc}(y). \tag{78}$$

The surface potential $\psi_s(y)$ at inversion can be approximated by $2\psi_B + V(y)$, where $V(y)$ as shown in Fig. 36c is the reverse bias between the point y and the source electrode (which is assumed to be grounded). The charge within the surface depletion region $Q_{sc}(y)$ was given previously as

$$Q_{sc}(y) = -qN_A W_m \simeq -\sqrt{2\epsilon_s qN_A[V(y) + 2\psi_B]}. \tag{79}$$

Substituting Eq. 79 in 78 yields

$$Q_n(y) \simeq - [V_G - V(y) - 2\psi_B]C_o + \sqrt{2\epsilon_s qN_A[V(y) + 2\psi_B]} . \quad (80)$$

The conductivity of the channel at position y can be approximated by

$$\sigma(x) = qn(x)\mu_n(x) . \quad (81)$$

For a constant mobility the channel conductance is then given by

$$g = \frac{Z}{L} \int_0^{x_i} \sigma(x) \, dx = \frac{Z\mu_n}{L} \int_0^{x_i} qn(x) \, dx . \quad (82)$$

The integral $\int_0^{x_i} qn(x) \, dx$ corresponds to the total charge per unit area in the inversion layer and is therefore equal to $|Q_n|$, or

$$g = \frac{Z\mu_n}{L} |Q_n| . \quad (83)$$

The channel resistance of an elemental section dy (Fig. 36b) is

$$dR = \frac{dy}{gL} = \frac{dy}{Z\mu_n |Q_n(y)|} \quad (84)$$

and the voltage drop across this elemental section is

$$dV = I_D \, dR = \frac{I_D \, dy}{Z\mu_n |Q_n(y)|} \quad (85)$$

where I_D is the drain current which is independent of y. Substituting Eq. 80 into Eq. 85 and integrating from the source ($y = 0$, $V = 0$) to the drain

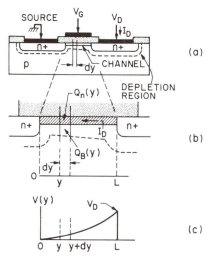

Fig. 36 (a) MOSFET operated in the linear region. (b) Enlarged view of the channel region. (c) Drain voltage drop along the channel.

$(y = L, V = V_D)$ yield

$$I_D \simeq \frac{Z}{L} \mu_n C_o \left\{ \left[V_G - 2\psi_B - \frac{V_D}{2} \right] V_D \right.$$

$$\left. - \frac{2}{3} \frac{\sqrt{2\epsilon_s q N_A}}{C_o} \left[(V_D + 2\psi_B)^{3/2} - (2\psi_B)^{3/2} \right] \right\}. \tag{86}$$

Figure 37 shows the current–voltage characteristics of an idealized MOS-FET based on Eq. 86. For a given V_G, the drain current first increases linearly with drain voltage (the linear region), then gradually levels off, approaching a saturated value (the saturation region). The dashed line indicates the locus of the drain voltage ($V_{D\,\text{sat}}$) at which the current reaches a maximum value.

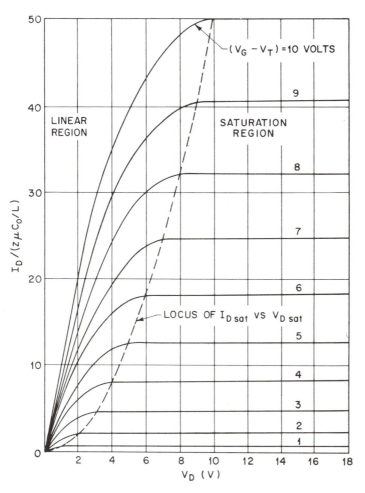

Fig. 37 Idealized drain characteristics of an MOSFET. For $V_D \geqslant V_{D\,\text{sat}}$, the drain current remains constant.

We shall consider the linear and saturation regions. For the case of small V_D, Eq. 86 reduces to

$$I_D \simeq \frac{Z}{L} \mu_n C_o (V_G - V_T) V_D \qquad \text{for} \quad V_D \ll (V_G - V_T) \qquad (87)$$

where V_T is the threshold voltage given previously in Eq. 69:

$$V_T \simeq \frac{\sqrt{2\epsilon_s q N_A (2\psi_B)}}{C_o} + 2\psi_B . \qquad (88)$$

By plotting I_D versus V_G (for a given small V_D), the threshold voltage can be deduced from the linearly extrapolated value at the V_G axis. In the linear region, Eq. 87, the channel conductance g_D and the transconductance g_m are given as

$$g_D \equiv \left. \frac{\partial I_D}{\partial V_D} \right|_{V_G = \text{const}} \simeq \frac{Z}{L} \mu_n C_o (V_G - V_T) \qquad (89)$$

$$g_m \equiv \left. \frac{\partial I_D}{\partial V_G} \right|_{V_D = \text{const}} \simeq \frac{Z}{L} \mu_n C_o V_D . \qquad (90)$$

When the drain voltage is increased to such a point that the charge in the inversion layer $Q_n(y)$ at $y = L$ becomes zero, the number of mobile electrons at the drain are reduced drastically. This point, called pinch-off, is analogous to the pinch-off point in the JFET. The drain voltage and the drain current at this point are designated as $V_{D\,\text{sat}}$ and $I_{D\,\text{sat}}$, respectively. For drain voltages larger than $V_{D\,\text{sat}}$, we have the saturation region. We can obtain the value of $V_{D\,\text{sat}}$ from Eq. 80 under the condition $Q_n(L) = 0$:

$$V_{D\text{sat}} \simeq V_G - 2\psi_B + K^2 \left[1 - \sqrt{1 + 2V_G/K^2} \right] \qquad (91)$$

where $K \equiv \sqrt{\epsilon_s q N_A}/C_o$. The saturation current $I_{D\,\text{sat}}$ can be obtained by substituting Eq. 91 into Eq. 86:

$$I_{D\,\text{sat}} \simeq \frac{Z \mu_n \epsilon_{ox}}{2dL} (V_G - V_T)^2 . \qquad (92)$$

It is interesting to note that Eq. 92 is identical to Eq. 51 for the JFET, except that the oxide thickness d replaces the channel depth a and that ϵ_{ox} replaces ϵ_s. The threshold voltage V_T in the saturation region for low substrate doping and thin oxide layers is the same as that from Eq. 88. At higher doping levels, V_T becomes V_G dependent. For the idealized MOSFET in the saturation region, the channel conductance is zero, and the transconductance can be obtained from Eq. 92:

$$g_m \equiv \left. \frac{\partial I_D}{\partial V_G} \right|_{V_D = \text{const}} = \frac{Z \mu_n \epsilon_{ox}}{dL} (V_G - V_T) . \qquad (93)$$

As in the case of the JFET, the g_m in the saturation region is equal to g_D in the linear region.

5.5.2 Equivalent Circuit and Frequency Response

The equivalent circuit for the MOSFET is essentially the same as that for the JFET. The low-frequency equivalent circuit is the same as that shown in Fig. 15a, where the transconductance g_m is given by Eq. 90 for the linear region and by Eq. 93 for the saturation region and the channel conductance g_D is given by Eq. 89 for the linear region and is zero for the saturation region. At higher frequencies, we have to take into account the capacitive coupling between device terminals. The modified equivalent circuit is shown in Fig 38. The capacitance C_{GS} is associated with the MOS gate and can be given by ZLC, where C is the MOS capacitance from Eq. 67. The capacitance C_{GD} is the gate-to-drain capacitance, which provides undesirable feedback between the input and output and is associated with the overlap capacitance that results from the portion of the gate that overlaps the drain region.

From the equivalent circuit, Fig. 38, we can estimate the maximum operating frequency f_T of a MOSFET. Let f_T be the frequency at which the ratio of the output current to input current becomes unity, that is, the device can no longer amplify the input signal when the output of the device is short-circuited. By inspection, the input current with the output short-circuited is

$$\tilde{i}_{in} = j\omega \, (C_{GS} + C_{GD})\tilde{v}_G \simeq j\omega \, (C_oZL)\tilde{v}_G \, . \tag{94}$$

The output current is

$$\tilde{i}_{out} = g_m \tilde{v}_G \, . \tag{95}$$

By letting $\tilde{i}_{out}/\tilde{i}_{in} = 1$, we obtain f_T:

$$f_T = \frac{g_m}{2\pi C_{in}} = \frac{\mu_n V_D}{2\pi L^2} \qquad \text{for} \quad V_D \leqslant V_{Dsat} \, . \tag{96}$$

To obtain high-frequency or high-speed operation, it is thus necessary to have a short channel length and high carrier mobility.

5.5.3 The Subthreshold Region

When the gate voltage is below the threshold voltage and the semiconductor surface is only weakly inverted, the corresponding drain current is

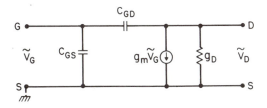

Fig. 38 Small-signal equivalent circuit of the MOSFET.

called the *subthreshold current*. The subthreshold region is particularly important when the MOSFET is used as a low-voltage, low-power device such as a
switch in digital logic and memory applications, because the subthreshold
region describes how the switch turns on and off.

In the subthreshold region, the drain current is dominated by diffusion
instead of drift, and is derived in the same way as the collector current is in a
bipolar transistor with homogeneous base doping. If we consider the MOS-
FET as an n–p–n (source–substrate–drain) bipolar transistor, we have

$$I_D = -qAD_n \frac{dn}{dy} = qAD_n \frac{n(0) - n(L)}{L} \tag{97}$$

where A is the channel cross section of the current flow and $n(0)$ and $n(L)$ are
the electron densities in the channel at the source and the drain, respectively
(Fig. 36b). These electron densities are given by Eq. 57a:

$$n(0) = n_i e^{q(\psi_s - \psi_B)/kT} \tag{98a}$$

$$n(L) = n_i e^{q(\psi_s - \psi_B - V_D)/kT} \tag{98b}$$

where ψ_s is the surface potential at the source. Substituting Eq. 98 into Eq. 97
gives

$$I_D = \frac{qAD_n n_i e^{-q\psi_B/kT}}{L} (1 - e^{-qV_D/kT}) e^{q\psi_s/kT} . \tag{99}$$

The surface potential ψ_s is approximately $V_G - V_T$. Therefore, the drain
current will decrease exponentially when V_G becomes less than V_T:

$$I_D \sim e^{q(V_G - V_T)/kT} . \tag{100}$$

A typical measured curve for the subthreshold region is shown in Fig. 39.
Note the exponential dependence of I_D on $V_G - V_T$ for $V_G < V_T$. To
reduce the subthreshold current to a negligible value, we must bias the MOS-
FET a half-volt or more below V_T.

5.5.4 Types of MOSFETs

There are basically four different types of MOSFETs, depending on the
type of inversion layer. If, at zero gate bias, the channel conductance is very
low and we must apply positive voltage to the gate to form the n-channel, then
the device is a normally-off (enhancement) n-channel MOSFET. If an n-
channel exists at zero bias and we must apply a negative voltage to the gate to
deplete carriers in the channel to reduce the channel conductance, then the
device is a normally-on (depletion) n-channel MOSFET. Similarly, we have
the p-channel normally-off (enhancement) and normally-on (depletion)
MOSFETs.

The device cross sections, output characteristics (i.e., I_D versus V_D), and
transfer characteristics (i.e., I_D versus V_G) of the four types are shown in
Fig. 40. Note that for the normally-off n-channel device, a positive gate bias

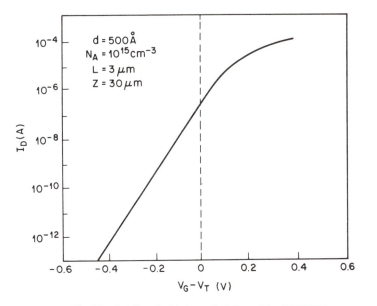

Fig. 39 Subthreshold characteristics of the MOSFET.

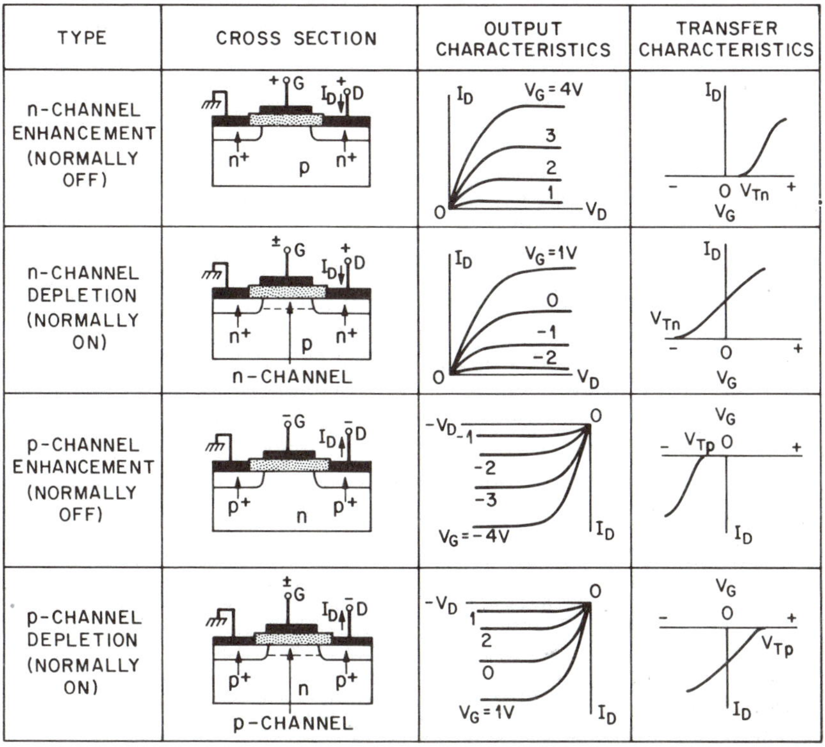

TYPE	CROSS SECTION	OUTPUT CHARACTERISTICS	TRANSFER CHARACTERISTICS
n-CHANNEL ENHANCEMENT (NORMALLY OFF)		$V_G = 4V$, 3, 2, 1	
n-CHANNEL DEPLETION (NORMALLY ON)	n-CHANNEL	$V_G = 1V$, 0, -1, -2	V_{Tn}
p-CHANNEL ENHANCEMENT (NORMALLY OFF)		-1, -2, -3, $V_G = -4V$	V_{Tp}
p-CHANNEL DEPLETION (NORMALLY ON)	p-CHANNEL	1, 2, 0, $V_G = 1V$	V_{Tp}

Fig. 40 Cross sections and output and transfer characteristics of four types of MOSFETs.

larger than the threshold voltage V_T must be applied before a substantial drain current flows. For the normally-on n-channel device, a large current can flow at $V_G = 0$, and the current can be increased or decreased by varying the gate voltage. This discussion can be readily extended to p-channel device by changing polarities.

5.6 THE MOSFET: THRESHOLD VOLTAGE AND DEVICE SCALING

One of the most important parameters of MOSFETs is the threshold voltage. In this section we shall consider various ways to control the threshold voltage. Since the beginning of the integrated circuit era in 1959, minimum device dimensions have been reduced by two orders of magnitude. The reduction in device dimensions is motivated by the need to make highly complex integrated circuits comprising as many as hundreds of thousands of transistors on a single semiconductor chip. At the present time, the minimum channel length is about 1 μm. We shall consider the device scaling and the associated "short-channel" effects.

5.6.1 Threshold Voltage

The ideal threshold voltage is given in Eq. 88. However, when we incorporate the effects of the fixed-oxide charge and the difference in work functions, there is a flat-band voltage shift. This in turn causes a change in the threshold voltage:

$$
\begin{aligned}
V_T &\simeq V_{FB} + 2\psi_B + \frac{\sqrt{2\epsilon_s q N_A (2\psi_B)}}{C_o} \\
&= \left[\phi_{ms} - \frac{Q_f}{C_o} \right] + 2\psi_B + \frac{\sqrt{4\epsilon_s q N_A \psi_B}}{C_o}
\end{aligned}
\tag{101}
$$

Figure 41 shows the calculated threshold voltage of n-channel (V_{Tn}) and p-channel (V_{Tp}) MOSFETs for two oxide thicknesses as a function of their substrate doping and assuming an n^+-polysilicon gate and $Q_f = 0$. The values of ϕ_{ms} were given previously in Fig. 28. For n-channel devices, V_{Tn} is negative at low dopings and becomes positive at high dopings. For p-channel devices, V_{Tp} is always negative.[13]

To control the threshold voltage, one of the most valuable tools is ion implantation (see Chapter 10). It is possible to obtain close control of V_T by ion implantation, because very precise quantities of impurity can be introduced. Figure 42 illustrates the result of a boron implantation through the gate oxide of an n-channel MOSFET (with p-type substrate) such that the implant dose peak occurs at the Si–SiO$_2$ interface. The negatively charged boron acceptors increase the doping level of the channel. As a result, V_T increases. Similarly, a shallow boron implant into a p-channel MOSFET can reduce V_T.

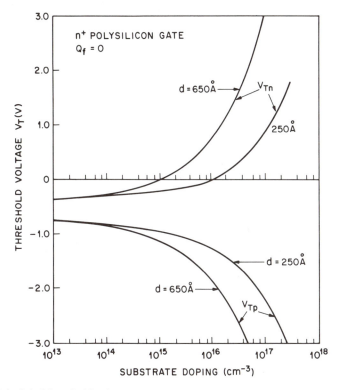

Fig. 41 Calculated threshold voltage of n-channel (V_{Tn}) and p-channel (V_{Tp}) MOSFETs as a function of impurity concentration, assuming an n^+-polysilicon gate and zero fixed oxide charge.[13]

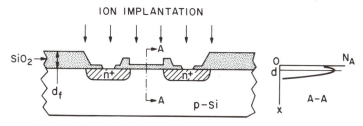

Fig. 42 Threshold voltage adjustment using ion implantation.

Problem

For an n-channel n^+-polysilicon–SiO$_2$–Si MOSFET with $N_A = 10^{16}$ cm^{-3} and $Q_f/q = 2 \times 10^{10}$ cm^{-2}, calculate V_T for a gate oxide of 250 Å. What is the boron ion dose required to increase V_T to 1 V? Assume that the implanted acceptors form a sheet of negative charge at the Si–SiO$_2$ interface.

Solution

From the example in Section 5.4.1 for an ideal device, we have $C_o = 1.38 \times 10^{-7}$ F/cm^2 and V_T(ideal MOS) = 1.04 V. For the practical devices, $\phi_{ms} = -1.08$ V (Fig. 28). Therefore, from Eq. 101,

$$V_T = \left[-1.08 - \frac{2\times10^{10}\times1.6\times10^{-19}}{1.38\times10^{-7}} \right] + V_T(\text{ideal MOS})$$

$$= -0.063 \text{ V} .$$

The boron charge causes a flat-band shift of qF_B/C_o. Thus,

$$1 = -0.063 + \frac{qF_B}{1.38\times10^{-7}}$$

$$F_B = \left[\frac{1.38\times10^{-7}}{1.6\times10^{-19}} \right] 1.063 = 9.1\times10^{11} \text{ cm}^{-2} .$$

We can also control V_T by varying the oxide thickness. This approach is used extensively for isolation of MOSFETs, as shown in Fig. 42, where the gate oxide is much thinner than the oxide outside the drain or source regions (the field oxide). The V_T for the field oxide is larger than that for the thin gate oxide. For example, for the device in the previous problem, the ideal V_T with a field oxide d_f of 1 μm is 14.7 V, which is an order of magnitude larger than the V_T (1.04 V) for the gate oxide. Therefore, if we apply a voltage of 5 V to both the gate and field oxides, an inversion channel forms under the gate, while the semiconductor surface under the field oxide remains depleted.

Substrate bias can also influence the threshold voltages. When a reverse bias is applied between the substrate and the source (i.e., negative voltage V_{BS} on the p-type substrate with respect to the source for an n-channel devices), the depletion region is widened and the threshold voltage required to achieve inversion must be increased to accommodate the larger Q_{sc}. A simplified view is that W is widened uniformly along the channel, so that Eq. 64a should be modified to

$$Q_B' = -\sqrt{2q\,\epsilon_s N_A(2\psi_B + V_{BS})} . \tag{102}$$

The change in threshold voltage due to the substrate bias is

$$\Delta V_T = \frac{\sqrt{2q\,\epsilon_s N_A}}{C_o} \left(\sqrt{2\psi_B + V_{BS}} - \sqrt{2\psi_B} \right) . \tag{103}$$

If we plot the channel conductance versus V_G, the intercept at the V_G-axis corresponds to the threshold voltage, Eq. 89. Such a plot is shown in Fig. 43 for four different substrate biases. As the magnitude of the substrate bias V_{BS} increases from 1 V to 16 V, the threshold voltage also increases from 1.25 V to 3.25 V. The substrate effect can be used to raise the threshold voltage of a marginal enhancement device ($V_T \simeq 0$) to a larger value.

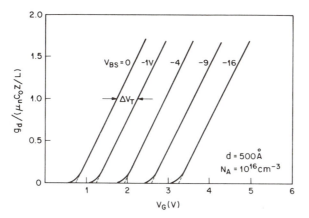

Fig. 43 Threshold voltage adjustment using substrate bias.

5.6.2 Device Scaling

To increase the number of components per integrated circuit chip, the device dimensions must be scaled down. Commercial devices with minimum feature length (e.g., gate length) of 1 to 2 μm are now available, and we expect that in the foreseeable future the minimum dimension will continue to shrink into the submicron region. As the channels become shorter, many undesirable effects, the so-called *short-channel effects*, will arise. We first consider these effects; then we shall consider how to scale the device dimensions so that short-channel effects can be minimized.

Short-Channel Effects As the channel length L is reduced, the depletion layer widths of the source and drain junction become comparable to the channel length. If we use an abrupt one-dimensional approximation, the width of the source junction W_S and that of the drain junction W_D are

$$W_S = \sqrt{\frac{2\epsilon_s}{qN_A} (V_{bi} + V_{BS})} \tag{104a}$$

$$W_D = \sqrt{\frac{2\epsilon_s}{qN_A} (V_D + V_{bi} + V_{BS})} \tag{104b}$$

where V_{BS} is the magnitude of the substrate bias. When $W_S + W_D = L$, punch-through will occur. At punch-through, the two depletion layers merge and the gate loses control of the current. Therefore, punch-through is a major limitation of device operation for short-channel MOSFETs.

In a typical circuit application, the biasing voltage V_D is usually kept at a constant value (say, 5 V). As the channel length becomes smaller, the longitudinal electrical field $\mathscr{E}_y$ will increase; the channel mobility becomes field

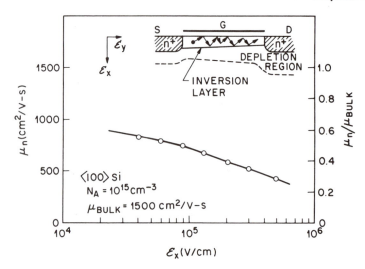

Fig. 44 Mobility versus longitudinal electric field in the inversion channel of an n-channel MOSFET.[14]

dependent, and eventually velocity saturation occurs. Even at very small $\mathscr{E}_y$, the mobility in the channel is lower than the mobility in the bulk. A measured result is shown[14] in Fig. 44 as a function of the transverse field $\mathscr{E}_x$. The mobility for bulk silicon with $N_A = 10^{15}$ cm^{-3} is 1500 cm²/V-s. In a MOSFET, the carrier transport is confined within the narrow inversion region (the inversion layer is about 10 to 100 Å), and surface scattering, shown in the insert of Fig. 44, causes some reduction of the mobility. The average channel mobility is approximately half that of the bulk mobility.

Figure 45 shows the measured drift velocity of electrons in an n-channel device for a selected number of transverse field $\mathscr{E}_x$ as a function of the longitudinal field $\mathscr{E}_y$. At low $\mathscr{E}_y$ ($\sim 10^3$ V/cm), the drift velocity varies linearly with $\mathscr{E}_y$; this results in a constant mobility. The low-field mobility corresponds to that shown in Fig. 44. As $\mathscr{E}_y$ increases, the drift velocity tends to increase more slowly; and when $\mathscr{E}_y$ reaches 10^5 V/cm, the drift velocity approaches a saturation velocity $v_s = 9 \times 10^6$ cm/s.

The drain current $I_{D\,\text{sat}}$ is the number of carriers times their velocity, multiplied by q. If $v = v_s$, we have

$$I_{D\,\text{sat}} = Zqv_s \int_0^{x_i} n(x)\, dx \qquad (105)$$

where x_i is the width of the inversion layer. The integral equals the inversion charge per unit area Q_n. At the source end of the channel, this charge can be expressed as $(V_G - V_T)C_o$. Therefore, Eq. 105 becomes

$$I_{D\,\text{sat}} \simeq ZC_o v_s (V_G - V_T) \qquad (106)$$

where the approximation introduced is due to the neglect of the variation in

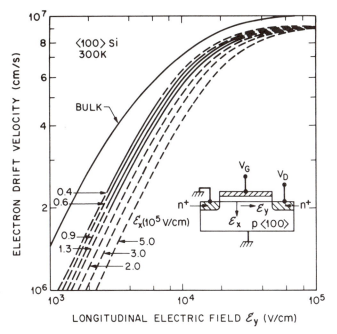

Fig. 45 Electron drift velocity in the inversion channel as a function of longitudinal electric field for a selected transverse field.[14]

Q_n along the channel. The transconductance becomes a constant,

$$g_m = \frac{\partial I_{D\,sat}}{\partial V_G} = Z C_o v_s \qquad (107)$$

and the expression for maximum operating frequency (Eq. 96) becomes

$$f_T = \frac{\mu_n \mathscr{E}_x}{2\pi L} = \frac{v_s}{2\pi L}. \qquad (108)$$

Problem

  For an n-channel MOSFET with $Z = 30\ \mu m$, $L = 1\ \mu m$, $\mu_n = 750\ cm^2/V\text{-}s$, $C_o = 1.5 \times 10^{-7}\ F/cm^2$, and $V_T = 1$ V, find the $I_{D\,sat}$ for an applied V_G of 5 V for the long-channel case. What is $I_{D\,sat}$ at velocity saturation? Also, find the transconductances for each of these cases.

Solution

  For the long-channel case,

$$I_{D\,sat} = \frac{Z \mu_n C_o}{2L} (V_G - V_T)^2 = 2.7 \times 10^{-2} = 27\ mA$$

$$g_m = \frac{Z \mu_n C_o}{L} (V_G - V_T) = 1.35 \times 10^{-2}\ S\,.$$

For velocity saturation with $v_s = 9 \times 10^6$ cm/s,

$$I_{D\,sat} = ZC_o v_s (V_G - V_T) = 1.6 \times 10^{-2} = 16 \text{ mA}$$

$$g_m = ZC_o v_s = 4.05 \times 10^{-3} \text{ S} .$$

From the above example we note that both the current and the transconductance are substantially reduced when velocity saturation occurs. When the field is increased further, carrier multiplication occurs near the drain. The generated electrons flow into the drain, and the generated holes flow into the substrate, giving rise to a substrate current. High fields also may cause some high-energy electrons to be injected into the oxide. These electrons then act as fixed oxide charges, causing a shift in threshold voltage.

Because these short-channel effects complicate device operation and degrade device performance, they should be eliminated or at least minimized so that physical short-channel devices can preserve the electrical characteristics of long-channel devices.

Device Miniaturization   One elegant approach to minimize the short-channel effects is to maintain the long-channel behavior by simply reducing all dimensions and voltages by a scaling factor κ (> 1), so that the internal electric fields are the same as those of a long-channel MOSFET.[15] The new dimensions are

$$L' = \frac{L}{\kappa}, \quad d' = \frac{d}{\kappa}, \quad Z' = \frac{Z}{\kappa} . \tag{109}$$

For a constant field, the operating voltages vary as

$$V' = \frac{V}{\kappa} . \tag{110}$$

For Eqs. 106, 108, 109, and 110, the physical quantities are scaled as follows:

$$C_o' = \frac{\epsilon_{ox}}{d/\kappa} = \kappa \frac{\epsilon_{ox}}{d} = \kappa C_o \quad \text{F/cm}^2 \tag{111a}$$

$$(C_o A)' = \left[\kappa C_o \right] \left[\frac{Z}{\kappa} \right] \left[\frac{L}{\kappa} \right] = \frac{C_o A}{\kappa} \quad \text{F} \tag{111b}$$

$$I_{D\,sat}' = \left[\frac{Z}{\kappa} \right] \frac{(\kappa C_o) v_s (V_G - V_T)}{\kappa} = \frac{I_{D\,sat}}{\kappa} \quad \text{A} \tag{111c}$$

$$J_{D\,sat}' = \frac{I_{D\,sat}'}{A'} = \left[\frac{I_{D\,sat}}{\kappa} \right] \frac{\kappa^2}{A} = \kappa J_{D\,sat} \quad \text{A/cm}^2 \tag{111d}$$

$$f_T' = \frac{v_s}{2\pi(L/\kappa)} = \kappa f_T \quad \text{Hz} . \tag{111e}$$

The switching power P_{ac} and dc power P_{dc} are scaled as

$$P'_{ac} = (C'_o A')V'^2 f'_T = \frac{C_o A}{\kappa} \left[\frac{V}{\kappa}\right]^2 \kappa f_T = \frac{P_{ac}}{\kappa^2} \quad \text{watts} \quad (112a)$$

and

$$P'_{dc} = I'V' = \left[\frac{I}{\kappa}\right]\left[\frac{V}{\kappa}\right] = \frac{P_{dc}}{\kappa^2} \quad \text{watts} . \quad (112b)$$

The switching energy is scaled as

$$E' = \frac{1}{2} (C_o A)'V'^2 = \frac{1}{2} \frac{C_o A}{\kappa} \left[\frac{V}{\kappa}\right]^2 = \frac{E}{\kappa^3} \quad \text{joules} . \quad (113)$$

Therefore, as the device is scaled down, all but one of these variables change favorably. The operating speed increases, the component density increases, and the power density remains constant. However, the current density is increased by the scaling factor. Metal conductors have an upper current density limit imposed by electromigration, the movement of atoms from one place to another under the influence of electrical force. This effect limits the maximum current density for an aluminum conductor to about 10^5 A/cm^2.

Figure 46 shows the traditional large device (right-hand insert), the scaled-down device (left-hand insert), and their corresponding output characteristics. Note that the threshold voltage is also scaled down by the same factor.

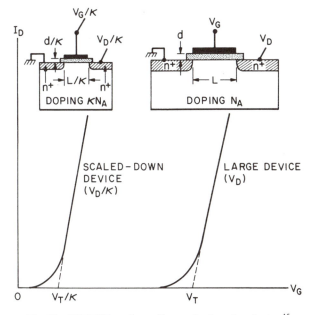

Fig. 46 MOSFET scaling with constant scaling factor.[15]

Another scaling approach is shown in Fig. 47, where L_{min} is plotted against a parameter γ. Here, L_{min} is the minimum channel length for which long-channel behavior can be observed, and γ is given by $r_j d(W_S + W_D)^2$ where r_j is the junction depth (in μm), d is the oxide thickness (in Å), and $W_S + W_D$ is the sum of source and drain depletion layer width (in μm) as given by Eq. 104. The result in Fig. 47, which is obtained from extensive experimental study and two-dimensional computer simulation, can be expressed by an empirical equation[16]:

$$L_{min} \simeq 0.4 \, [r_j d(W_S + W_D)^2]^{1/3} \equiv 0.4\gamma^{1/3} . \tag{114}$$

For example, in order to design a 0.5-μm-channel MOSFET with adequate long-channel behavior, the parameter γ should be 2. Once γ is determined, we can choose r_j, d, W_S, and W_D such that the resulting γ is not larger than 2.

The expression for the minimum channel length, Eq. 114, can be used for a more flexible scaling approach, because it allows the various device parameters to be adjusted independently as long as the value of γ remains the same. Therefore, all device parameters do not have to be scaled by the same factor κ. This flexibility allows us to choose new geometries that are easier to make or that optimize other aspects of device operation, rather than strictly scaled geometries. Table 1 lists a set of device parameters for submicron-channel MOSFETs based on Eq. 114. Note that as the channel length is reduced, all device parameters and the biasing voltages must be reduced accordingly to maintain long-channel characteristics.

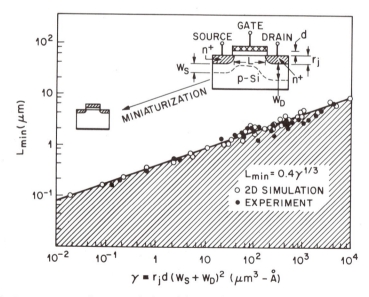

Fig. 47 L_{min} versus γ, where L_{min} is the minimum channel length for which long-channel behavior can be observed.[16]

Table 1 MOSFET Miniaturization

Channel Length (μm)	Junction Depth (μm)	V_D (V)	V_T (V)	Gate Oxide (Å)
0.75	0.30	5.0	1.0	400
0.5	0.20	4.0	0.7	250
0.25	0.10	2.5	0.5	160
0.1	0.05	1.0	0.2	100

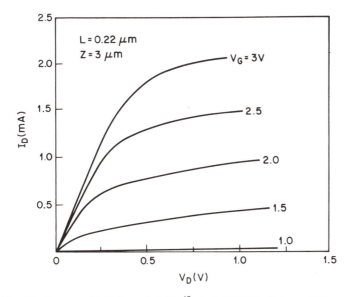

Fig. 48 Measured *I–V* characteristics[17] of a MOSFET with $L = 0.22$ μm.

Figure 48 shows the characteristics of a submicron-channel MOSFET.[17] The channel length is 0.22 μm. The other device parameters are chosen as follows: Z is 3 μm, d is 80 Å, r_j is 0.09 μm, N_A is about 10^{18} cm^{-3}, and V_T is 0.5 V. To obtain long-channel behavior for a MOSFET with $L_{min} = 0.22$ μm, the corresponding γ found from Fig. 47 is 0.2. The device parameters mentioned above, together with a maximum drain voltage of 2 V, can ensure that γ is not larger than 0.2. As can be seen, the device indeed exhibits reasonably good long-channel behavior.

REFERENCES

1 S. M. Sze, *Physics of Semiconductor Devices*, 2nd ed., Wiley, New York, 1981.

2 A. M. Cowley and S. M. Sze, "Surface States and Barrier Height of Metal Semiconductor System," *J. Appl. Phys.*, **36**, 3212 (1965).

3 C. R. Crowell, J. C. Sarace, and S. M. Sze, "Tungsten–Semiconductor Schottky-Barrier Diodes," *Trans. Met. Soc. AIME,* **233**, 478 (1965).

4 V. L. Rideout, "A Review of the Theory, Technology and Applications of Metal–Semiconductor Rectifiers," *Thin Solid Films,* **48**, 261 (1978).

5 W. Shockley, "A Unipolar Field-Effect Transistor," *Proc. IRE,* **40**, 1365 (1952).

6 C. A. Mead, "Schottky Barrier Gate Field-Effect Transistor," *Proc. IEEE,* **54**, 307 (1966).

7 J. Barnard, H. Ohno, C. E. C. Wood, and L. F. Eastman, "Double Heterostructure GaInAs MESFETs with Submicron Gates," *IEEE Electron Device Lett.,* **EDL-1**, 174 (1980).

8 T. Mimura, K. Joshin, S. Hiyamizu, K. Hikosala and M. Abe, *Jpn. J. Appl. Phys.,* **20**, L598 (1981).

9 E. H. Nicollian and J. R. Brews, *MOS Physics and Technology,* Wiley, New York, 1982.

10 A. S. Grove, *Physics and Technology of Semiconductor Devices,* Wiley, New York, 1967.

11 B. E. Deal, "Standardized Terminology for Oxide Charge Associated with Thermally Oxidized Silicon," *IEEE Trans. Electron Devices,* **ED-27**, 606 (1980).

12 D. Kahng and M. M. Atalla, "Silicon–Silicon Dioxide Field Induced Surface Devices," IRE Solid State Device Res. Conf., Pittsburgh, Pa., 1960. D. Kahng, "A Historical Perspective on the Development of MOS Transistors and Related Devices," *IEEE Trans. Electron Devices,* **ED-23**, 655 (1976).

13 L. C. Parrillo, "VLSI Process Integration," in S. M. Sze, Ed., *VLSI Technology,* McGraw-Hill, New York, 1983.

14 J. A. Cooper and D. F. Nelson, "High-Field Drift Velocity of Electrons at the Si–SiO$_2$ Interface as Determined by a Time-of-Flight Technique," *J. Appl. Phys.,* **54**, 1445 (1983).

15 R. H. Dennard, F. H. Gaensslen, H. Yu, V. L. Rideout, E. Bassons, and A. R. LeBlance, "Design of Ion Implanted MOSFET's with Very Small Physical Dimensions," *IEEE J. Solid State Circuits,* **SC-9**, 256 (1974).

16 J. R. Brews, W. Fichtner, E. H. Nicollain, and S. M. Sze, "Generalized Guide for MOSFET Miniaturization," *IEEE Electron Devices Lett.,* **EDL-1**, 2 (1980).

17 W. Fichtner, E. N. Fuls, R. L. Johnston, R. K. Watts, and W. W. Weick, "Optimized MOSFETs with Subquartermicron Channel Length," *IEEE Int. Electron Devices Meeting Digest* (1983), p. 384.

PROBLEMS

1 (*a*) Find the donor concentration and the barrier height of the W–GaAs Schottky barrier diode shown in Fig. 5. (*b*) Compare the barrier height with that obtained from the saturation current density of 5×10^{-7} A/cm^2 shown in Fig. 7. (*c*) For a reverse bias of -1 V, calculate the depletion layer width W, the maximum field, and the capacitance.

2 Find the ratio of hole to electron current for a Au–Si Schottky diode with barrier height of 0.8 V. The silicon is 1 Ω-cm, n-type with $\tau_p = 100$ μs and $\mu_p = 400$ cm^2/V-s.

3 An ohmic contact has an area of 10^{-5} cm^2 and a specific contact resistance of 10^{-6} Ω-cm^2. The ohmic contact is formed in an n-type silicon. If $N_D = 5 \times 10^{19}$ cm^{-3}, and $\phi_{Bn} = 0.8$ V and the electron effective mass is $0.26m_0$, find the voltage drop across the contact when a forward current of 1 A flows through it.

4 An n-channel GaAs MESFET is shown in Fig. 17, with $\phi_{Bn} = 0.9$ V, $N_D = 10^{17}$ cm^{-3}, $a = 0.2$ μm, $L = 1$ μm, and $Z = 10$ μm. (a) Is this an enhancement or depletion mode device? (b) Find the threshold voltage. (c) Find the saturation current at $V_G = 0$. (d) Calculate the cutoff frequency.

5 (a) Plot an ideal C–V curve of a Si–SiO$_2$ MOS diode at 300 K with $d = 300$ Å, $N_A = 5 \times 10^{15}$ cm^{-3}, and a metal plate area of 5×10^{-4} cm^2. (b) If the metal work function is 3 eV, $q\chi = 4.05$ eV, $Q_f/q = 10^{11}$ cm^{-2}, $Q_m/q = 10^{10}$ cm^{-2}, $Q_{ot}/q = 5 \times 10^{10}$ cm^{-2}, and $Q_{it} = 0$, plot the corresponding C–V curve.

6 For an ideal Si–SiO$_2$ MOS diode with $d = 300$ Å and $N_A = 5 \times 10^{15}$ cm^{-3}, find the applied voltage and the electric field at the interface required (a) to make the silicon surface intrinsic and (b) to bring about strong inversion.

7 Derive the I–V characteristics of a MOSFET with the drain and gate connected together and the source and substrate grounded. Can one obtain the threshold voltage from these characteristics?

8 Consider a long-channel MOSFET with $L = 3$ μm, $Z = 21$ μm, $N_A = 5 \times 10^{15}$ cm^{-3}, $C_o = 1.5 \times 10^{-7}$ F/cm^2, and $V_T = 1.5$ V. Find $V_{D\,sat}$ for $V_G = 4$ V. If a constant scaling factor is used to reduce the channel length to 1 μm, find the following scaled-down parameters: Z, C_o, $I_{D\,sat}$, and f_T.

9 Design a submicron MOSFET with a gate length of 0.75 μm. (The gate length is the channel length plus twice the junction depth.) If the junction depth is 0.2 μm, the gate oxide thickness is 200 Å, and the maximum drain voltage is limited to 2.5 V, find the required channel doping so that the MOSFET can maintain its long-channel characteristics.

10 For the submicron MOSFET shown in Fig. 48, (a) find the channel conductance for $V_G = 2$ V and 0.5V $< V_D <$ 1 V, (b) find the transconductance for $V_D = 0.75$ V and 2V $< V_G <$ 2.5 V, (c) explain why the current does not vary as $(V_G - V_T)^2$, and (d) estimate the saturation velocity from the transconductance.

6

Microwave Devices

The microwave frequencies cover the range from about 1 GHz (10^9 Hz) to 1000 GHz, with corresponding wavelengths from 30 to 0.03 cm. The frequencies from 30 to 300 GHz are called the millimeter wave band, because the wavelength is between 10 and 1 mm; higher frequencies are called the submillimeter wave band.

Many semiconductor devices discussed in the previous chapters can be operated in the microwave region. However, to achieve microwave capability, the device dimensions must be reduced and the parasitic capacitance and resistance must be minimized. For example, in a MESFET the gate length must be reduced to increase the cutoff frequency, and the contact resistance to the source and drain must be lowered so that the frequency response is not limited by the RC product. Table 1 summarizes the representative microwave devices and their operational principles. In this chapter, we consider four special microwave devices: the tunnel diode, the IMPATT diode, the BARITT diode, and the transferred-electron device. We shall investigate the operational principles of these devices and how they can be used in microwave applications.

6.1 TUNNEL DIODE

The tunnel diode is associated with the quantum tunneling phenomena.[1] The tunneling time across the device is very short, permitting its use well into the millimeter wave region. Because of its mature technology, the tunnel diode is used in special low-power microwave applications, such as local oscillator and frequency locking circuit.

6.1.1 Transmission Coefficient*

To understand the operation of the tunnel diode, we shall first consider the transmission (or tunneling) coefficient of a particle through a one-dimensional potential barrier shown in Fig. 1a. In the corresponding classical case, the particle is always reflected if its energy E is less than the potential barrier height qV_o. We shall see that in the quantum case, the particle has a finite probability to transmit or "tunnel" through the potential barrier.

*See footnote to Section 2.4.2, p. 48.

Table 1 Microwave Semiconductor Devices

Device	Chapter	Operational Principle
Varactor diode	3	Reactance varies with bias voltage
$p-i-n$ diode	3	Nearly constant capacitance, high breakdown voltage
Bipolar transistor	4	Electrons and holes participate in transport processes
Point contact diode	5	Small area, small capacitance
Schottky diode	5	Majority carrier transport, thermionic injection
JFET	5	Majority carrier, current modulated by junction gate bias
MESFET	5	Majority carrier, current modulated by Schottky gate bias
MOSFET	5	Minority carrier transport in surface inversion channel
Tunnel diode	6	Tunneling in forward-biased p^+-n^+ junction, negative differential resistance
Backward diode	6	Tunneling in reverse-biased junction or near-zero bias
IMPATT diode	6	Avalanche and transit time effects to generate high power
BARITT diode	6	Barrier injection and transit time effects
TED	6	Electrons transferred from low-energy high-mobility valley to high-energy low-mobility valleys

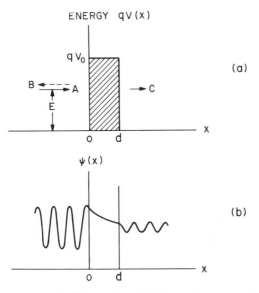

Fig. 1 (a) One-dimensional potential barrier. (b) Schematic representation of the wave functions across the potential barrier.

223

The behavior of a particle (e.g., a conduction electron) in the region where $qV(x) = 0$ can be described by a wave equation:

$$-\frac{\hbar^2}{2m_n} \frac{d^2\psi}{dx^2} = E\psi \tag{1}$$

or

$$\frac{d^2\psi}{dx^2} = -\frac{2m_n E}{\hbar^2} \psi \tag{2}$$

where m_n is the effective mass, $\hbar$ is the reduced Planck constant, E is the kinetic energy, and ψ is the wave function of the particle. The solutions are

$$\psi(x) = Ae^{jkx} + Be^{-jkx} \qquad x \leqslant 0 \tag{3}$$

$$\psi(x) = Ce^{jkx} \qquad x \geqslant d \tag{4}$$

where $k \equiv \sqrt{2m_n E/\hbar^2}$. For $x \leqslant 0$, we have an incident-particle wave function (with amplitude A) and a reflected wave function (with amplitude B); for $x \geqslant d$, we have a transmitted wave function (with amplitude C).

Inside the potential barrier, the wave equation is given by

$$-\frac{\hbar^2}{2m_n} \frac{d^2\psi}{dx^2} + qV_o\psi = E\psi \tag{5}$$

or

$$\frac{d^2\psi}{dx^2} = \frac{-2m_n(qV_o - E)}{\hbar^2}\psi. \tag{6}$$

The solution for $E < qV_o$ is

$$\psi(x) = Fe^{\beta x} + Ge^{-\beta x} \tag{7}$$

where $\beta \equiv \sqrt{2m_n(qV_o - E)/\hbar^2}$. A schematic representation of the wave functions across the barrier is shown in Fig. 1b. The continuity of ψ and $d\psi/dx$ at $x = 0$ and $x = d$, which is required by the boundary conditions, provides four relations between the five coefficients (A, B, C, F, and G). We can solve for $(C/A)^2$, which is the *transmission coefficient*:

$$\left[\frac{C}{A}\right]^2 = \left[1 + \frac{(qV_o \sinh \beta d)^2}{4E(qV_o - E)}\right]^{-1} \tag{8}$$

The transmission coefficient decreases monotonically as E decreases. When $\beta d \gg 1$, the transmission coefficient becomes quite small and varies as

$$\left[\frac{C}{A}\right]^2 \sim \exp(-2\beta d) = \exp\left[-2d\sqrt{2m_n(qV_o - E)/\hbar^2}\right]. \tag{9}$$

To have a finite transmission coefficient, we require a small tunneling distance d, a low potential barrier qV_o, and a small effective mass.

6.1.2   Current–Voltage Characteristics

A tunnel diode consists of a simple $p-n$ junction in which both the p- and n-sides are degenerate (i.e., very heavily doped with impurities). Figure 2 shows a schematic energy diagram of a tunnel diode in thermal equilibrium. Because of the high dopings, the depletion region is very narrow and the tunneling distance d is quite small (of the order of 50 to 100 Å). The dopings also cause the Fermi levels to be located within the allowed bands. The amount of degeneracies, $V_p \equiv (E_V - E_{Fp})/q$ and $V_n \equiv (E_{Fn} - E_C)/q$, are typically 50 to 100 mV.

Figure 3 shows a representative current–voltage characteristic of a tunnel diode. In the reverse direction (p-side negative with respect to n-side), the current increases monotonically. In the forward direction, the current first increases to a maximum value (peak current I_P) at a voltage V_P, then decreases to a minimum value I_V at a voltage V_V. For voltages larger than V_V, the current increases exponentially with the voltage. The I–V characteristic is the result of two current components: tunneling current and thermal current as indicated in Fig. 3.

We now consider the tunneling processes. We note that in thermal equilibrium, Fig. 2, the Fermi level ($E_{Fn} = E_{Fp}$) is constant across the junction. Assume that the device is operated at low temperatures so that there are no filled states above the Fermi level and no empty states below the Fermi level on either side of the junction. When a biasing voltage is applied, the electrons may tunnel from the valence band to the conduction band, or vice versa. However, there are three conditions necessary for tunneling: (1) occupied energy states on the side from which the electron tunnels, (2) unoccupied energy states at the same energy levels as in (1) on the side to which the electron can tunnel, and (3) a low tunneling potential barrier height and a small tunneling distance so that there is a finite transmission coefficient.

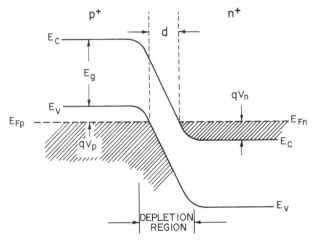

Fig. 2   Energy band diagram of a tunnel diode in thermal equilibrium.[1]

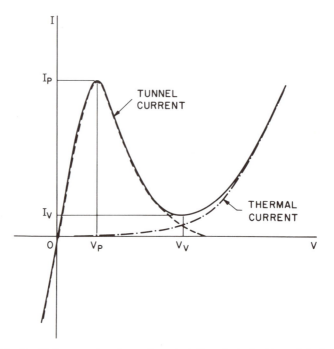

Fig. 3 Static current–voltage characteristics of a typical tunnel diode.

When a small forward voltage is applied (Fig. 4a), a band of energies exists in which filled states on the n-side correspond to unoccupied states on the p-side that are available to accept electrons. The electrons can thus tunnel from the n-side to the p-side. The corresponding current is designated by the dot on the $I–V$ curve. As the voltage increases, the current will reach a maximum (peak current) corresponding to the maximum crossing of electrons from the filled states on the n-side to the unoccupied states on the p-sides. When the forward voltage is further increased, there are fewer available unoccupied states in the p-side (Fig. 4b), and the current decreases. If forward voltage is applied such that the band is "uncrossed," that is, the bottom of the conduction band is exactly opposite the top of the valence band, there are no available energy states that are opposite filled states. Thus, at this point the tunneling current can no longer flow.

With still further increases of the voltage, the normal thermal current, as discussed in Chapter 3, will flow (Fig. 4c) and will increase exponentially with the applied voltage. Figure 4d shows electrons tunneling from the valence band into the conduction band when a reverse voltage is applied. The reverse current will increase monotonically, because more occupied states on the p-side and more unoccupied states on the n-side will become available as the reverse voltage increases.

From this discussion we expect that in the forward direction the tunneling current increases from zero to a peak current I_P as the voltage increases. With

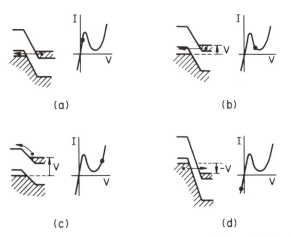

Fig. 4 Simplified energy band diagrams of the tunnel diode at (a) low forward bias, (b) forward bias such that the valley current is approached, (c) forward bias with thermal current flowing, and (d) reverse bias.

a further increase in voltage, the current then decreases to zero when $V = V_n + V_p$, where V is the applied forward voltage and V_n and V_p are the amount of degeneracy on the n- and p-side, respectively (Fig. 2). The decreasing portion after the peak current in Fig. 3 is the negative differential resistance region. The values of the peak current I_P and the valley current I_V determine the magnitude of the negative resistance. For this reason their ratio I_P/I_V is used as a figure of merit for tunnel diode.

An empirical form for the $I-V$ characteristics is given by

$$I = I_P \left[\frac{V}{V_P} \right] \exp \left[1 - \frac{V}{V_P} \right] + I_o \exp \left[\frac{qV}{kT} \right] \quad (10)$$

where the first term is the tunnel current and I_P and V_P are the peak current and peak voltage, respectively, as shown in Fig. 3. The second term is the normal thermal current. The negative differential resistance can be obtained from the first term in Eq. 10:

$$R = \left[\frac{dI}{dV} \right]^{-1} = - \left[\left[\frac{V}{V_P} - 1 \right] \frac{I_P}{V_P} \exp \left[1 - \frac{V}{V_P} \right] \right]^{-1}. \quad (11)$$

Figure 5 shows a comparison of the typical current–voltage characteristics of germanium, gallium antimonide, and gallium arsenide tunnel diodes at room temperature. The current ratios of I_P/I_V are 8:1 for germanium and 12:1 for gallium antimonide and gallium arsenide. Because of its smaller effective mass ($0.042m_0$) and small bandgap (0.72 eV), the gallium antimonide tunnel diode has the largest negative resistance among the three devices.

A related tunnel device is the *backward diode* as shown in the insert of Fig. 6. The doping concentration on one side of the device is nearly or not quite

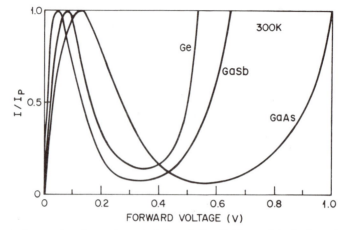

Fig. 5 Typical current–voltage characteristics of Ge, GaSb, and GaAs tunnel diodes at room temperature.

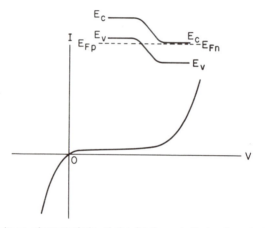

Fig. 6 Current–voltage characteristic of the backward diode. Insert shows the energy band diagram of a backward diode in thermal equilibrium.

degenerate. The other side is usually degenerate. Figure 6 shows the current–voltage characteristics of a backward diode. The current in the "reverse" direction for small bias is larger than the current in the "forward" direction—hence the name backward diode. When we apply a small reverse voltage, the energy band diagram is similar to Fig. 4d except that there is no degeneracy on one side (e.g., $V_n \simeq 0$). Under reverse bias, electrons can readily tunnel from the valence band into the conduction band to produce a tunneling current. In the forward direction, however, the device has a very small or no tunnel current component, because there are no occupied states on the n–side (Fig. 6 insert).

The backward diode can be used for rectification of small signals, microwave detection, and mixing. It has good frequency response, similar to that of a tunnel diode, because there is no minority carrier storage effect.

6.2 IMPATT DIODE

The word IMPATT stands for "impact ionization avalanche transit time." IMPATT diodes employ impact ionization and transit time properties of semiconductor devices to produce negative resistance at microwave frequencies. The IMPATT diode is one of the most powerful solid-state sources of microwave power. At present, the IMPATT diode can generate the highest cw (continuous wave) power output of all solid-state devices at millimeter wave frequencies (i.e., about 30 GHz). But there is one noteworthy difficulty in IMPATT applications: the noise is high due to random fluctuations of the avalanche multiplication processes. A discussion of the power and noise performances for various microwave devices will be presented in Section 6.5.

6.2.1 Device Structure

The IMPATT diode family includes many different junction and metal–semiconductor devices. The first IMPATT oscillation was obtained from a simple silicon $p-n$ junction diode biased into reverse avalanche breakdown and mounted in a microwave cavity.[2] The doping profile and the electric-field distribution at breakdown for the $p-n$ junction diode are shown in Fig. 7a.

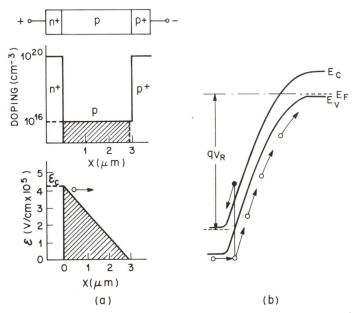

Fig. 7 (a) Doping profile and electric-field distribution for a one-sided abrupt n^+-p diode at avalanche breakdown. (b) Energy band diagram of the diode.[2]

Because of the strong dependence of the ionization coefficient on the electric field, most of the electron–hole pairs are generated in the high-field region (i.e., in the region $0 \leqslant x < 0.5$ μm where $\mathscr{E} > 3 \times 10^5$ V/cm). The corresponding energy band diagram at breakdown is shown in Fig. 7b. The generated electrons immediately move into the n^+-region, while the generated holes drift across the p-region. The time required for the holes to reach the p^+-contact constitutes the transit time delay.

The original proposal for a microwave device of the IMPATT type was made by Read and involved an $n^+ - p - i - p^+$ or an $n^+ - p - \pi - p^+$ structure as shown[3] in Fig. 8a. The Read diode consists of two regions as illustrated in Figs. 8b and 8c: (1) the avalanche region (p_1-region with relatively high doping and high field, $0 \leqslant x \leqslant b$), in which avalanche multiplication occurs; and (2) the drift region (p_2-region with essentially intrinsic doping and constant field, $b \geqslant x \geqslant b + W$), in which the generated holes drift toward the p^+-contact. Of course, a similar device can be built with the $p^+ - n - i - n^+$ configuration, in which electrons generated from the avalanche multiplication drift through the intrinsic region.

Figures 9a and b show a modified Read diode called a *lo–hi–lo structure* in which a "clump" of charge Q is located at $x = b$ and the dopings p_1 and p_2 are

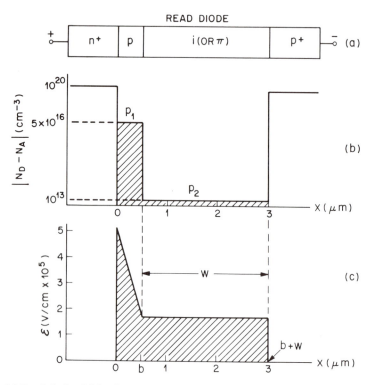

Fig. 8 (a) Read diode. (b) Doping profile. (c) Electric field distribution at avalanche breakdown.[3]

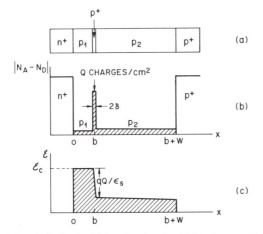

Fig. 9 (a) Modified Read diode (lo–hi–lo structure). (b) Doping profile. (c) Electric-field distribution at avalanche breakdown.

nearly intrinsic. This device can be made by epitaxial techniques such as molecular-beam epitaxy (see Chapter 8). The electric field across the lo–hi–lo device at breakdown (Fig. 9c) is

$$\mathscr{E} = \mathscr{E}_c - \frac{1}{\epsilon_s} \int_0^x \rho_s \, dx \tag{12}$$

where $\mathscr{E}_c$ is the critical field, ϵ_s is the dielectric permittivity, and ρ_s is the space charge density. Because p_1 is nearly intrinsic, the field remains essentially constant from $x = 0$ to $x = b$. At $x = b$, the "clump" of charge causes a field reduction by an amount of qQ/ϵ_s (since $\int_{b-\delta}^{b+\delta} \rho_s \, dx = qQ$). The field then remains at an essentially constant value of $(\mathscr{E}_c - qQ/\epsilon_s)$ from $x = b$ to the p^+-contact. The breakdown voltage is given by

$$V_B = \mathscr{E}_c b + \left[\mathscr{E}_c - \frac{qQ}{\epsilon_s} \right] W . \tag{13}$$

Because of the nearly uniform high-field region ($0 \leqslant x \leqslant b$), the value of the critical field and thus the junction temperature can be kept much lower than that for a Read diode.

A fabricated IMPATT diode generally is mounted in a microwave package. A typical microwave package is shown in Fig. 10. The diode is mounted with its high-field region close to the copper heat sink so that the heat generated at the junction can be conducted away readily. Similar microwave packages are used to house other microwave devices considered in this chapter.

6.2.2 Dynamic Characteristics

We shall now use the lo–hi–lo structure to discuss the IMPATT operational principles. When a reverse dc voltage V_B is applied to the diode so that the

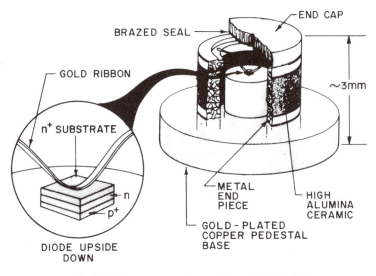

Fig. 10 Microwave package with an IMPATT diode.

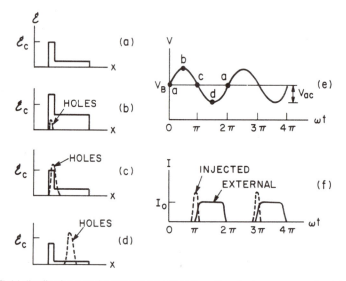

Fig. 11 Field distributions and generated-carrier densities during an ac cycle at four intervals of time, (a) to (d). (e) ac Voltage. (f) Injected and external current.[2, 3]

critical field for avalanche $\mathscr{E}_c$ is just reached (Fig. 11a), avalanche multiplication will begin. An ac voltage is superimposed onto this dc voltage at $t = 0$. This is indicated as a in Fig. 11e. Electrons generated in the avalanche region move to the n^+-region, and holes enter the drift region. As the applied ac voltage goes positive, more holes are generated in the avalanche region as shown by the dotted line in Fig. 11b. The hole pulse keeps increasing as long as the electrical field is above $\mathscr{E}_c$. Therefore, the hole pulse reaches its peak value not at $\pi/2$ when the voltage is maximum, but at π (Fig. 11c). The important consequence is that there is a $\pi/2$ phase delay inherent in the avalanche process itself, that is, the injected-carrier density (hole pulse) lags the ac voltage by 90°.

An additional delay is provided by the drift region. Once the applied voltage drops below V_B ($\pi \leqslant \omega t \leqslant 2\pi$), the injected holes will drift toward the p^+-contact (Fig. 11d) with a saturation velocity, provided that the field across the drift region is sufficiently high (e.g., larger than 8×10^4 V/cm for holes in silicon).

The situation described above is illustrated by the injected carriers in Fig. 11f. By comparing Figs. 11e and 11f, we note that the peak value of the ac field (or voltage) occurs at $\pi/2$, but the peak of the injected carrier density occurs at π. The injected carriers then enter and traverse the drift region at saturation velocity, thereby introducing the transit time delay. The induced external current is also shown in Fig. 11f. Comparing the ac voltage and the external current clearly shows that the diode exhibits a negative resistance characteristic.

The injected carriers (hole pulse) will traverse the length W of the drift region during the negative half-cycle if we choose the transit time to be one half the oscillation period, that is,

$$\frac{W}{v_s} = \frac{1}{2}\left[\frac{1}{f}\right] \tag{14}$$

or

$$f = \frac{v_s}{2W} \tag{15}$$

where v_s is the saturation velocity (10^7 cm/s for silicon at 300 K).

Problem

Consider a lo–hi–lo silicon structure ($p^+ - i - n^+ - i - n^+$) having $b = 1$ μm and $W = 5\mu$m. If $Q = 2.0 \times 10^{12}$ charges/cm^2, find the dc breakdown voltage and the IMPATT operating frequency.

Solution

For a 1-μm uniform avalanche region, the avalanche breakdown voltage will be the same as for a $p^+ - i - n^+$ diode with 1-μm intrinsic region. From Fig. 27 in Chapter 3, we find the breakdown voltage to be 33 V. Therefore, the field at breakdown is

$$\mathscr{E}_c = \frac{33 \text{ V}}{1 \times 10^{-4} \text{ cm}} = 3.3 \times 10^5 \text{ V/cm} .$$

From Eq. 13, we find the dc breakdown voltage for the lo–hi–lo structure to be

$$V_B = 3.3 \times 10^5 \times 10^{-4} + \left[3.3 \times 10^5 - \frac{1.6 \times 10^{-19} \times 2.0 \times 10^{12}}{8.85 \times 10^{-14} \times 11.9}\right] \times 5 \times 10^{-4}$$

$$= 33 + 13 = 46 \text{ V} .$$

The field in the drift region is

$$\frac{13}{5 \times 10^{-4}} = 2.6 \times 10^4 \text{ V/cm} .$$

The drift field is high enough for the injected electrons to maintain their saturation velocity. Therefore,

$$f = \frac{v_s}{2W} = \frac{10^7 \text{ cm/s}}{2 \times 5 \times 10^{-4} \text{ cm}} = 10^{10} \text{ Hz} = 10 \text{ GHz} .$$

We can also estimate the dc-to-ac power conversion efficiency of the IMPATT diode using Figs. 11e and 11f. The dc power input is the product of the average dc voltage and the average dc current, that is, $V_B(I_o/2)$. The ac power output can be estimated by assuming that the maximum ac voltage swing to be $\frac{3}{4}V_B$, that is, $V_{ac} = 3V_B/4$, and that the external current is zero between $0 \leqslant \omega t \leqslant \pi$ and is I_o between $\pi \leqslant \omega t \leqslant 2\pi$. Therefore, the microwave power-generating efficiency η is

$$\eta = \frac{\text{ac power output}}{\text{dc power input}} = \frac{\int_0^{2\pi} (V_{ac} \sin \omega t)I \ d(\omega t)}{\left[V_B \dfrac{I_o}{2} \right] 2\pi}$$

$$= \frac{\int_{\pi}^{2\pi} \left[\dfrac{3V_B}{4} \sin \omega t \right] I_o \ d(\omega t)}{V_B I_o \pi} = \frac{3}{2\pi} = 48\% . \tag{16}$$

State-of-the-art IMPATT diodes have cw power capabilities up to 10 W at 10 GHz with over 30% efficiency, up to 0.5 W at 100 GHz with 10% efficiency, and 5 mW at 300 GHz with 0.1% efficiency. The substantial reduction in power and efficiency in the millimeter wave region is partly due to difficulties in device fabrication and circuit optimization and partly due to additional delays introduced by energy relaxation (i.e., transfer of energy to the carriers) and tunneling.

6.3 BARITT DIODE

The word BARITT stands for "barrier injection transit time." The BARITT diode belongs to the transit time microwave diode family. The mechanisms responsible for the microwave oscillation are the thermionic injection and diffusion of minority carriers across a forward-biased barrier and a transit time delay of the injected carriers traversing the drift region. Because it has no avalanche delay, we expect the BARITT diode to operate at lower power and lower efficiency than the IMPATT diode. On the other hand, the noise associated with carrier injection across the barrier is much smaller than the avalanche noise in an IMPATT diode. The low-noise property and the stability of the device make the BARITT diode suitable for many low-power applications such as local oscillators and Doppler detectors.

6.3.1 Device Structures

The first BARITT operation was obtained from a metal–semiconduc or–metal reach–through diode[4] shown in Fig. 12a. It is basically two Schottky diodes connected back-to-back. The field distribution at thermal equilibrium is shown in Fig. 12b. When a sufficiently large bias is applied to the device, the electric field will reach through the entire device (Fig. 12c). The corresponding band diagram is shown in Fig. 12d. Under this condition, thermionic injection of holes across the barrier (located at x_R) occurs. The injected holes then traverse the drift region from x_R to the right-side metal contact. The time required to reach the metal contact is the transit time delay.

An optimum BARITT structure[5] is shown in Fig. 13a. A "clump" of charge is located at the injection point (Fig. 13b). By proper choice of the amount and the location of the charge, we can minimize the voltage we must apply to the device and still maintain a sufficiently large field in the drift region for velocity saturation. Note that in the nonoptimum structure (Fig. 12c), the field near x_R, is very low and the injected carriers will drift at a much lower speed than the saturation velocity.

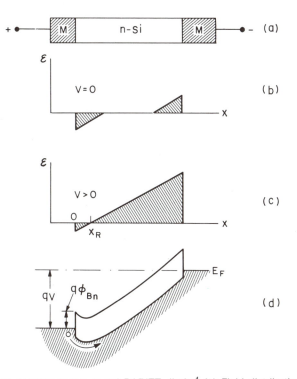

Fig. 12 (a) Metal–semiconductor–metal BARITT diode.[4] (b) Field distribution at thermal equilibrium. (c) Field distribution after reach through. (d) Energy band diagram showing hole injection.

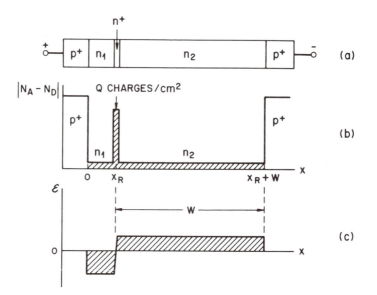

Fig. 13 (a) Optimum BARITT diode structure.[5] (b) Doping profile for a lo–hi–lo structure. (c) Field distribution after reach through.

6.3.2 Dynamic Operation

The basic operational principle of BARITT diode is similar to that of IMPATT diode. However, there is one major difference; that is, in BARITT operation we do not have the avalanche delays because the thermionically injected carriers are in phase with the ac voltage swings.

The large-signal BARITT operation is shown in Fig. 14. The field $\mathscr{E}_{RT}$ corresponds to the reach-through condition. We assume that the device is biased with a dc voltage V_{RT} corresponding to the reach-through field $\mathscr{E}_{RT}$ (Fig. 14a). An ac voltage is applied at $t = 0$. This is indicated as a in Fig. 14e. As the ac voltage goes positive, the injected holes increase with the ac voltage, and the peak value of injected holes is reached when the voltage is a maximum (Fig. 14b). In other words, the injected carrier density is in phase with the ac voltage. Once injected, the carriers will traverse the drift region (Figs. 14c and 14d). Figure 14f shows the injected hole pulse at $\pi/2$ and the corresponding induced external current, which travels three fourths of a cycle to reach the negative terminal:

$$\frac{W}{v_s} = \frac{3}{4}\left[\frac{1}{f}\right] \tag{17}$$

or

$$f = \frac{3v_s}{4W}. \tag{18}$$

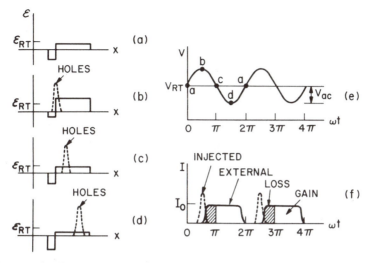

Fig. 14 Field distribution and injected carrier of a BARITT diode during one ac cycle, at four intervals of time, (a) to (d). (e) ac Voltage. (f) Injected and external current.

The efficiency of a BARITT diode is substantially lower than that of an IMPATT diode. This is because in the period $\pi/2 \leqslant \omega t \leqslant \pi$, both the ac voltage and the external current are positive (Figs. 14e and 14f); thus, ac power is dissipated in the device. This power will cancel some of the ac power generated by the device in the period $\pi \leqslant \omega t \leqslant 2\pi$. Therefore, the net ac power generated is smaller than the ac power of the IMPATT diode (Fig. 11). State-of-the-art BARITT diodes have cw power capabilities of about 100 mW at 10 GHz with 2% efficiency.

6.4 TRANSFERRED-ELECTRON DEVICE

The transferred-electron effect was first observed in 1963. In the first experiment, a microwave output was generated when a dc electric field that exceeded a critical threshold value of several thousand volts per centimeter was applied across a short n-type sample of gallium arsenide or indium phosphide.[6] The transferred-electron device (TED) is an important microwave device. It is used extensively as a local oscillator and power amplifier covering the microwave frequency range from 1 to 100 GHz. The TEDs have matured to become important solid-state microwave sources used in radars, intrusion alarms, and microwave test instruments.

6.4.1 Negative Differential Resistance

In Section 2.6 we considered the transferred-electron effect, that is, the transfer of conduction electrons from a high–mobility energy valley to low-mobility higher-energy satellite valleys. Using the relations given by Eqs. 98

and 99 in Chapter 2, the current density takes on asymptotic values

$$J \simeq qn\,\mu_1 \mathscr{E} \qquad \text{for} \quad 0 < \mathscr{E} < \mathscr{E}_a \tag{19}$$

$$J \simeq qn\,\mu_2 \mathscr{E} \qquad \text{for} \quad \mathscr{E} > \mathscr{E}_b \,. \tag{20}$$

Now, if $\mu_1 \mathscr{E}_a$ is larger than $\mu_2 \mathscr{E}_b$, there will exist a region of negative differential resistance (NDR)[7] between $\mathscr{E}_a$ and $\mathscr{E}_b$ as shown in Fig. 15. Also shown are the threshold field $\mathscr{E}_T$ corresponding to the onset of the NDR, the threshold current density J_T, the valley field $\mathscr{E}_V$, and the valley current density J_V. The NDR exists between $\mathscr{E}_T$ and $\mathscr{E}_V$.

For the transferred-electron mechanism to give rise to the NDR, certain requirements must be met: (1) The lattice temperature must be low enough that, in the absence of an electric field, most of the electrons are in the lower valley (the conduction band minimum), that is, $\Delta E > kT$. (2) In the lower valley the electrons must have high mobility and small effective mass, while in the upper satellite valleys the electrons must have low mobility and large effective mass. (3) The energy separation between the two valleys must be smaller than the semiconductor bandgap (i.e., $\Delta E < E_g$) so that avalanche breakdown does not set in before the transfer of electrons into the upper valleys.

Of the semiconductors satisfying these requirements, n-type gallium arsenide and n-type indium phosphide are the most widely studied and used. The measured room temperature velocity field characteristics for these semiconductors are shown in Fig. 16. The threshold field $\mathscr{E}_T$ is 3.2 kV/cm for gallium arsenide and 10.5 kV/cm for indium phosphide. The peak velocity v_p is about 2.2×10^7 cm/s for gallium arsenide and 2.5×10^7 cm/s for indium phosphide. The maximum negative differential mobility (i.e., $dv/d\mathscr{E}$) is about -2400 cm^2/V-s for gallium arsenide and -2000 cm^2/V-s for indium phosphide.

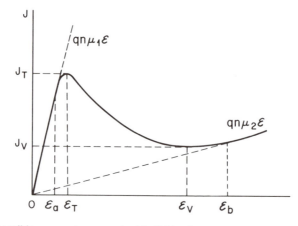

Fig. 15 A possible current-versus-electric-field characteristic of a two-valley semiconductor. $\mathscr{E}_T$ is the threshold field and $\mathscr{E}_V$ is the valley field.

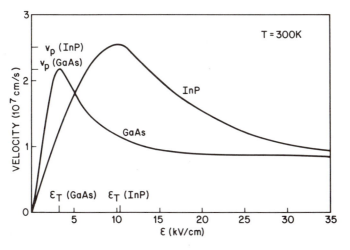

Fig. 16 Measured velocity–field characteristics for GaAs and InP.

A semiconductor exhibiting NDR is inherently unstable, because a random fluctuation of carrier density at any point in the semiconductor produces a momentary space charge that will grow exponentially with time. The one–dimensional continuity equation is given by

$$\frac{\partial n}{\partial t} = \frac{1}{q}\frac{\partial J}{\partial x} . \qquad (21)$$

If there is a small local fluctuation of the majority carriers from the uniform equilibrium concentration n_o, the locally created space charge density is $n - n_o$. Poisson's equation and the current density equation are

$$\frac{\partial \mathscr{E}}{\partial x} = \frac{-q(n - n_o)}{\epsilon_s} \qquad (22)$$

$$J = qn_o\bar{\mu}\mathscr{E} + qD\frac{\partial n}{\partial x} \qquad (23)$$

where $\bar{\mu}$ is the average mobility (defined by Eq. 96 in Chapter 2), ϵ_s is the dielectric permittivity, and D is the diffusion constant. Differentiating Eq. 23 with respect to x and inserting Poisson's equation yield

$$\frac{1}{q}\frac{\partial J}{\partial x} = -\frac{n - n_o}{\epsilon_s/qn_o\bar{\mu}} + D\frac{\partial^2 n}{\partial x^2} . \qquad (24)$$

Substituting this expression into Eq. 21 gives

$$\frac{\partial n}{\partial t} = -\frac{n - n_o}{\epsilon_s/qn_o\bar{\mu}} + D\frac{\partial^2 n}{\partial x^2} . \qquad (25)$$

We can solve Eq. 25 by separation of variables, that is, let $n(x, t) =$

$n_1(x)\, n_2(t)$. For the temporal response, the solution of Eq. 25 is

$$n - n_o = (n - n_o)_{t=0} \exp\left[\frac{-t}{\tau_R}\right] \qquad (26)$$

where τ_R is the *dielectric relaxation time* given by

$$\tau_R = \frac{\epsilon_s}{qn_o\bar{\mu}}. \qquad (27)$$

τ_R represents the time constant for the decay of the space charge to neutrality if the mobility $\bar{\mu}$ is positive. However, if the semiconductor exhibits NDR, any charge imbalance will grow with a time constant equal to $|\tau_R|$.

6.4.2 Device Operation

The TEDs require very pure and uniform materials with a minimum of deep impurity levels and traps. Modern TEDs almost always have epitaxial layers on n^+-substrates deposited by various epitaxial techniques. Typical donor concentrations range from 10^{14} to 10^{16} cm^{-3}, and typical device lengths range from a few microns to several hundred microns. A TED having an epitaxial n-layer on n^+-substrate and an ohmic n^+-contact to the cathode electrode is shown in Fig. 17a. For such an ohmic contact there is always a low-field region near the cathode, and the field is nonuniform across the device length resulting in a space charge build-up at the cathode as discussed below.

To improve device performance, we use the two-zone cathode contact instead of the n^+-ohmic contact. The two-zone cathode contact consists of a high-field zone and an n^+-zone (Fig. 17b). This configuration is similar to that of a lo–hi–lo IMPATT diode. Electrons are "heated" in the high-field zone and subsequently injected into the active region, which has a uniform field. This structure has been used successfully over a wide temperature range with high efficiency and high power output.

The operational characteristics of a TED depend on five factors: (1) doping concentration and doping uniformity in the device, (2) length of the active region, (3) cathode contact characteristics, (4) type of circuit, and (5) operating bias voltage.

We have shown that for a device with NDR, the initial space charge will grow exponentially with time (Eq. 26) and that the time constant is given by Eq. 27:

$$|\tau_R| = \frac{\epsilon_s}{qn_o\,|\mu_-|} \qquad (28)$$

where μ_- is the negative differential mobility. If Eq. 26 remains valid throughout the entire transit time of the space charge layer, the maximum growth factor would be $\exp(L/v\,|\tau_R|)$, where L is the device length and v is the average drift velocity of the space charge layer. For large space charge growth, this growth factor must be greater than unity, making $L/$

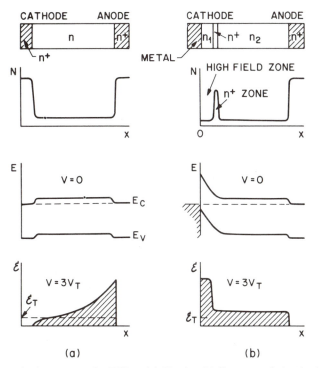

Fig. 17 Two cathode contacts for TEDs. (a) Ohmic. (b) Two-zone Schottky barrier contact.

$$v \mid \tau_R \mid \; > 1, \text{ or}$$

$$n_o L \; > \; \epsilon_s v / q \mid \mu_- \mid \; \approx 10^{12} \text{ cm}^{-2} \qquad (29)$$

for n-type gallium arsenide and indium phosphide. The TEDs with $n_o L$ products smaller than 10^{12} cm^{-2} exhibit a stable field distribution. Therefore, an important boundary that separates the various modes of operation is the product of the carrier concentration and the device length $n_o L \; = \; 10^{12}$ cm^{-2}.

We shall now consider a few important modes of operation in TEDs. The simplest form of space charge instability is an accumulation layer. Lightly doped or short samples ($n_o L < 10^{12}$ cm^{-2}) exhibit a stable field distribution when a constant voltage is applied. In a device with $n_o L > 10^{12}$ cm^{-2}, a traveling accumulation layer will be formed. Figure 18a shows a current–field ($J - \mathscr{E}$) plot, and Fig. 18b shows the profile of the device.[8] Assume that at point A an excess (or accumulation) of negative charges exists that could be due to a random–noise fluctuation or possibly to a permanent nonuniformity in doping (Fig. 18c). Integrating Poisson's equation once yields the electric-field distribution shown in Fig. 18d, where the field to the left of point A is lower than that to the right of A. If the device is biased at point $\mathscr{E}_A$ on the $J - \mathscr{E}$ curve (Fig. 18a), this condition implies that the carriers (or current)

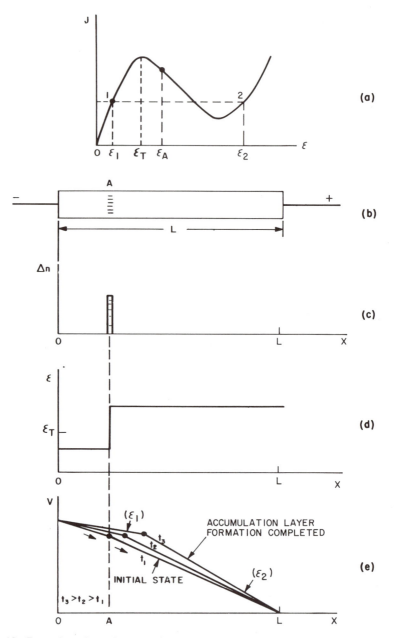

Fig. 18 Formation of an electron accumulation layer in a perturbed medium that has a negative differential resistivity.[8]

flowing into point A are greater than those flowing out of point A, thereby increasing the excess negative space charge at A. Now, the field to the left of A is lower than it was originally, and the field to the right is higher than it was originally, resulting in an even greater space charge accumulation. This process continues as illustrated in Fig. 18e until the high and low fields both obtain values outside the NDR region and settle at points $\mathscr{E}_1$ and $\mathscr{E}_2$ in Fig. 18a, where the currents in the two field regions are equal. As a result, a traveling space charge accumulation forms.

A more complicated and frequently occurring situation arises when there are positive and negative charges separated by a small distance. We have a dipole formation (also called a domain) as shown in Figs. 19a and 19b. The electric field inside the dipole would be greater than the fields on either side of it, Fig. 19c. The corresponding voltage variation across the device is shown in Fig. 19d. Because of the NDR, the current in the low-field region would be greater than that in the high-field region. The two field values tend toward the equilibrium level outside the NDR region, where the high and low currents are the same (Fig. 18a). The dipole has now reached a stable configuration. The dipole layer moves through the active region and disappears at the anode, at which time the field begins to rise uniformly across the device through the threshold (i.e., $\mathscr{E} > \mathscr{E}_T$), thus forming a new dipole, and the process repeats itself. The time required for the domain to travel from the cathode to anode is L/v, where L is the active device length and v is the average velocity. The corresponding frequency for the transit time domain mode is

$$f = \frac{v}{L}. \tag{30}$$

The field distributions across a 35-μm sample during one ac cycle at four intervals are shown in Figs. 20a through 20d. Also shown are the terminal voltage and current wave forms[9] (Fig. 20e). The $n_o L$ product is 2.1×10^{12} cm^{-2} for this device. The current waveform is reasonably close to a sinusoidal form and the dc- to ac–power conversion efficiency is about 10%.

Figure 21a shows a simulated time-dependent behavior of a domain in a gallium arsenide TED 100 μm long and has a doping of 5×10^{14} cm^{-3} ($n_o L = 5 \times 10^{12}$ cm^{-2}).[10] The time between successive vertical displays of $\mathscr{E}(x, t)$ is 16 τ_R, where τ_R is the low-field dielectric relaxation time from Eq. 27 ($\tau_R = 1.5$ ps for this device). Each time a domain is absorbed at the anode, the current in the external circuit increases and a new domain is nucleated at the cathode contact where the largest doping fluctuation and space charge perturbation usually exist.

A TED in a resonant circuit can operate at frequencies higher than the transit time frequency (Eq. 30) if the high-field domain can be quenched before it reaches the anode. The domain width generally decreases with decreasing biasing voltage. If we reduce the biasing voltage sufficiently below the threshold during an ac cycle, the domain will vanish (or is "quenched"). When the biasing voltage swings back above threshold, a new domain is

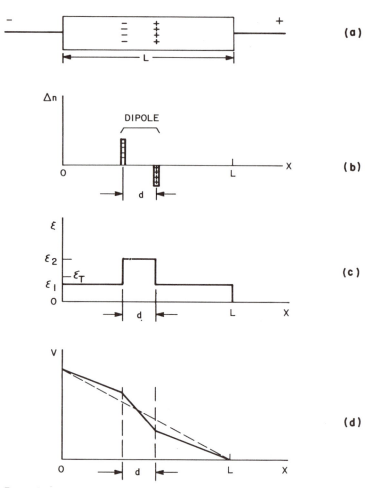

Fig. 19 Formation of a domain (dipole layer) in a perturbed medium that has a negative differential resistivity.[8]

nucleated and the process is repeated. Therefore, the oscillation occurs at the frequency of the resonant circuit rather than at the transit time frequency. Figure 21*b* shows an example of the quenched domain mode. The device has the identical length and doping as that of Fig. 21*a*. The domain is quenched at a distance of about $L/3$ from the cathode, and the operating frequency is about three times higher than the transit time domain mode shown in Fig. 21*a*.

6.5 COMPARISON OF MICROWAVE DEVICES

We shall now compare the various microwave devices in terms of their power output, maximum frequency, and noise performances.[11]

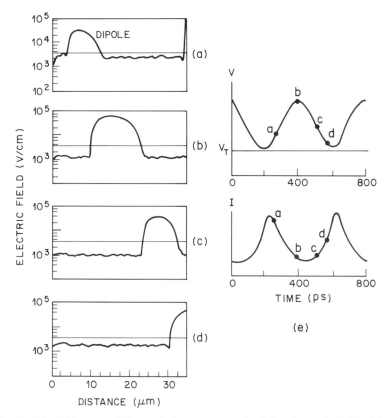

Fig. 20 Electric field versus distance during one ac cycle at four intervals, (a) to (d). (e) Voltage and current wave forms of a transit time domain mode.[9]

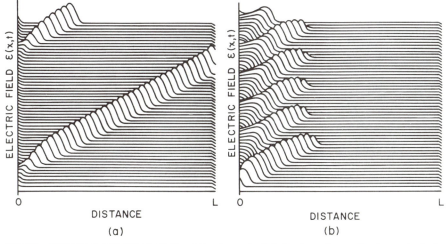

Fig. 21 Numerical simulation of the time-dependent behavior of a cathode–nucleated TED.[10] (a) Transit time domain mode. Each successive time is 24 ps. (b) Quenched domain mode.

One of the most important figure of merits for microwave devices is power output as a function of oscillation frequency. Because of the inherent limitations of semiconductor materials and the impedance levels attainable in microwave circuitry, the maximum output power of a single device at a given frequency is limited. There are two basic limitations of semiconductor materials: (1) the critical electrical field $\mathscr{E}_c$ at which avalanche breakdown occurs and (2) the saturation velocity v_s, which is the maximum attainable velocity in semiconductors.

The maximum voltage that can be applied across a semiconductor sample is limited by the breakdown voltage, which for a uniform avalanche is given by $V_m = \mathscr{E}_c W$, where W is the depletion layer width. The maximum current that can be carried by the semiconductor is also limited by the avalanche breakdown process, because the current in the space charge region causes an increase in the electric field. If we assume that the electrons travel at their saturation velocity v_s across the depletion region, the space charge current I is given by

$$I = v_s \rho_s A \tag{31}$$

where ρ_s is the space charge density and A is the area. The disturbance $\Delta\mathscr{E}(x)$ in the electric field due to the space charge is obtained from Eq. 31 and Poisson's equation:

$$\Delta\mathscr{E}(W) = \frac{\int_0^W \rho_s \, dx}{\epsilon_s} = \frac{IW}{A\,\epsilon_s v_s} . \tag{32}$$

Setting $\Delta\mathscr{E}(W) = \mathscr{E}_c$, we obtain I_m, the maximum current allowed:

$$I_m = \frac{\mathscr{E}_c \epsilon_s v_s A}{W} . \tag{33}$$

Therefore, the upper limit on the power input is given by the product of V_m and I_m:

$$P_m = V_m I_m = \mathscr{E}_c^2 \epsilon_s v_s A . \tag{34}$$

The transit time frequency is given by

$$f = \frac{\gamma v_s}{W} \tag{35}$$

where γ is ½ for the IMPATT diode, ¾ for the BARITT diode, and 1 for the TED operated under the transit time domain mode. Equation 34 can be written as

$$P_m f^2 = \frac{\gamma \mathscr{E}_c^2 v_s^2}{2\pi X_c} \tag{36}$$

where X_c is the device reactance $(2\pi f \epsilon_s A /W)^{-1}$. Assuming that we are limited to some minimum circuit impedance, Eq. 36 predicts that the maximum power decreases as $1/f^2$.

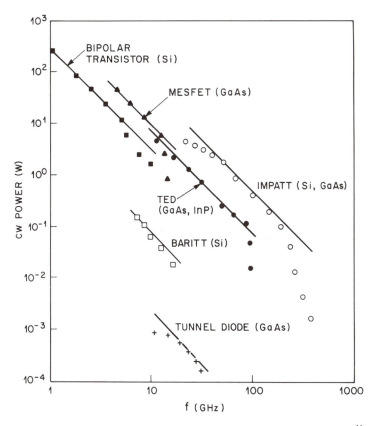

Fig. 22 Power output versus frequency for various microwave devices.[11]

Figure 22 shows the state-of-the-art cw power output versus frequency for six microwave device families. Note that the power output indeed varies approximately as $1/f^2$ for most devices. IMPATT diodes, because of their high operating fields (which are the critical fields at avalanche breakdown), have the highest $P_m f^2$ product and the highest power output in the millimeter wave region. Bipolar transistors, MESFETs, and TEDs have comparable $P_m f^2$ products. However, TEDs can extend substantially into the millimeter region, while the power output for both the bipolar transistor and the MESFET begin to drop rapidly beyond 10 GHz. As expected, the power output of BARITT diodes, which operate at about one tenth of the critical field and have about one tenth the efficiency of IMPATT diodes, is about three orders of magnitude below that of IMPATT diodes (because the power output is proportional to $\eta \mathscr{E}^2$). Tunnel diodes with a maximum voltage swing of only 1 V or less show the lowest power output.

In selecting a device for a particular microwave application, power output is not the only criterion; many other factors must be considered. One of the factors is noise. The term *noise* refers to spontaneous fluctuations either of current passing through a device or of voltage developed across the device.

The term originated with the study of high-gain audio-frequency amplifiers. When a fluctuating voltage or current generated in a device is amplified by an audio-frequency amplifier and the amplified signal is fed into a loudspeaker, the loudspeaker produces a hissing sound, hence the name noise. The descriptive term noise now refers to any spontaneous fluctuation, regardless of whether an audible sound is produced.

Because semiconductor devices are used mainly to measure small physical quantities or to amplify small signals, spontaneous fluctuations in current and voltage set a lower limit to the quantities to be measured or the signal to be amplified.

Noise is related to the discrete nature of charge carriers: electrons and holes. The macroscopic behavior of semiconductor devices, that is, their response to dc or ac voltages, is described in terms of the mean density of carriers, their drift velocity, and their mean lifetime. However, the instantaneous voltage V across a device fluctuates about a mean value $\overline{V}$. Consequently, there is an instantaneous fluctuation voltage $\Delta V \equiv V - \overline{V}$ called the noise voltage. A chart recording typical variations of the noise voltage with time is shown in Fig. 23. The noise wave form variation is completely random although the mean value of the noise voltage is zero.

In a $p - n$ junction, there are two fundamental sources of noise: (1) thermal noise resulting from the random motion of charge carriers in the neutral region (which is composed of resistive bulk materials), and (2) shot noise resulting from the random injection of charge carriers across the depletion region. The mean–square noise voltage $\overline{(\Delta V)^2}$ is found to be proportional to R for thermal noise, where R is the resistance of the neutral region, and to $1/I$ for shot noise, where I is the injected current.[12]

To compare the performance of microwave devices, an important figure of merit is the noise figure, which is the ratio of total mean-square noise voltage at the output of a device to mean-square noise voltage at the output resulting from thermal noise only. For example, if the total $\overline{(\Delta V)^2}$ at the output is

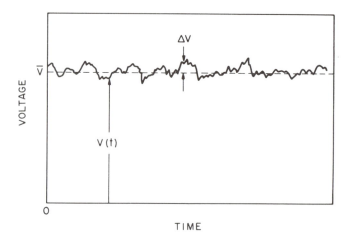

Fig. 23 Chart recording typical variations of fluctuating (noise) voltage with time.

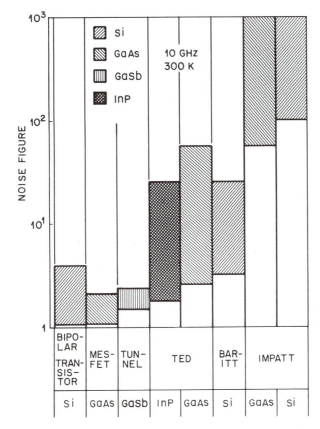

Fig. 24 Noise behavior of various microwave devices.

10^{-14} V^2 and the $\overline{(\Delta V)^2}$ from thermal noise is 10^{-16} V^2, the noise figure is $10^{-14}/10^{-16} = 100$. Figure 24 shows the measured noise figures of the six device families. Note that although IMPATT diodes have the highest power output, they also have the worst noise behavior due to the large random fluctuations of the avalanche processes. On the other hand, tunnel diodes have low power output and low noise. Therefore, there is a trade-off between power and noise.

The ease of fabrication is another important consideration. There is an inherent advantage to the structural simplicity of two-terminal devices compared to three-terminal transistors. Parasitic resistances and capacitances in three-terminal devices usually limit their upper operating frequencies to below the millimeter wave region.

REFERENCES

1   L. Esaki, "New Phenomenon in Narrow Ge $p-n$ Junction," *Phys. Rev.,* **109**, 603 (1958); "Discovery of the Tunnel Diode," *IEEE Trans. Electron Devices,* **ED-23**, 644 (1976).

2 B. C. DeLoach, Jr., "The IMPATT Story," *IEEE Trans. Electron Devices,* **ED-23**, 57 (1976); R. L. Johnston, B. C. DeLoach, Jr., and B. G. Cohen, "A Silicon Diode Oscillator," *Bell Syst. Tech. J.,* **44**, 369 (1965).

3 W. T. Read, "A Proposed High Frequency Negative Resistance Diode," *Bell Syst. Tech. J.,* **37**, 401 (1958).

4 D. J. Coleman, Jr., and S. M. Sze, "A Low–Noise Metal–Semiconductor–Metal (MSM) Microwave Oscillator," *Bell Syst. Tech. J.,* **50**, 1695 (1971).

5 S. Luryi and R. F. Kazarinov, "Optimum BARITT Structure," *Solid St. Electron.,* **25**, 943 (1982).

6 J. B. Gunn, "Microwave Oscillation of Current in III–V Semiconductors," *Solid State Comm.,* **1**, 88 (1963).

7 B. K. Ridley and T. B. Watkins, "The Possibility of Negative Resistance Effects in Semiconductors," *Proc. Phys. Soc. Lond.,* **78**, 293 (1961); C. Hilsum, "Transferred Electron Amplifiers and Oscillators," *Proc. IRE,* **50**, 185 (1962).

8 H. Kroemer, "Negative Conductance in Semiconductors," *IEEE Spectrum,* **5**, 47 (1968).

9 H. W. Thim, "Solid State Microwave Sources," in C. Hilsum, Ed., *Handbook on Semiconductor*, Vol. 4, *Device Physics*, North Holland, Amsterdam, 1980.

10 M. Shaw, H. L. Grubin, and P. R. Solomon, *The Gunn-Hilsum Effect*, Academic, New York, 1979.

11 S. M. Sze, *Physics of Semiconductor Devices*, 2nd ed., Wiley, New York, 1981, p 513–678.

12 A. Van der Ziel, *Noise in Measurements*, Wiley, New York, 1976.

PROBLEMS

1 (a) Find the depletion layer capacitance and depletion layer width at 0.25 V forward bias for a GaAs tunnel diode doped to 10^{19} cm^{-3} on both sides, using abrupt approximation and assuming $V_n = V_p = 0.03$ V. (b) If the tunneling distance is one half of the depletion layer width in (a), the potential barrier qV_o is 0.7 eV, and $m_n = 0.07m_0$, find the transmission coefficient for a conduction electron with $E = 0.25$ eV.

2 The current–voltage characteristic of a GaSb tunnel diode can be expressed by the empirical form Eq. 10 with $I_P = 10$ mA, $V_P = 0.1$ V, and $I_o = 0.1$ nA. (a) Find the largest negative differential resistance and the corresponding voltage. (b) Find the valley voltage V_V and the corresponding valley current I_V at 300 K.

3 Consider a silicon IMPATT diode shown in Fig. 9 with $b = 1.5$ μm, $W = 6$ μm, and a clump doped to 10^{18} cm^{-3} with a half-width $\delta = 70$ Å. (a) Find the breakdown voltage of the diode and the maximum field at breakdown. (b) Is the field in the drift region high enough to maintain the saturation velocity of holes? (c) Find the operating frequency.

4 The variation of electric field in the depletion region due to avalanche-generated space charge gives rise to an incremental resistance for abrupt p^+-n and Read diodes. The incremental resistance is called the space charge resistance R_{sc} and is given by $1/I \int_0^W \Delta\mathscr{E} \, dx$, where $\Delta\mathscr{E}$ is given by Eq. 32. (a) Find R_{sc} for a p^+-n Si IMPATT diode with $N_D = 10^{15} \, cm^{-3}$, $W = 12 \, \mu m$, and $A = 5 \times 10^{-4} \, cm^2$. (b) Find the total applied dc voltage for a current density of $10^3 \, A/cm^2$.

5 A GaAs IMPATT diode is operated at 10 GHz with a dc bias of 100 V and an average biasing current $(I_o/2)$ of 100 mA. (a) If the power-generating efficiency is 25% and the thermal resistance of the diode is $10°C/W$, find the junction temperature rise above the room temperature. (b) If the breakdown voltage increases with temperature at a rate of 60 mV/°C, find the breakdown voltage of the diode at room temperature.

6 For the BARITT diode shown in Fig. 12, the width of the n-type silicon is 5 μm and the doping is $5 \times 10^{15} \, cm^{-3}$. (a) Find the reach-through voltage at which the reverse-biased depletion region reaches through to the forward–biased depletion region. (b) Find the flat-band voltage at which the electric field is zero at the forward-biased metal–semiconductor contact. (c) Find the frequency of oscillation of the device.

7 Estimate the power-generating efficiency of a BARITT diode, assuming $V_{ac} = \frac{1}{2} V_{RT}$.

8 An important figure of merit for microwave application of $p-i-n$ diodes and Schottky diodes is the zero bias cutoff frequency f_c, which is defined as $(2\pi RC)^{-1}$, where R is the device series resistance and C is the junction capacitance at zero bias. (a) Find f_c for a Si $p-i-n$ diode with an area of $4 \times 10^{-4} \, cm^2$, an i-region width of 3 μm, and a series resistance of 0.2 Ω. (b) Find f_c for an Au-n-type GaAs Schottky diode with an area of $4 \times 10^{-4} \, cm^2$, an n-type region of 1 μm, and a doping concentration of $10^{17} \, cm^{-3}$.

9 (a) Find the effective density of states in the upper valley N_{CU} of the GaAs conduction band. The upper-valley effective mass is $1.2m_0$. (b) The ratio of electron concentrations between the upper and lower valleys is given by $(N_{CU}/N_{CL}) \exp(-\Delta E/kT_e)$, where N_{CL} is the effective density of states in the lower valley, $\Delta E = 0.31 \, eV$ is the energy difference, and T_e is the effective electron temperature. Find the ratio at $T_e = 300 \, K$. (c) When electrons gain kinetic energies from the electric field, T_e increases. Find the concentration ratio for $T_e = 1500 \, K$.

10 (a) A GaAs TED is 10 μm long and is operated in the transit time domain mode. Find the minimum electron density n_o required and the time between current pulses. (b) Calculate the power dissipated in the device, if it is biased at one half the threshold field. The device cross-sectional area is $10^{-2} \, cm^2$, and the electron density n_o is the same as in (a).

7

Photonic Devices

Photonic devices are devices in which the basic particle of light—the photon—plays a major role. In this chapter we consider four groups of photonic devices: *light-emitting diodes* (LEDs) and diode *lasers* (*l*ight *a*mplification by *s*timulated *e*mission of *r*adiation), which convert electrical energy to optical energy; *photodetectors*, which electronically detect optical signals; and *solar cells*, which convert optical energy into electrical energy.

7.1 RADIATIVE TRANSITIONS AND OPTICAL ABSORPTION

Figure 1 shows the electromagnetic spectrum of the optical region. The range of light detectable by the human eye extends only from approximately 0.4 μm to 0.7 μm. Figure 1 also shows the major color bands from violet to red in the expanded scale. The ultraviolet region includes wavelengths from 0.01 μm (i.e., 100 Å or 10 nm) to 0.4 μm, and the infrared region extends from 0.7 μm to 1000 μm. In this chapter we are primarily interested in the wavelength range from near-ultraviolet (~0.3 μm) to near-infrared (~1.5 μm).

Figure 1 also shows the photon energy on a separate horizontal scale. To

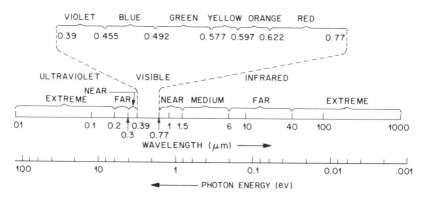

Fig. 1 Chart of the electromagnetic spectrum from the ultraviolet region to the infrared region.

convert the wavelength to photon energy, we used the relationship

$$\lambda' = \frac{c}{\nu} = \frac{hc}{h\nu} = \frac{1.24}{h\nu \text{ (eV)}} \ \mu m \tag{1}$$

where c is the speed of light in vacuum, ν is the light frequency, h is Planck's constant, and $h\nu$ is the energy of a photon, and is measured in electron volts. For example, a 0.5-μm green light corresponds to a photon energy of 2.48 eV.

7.1.1 Radiative Transitions

There are basically three processes for interaction between a photon and an electron in a solid: absorption, spontaneous emission, and stimulated emission.[1] We shall use a simple system to demonstrate these processes. Consider two energy levels E_1 and E_2 of an atom, where E_1 corresponds to the ground state and E_2 corresponds to an excited state (Fig. 2). Any transition between these states involves the emission or absorption of a photon with frequency ν_{12} given by $h\nu_{12} = E_2 - E_1$. At room temperature, most of the atoms in a solid are at the ground state. This situation is disturbed when a photon of energy exactly equal to $h\nu_{12}$ impinges on the system. An atom in state E_1 absorbs the photon and thereby goes to the excited state E_2. The change in

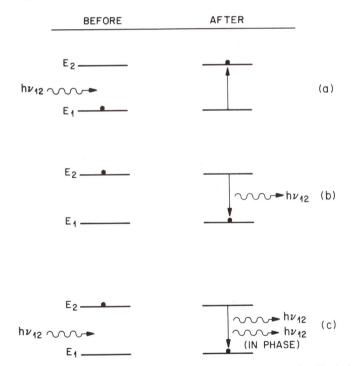

Fig. 2 The three basic transition processes between two energy levels. Black dots indicate the state of the atom. The initial state is at the left; the final state, after the transition process, is at the right. (a) Absorption. (b) Spontaneous emission. (c) Stimulated emission.

the energy state is the *absorption* process, shown in Fig. 2*a*. The excited state of the atom is unstable; and after a short time, without any external stimulus, it makes a transition to the ground state, giving off a photon of energy $h\nu_{12}$. This process is called *spontaneous emission* (Fig. 2*b*). When a photon of energy $h\nu_{12}$ impinges on an atom while it is in the excited state (Fig. 2*c*), the atom can be stimulated to make a transition to the ground state and gives off a photon of energy $h\nu_{12}$, which is in phase with the incident radiation. This process is called *stimulated emission*. The radiation from stimulated emission is monochromatic because each photon has an energy of precisely $h\nu_{12}$ and is coherent because all photons emitted are in phase.

The dominant operating process for the light-emitting diode (LED) is spontaneous emission; for the laser, it is stimulated emission; and for the photo-detector and the solar cell, it is absorption.

Let us assume that the instantaneous populations of E_1 and E_2 are n_1 and n_2, respectively. Under a thermal-equilibrium condition and for $(E_2 - E_1) > 3kT$, the population is given by the Boltzmann distribution:

$$\frac{n_2}{n_1} = e^{-(E_2-E_1)/kT} = e^{-h\nu_{12}/kT}. \tag{2}$$

The negative exponent indicates that n_2 is less than n_1 in thermal equilibrium; that is, most electrons are at the lower energy level.

In steady state, the stimulated-emission rate (i.e., the number of stimulated-emission transitions per unit time) and the spontaneous-emission rate must be balanced by the rate of absorption to maintain the populations n_1 and n_2 constant. The stimulated-emission rate is proportional to the photon field energy density $\rho(h\nu_{12})$, which is the total energy in the radiation field per unit volume per unit frequency. Thus, the stimulated-emission rate can be written as $B_{21}n_2\rho(h\nu_{12})$, where n_2 is the number of electrons in the upper level and B_{21} is a proportionality constant. The spontaneous-emission rate is proportional only to the population of the upper level and can be written as $A_{21}n_2$ where A_{21} is a constant. The absorption rate is proportional to the electron population at the lower level and to $\rho(h\nu_{12})$; this rate can be written as $B_{12}n_1\rho(h\nu_{12})$, where B_{12} is a proportionality constant. Therefore, we have at steady state

Stimulated-emission rate + spontaneous-emission rate = absorption rate

or

$$B_{21}n_2\rho(h\nu_{12}) + A_{21}n_2 = B_{12}n_1\rho(h\nu_{12}). \tag{3}$$

From Eq. 3 we observe that

$$\frac{\text{Stimulated-emission rate}}{\text{Spontaneous-emission rate}} = \frac{B_{21}}{A_{21}}\rho(h\nu_{12}). \tag{4}$$

To enhance stimulated emission over spontaneous emission, we must have a very large photon field energy density $\rho(h\nu_{12})$. To achieve this density, an optical resonant cavity is used (refer to Section 7.3) to increase the photon field. We also observe from Eq. 3 that

$$\frac{\text{Stimulated-emission rate}}{\text{Absorption rate}} = \frac{B_{21}}{B_{12}}\left[\frac{n_2}{n_1}\right]. \tag{5}$$

If the stimulated emission of photons is to dominate over the absorption of photons, we must have higher electron density in the upper level than in the lower level. This condition is called *population inversion*, since under an equilibrium condition the reverse is true. We shall consider various ways to have a large photon field energy density and achieve population inversion, so that the stimulated emission becomes dominant over both spontaneous emission and absorption.

7.1.2 Optical Absorption

Figure 3 shows the basic transitions in a semiconductor. When the semiconductor is illuminated, photons are absorbed to create electron–hole pairs as shown at (*a*) in Fig. 3 if the photon energy is equal to the bandgap energy, that is, $h\nu$ equals E_g. If $h\nu$ is greater than E_g, an electron–hole pair is generated and, in addition, the excess energy $(h\nu - E_g)$ is dissipated as heat as shown at (*b*) in Fig. 3. Both processes, (*a*) and (*b*), are called *intrinsic transitions* (or band-to-band transitions). On the other hand, for $h\nu$ less than E_g, a photon will be absorbed only if there are available energy states in the forbidden bandgap due to chemical impurities or physical defects as shown at (*c*) in Fig. 3. Process (*c*) is called *extrinsic transition*. This discussion also is generally true for the reverse situation. For example, an electron at the conduction band edge combining with a hole at the valence band edge will result in the emission of a photon with energy equal to that of the bandgap.

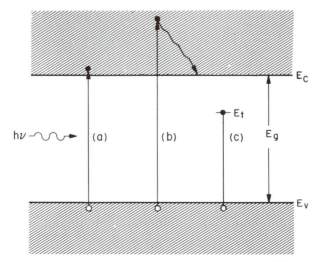

Fig. 3 Optical absorption for (a) $h\nu = E_g$, (b) $h\nu > E_g$, and (c) $h\nu < E_g$.

Assume that a semiconductor is illuminated from a light source with $h\nu$ greater than E_g and a photon flux of Φ_o (in units of photons per square centimeter per second). As the photon flux travels through the semiconductor, the fraction of the photons absorbed is proportional to the intensity of the flux. Therefore, the number of photons absorbed within an incremental distance Δx (Fig. 4a) is given by $\alpha\Phi(x)\Delta x$, where α is a proportionality constant defined as the *absorption coefficient*. From the continuity of photon flux as shown in Fig. 4a, we obtain

$$\Phi(x + \Delta x) - \Phi(x) = \frac{d\Phi(x)}{dx}\Delta x = -\alpha\Phi(x)\Delta x$$

or

$$\frac{d\Phi(x)}{dx} = -\alpha\Phi(x) . \tag{6}$$

The negative sign indicates decreasing intensity of the photon flux due to absorption. The solution of Eq. 6 with the boundary condition $\Phi(x) = \Phi_o$ at $x = 0$ is

$$\Phi(x) = \Phi_o e^{-\alpha x} . \tag{7}$$

The fraction of photon flux that exits from the other end of the semiconductor at $x = W$ (Fig. 4b) is

$$\Phi(W) = \Phi_o e^{-\alpha W} . \tag{8}$$

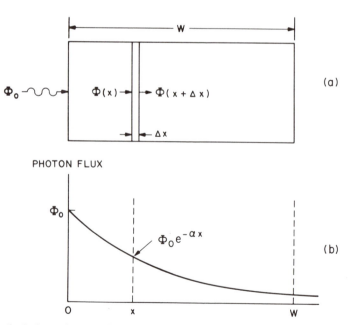

Fig. 4 Optical absorption. (a) Semiconductor under illumination. (b) Exponential decay of photon flux.

The absorption coefficient α is a function of $h\nu$. Figure 5 shows the measured absorption coefficient for some important semiconductors that are used for photonic devices.[2] Also shown is the absorption coefficient for amorphous silicon (dashed curve), which is an important material for solar cells. The absorption coefficient decreases rapidly at the cutoff wavelength λ_c; that is,

$$\lambda_c = \frac{1.24}{E_g} \ \mu m \tag{9}$$

because the optical band-to-band absorption becomes negligible for $h\nu < E_g$, or $\lambda > \lambda_c$.

Problem

A 0.25-μm-thick single-crystal silicon sample is illuminated with a monochromatic light (single frequency) having an $h\nu$ of 3 eV. The incident power is 10 mW. Find the total energy absorbed by the semiconductor per second, the rate of excess thermal energy dissipated to the lattice, and the number of photons per second given off from recombination by intrinsic transitions.

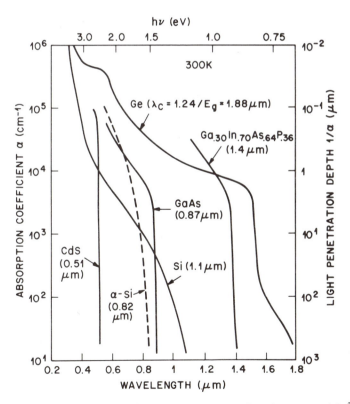

Fig. 5 Optical absorption coefficients for various semiconductor materials.[2]

Solution

From Fig. 5 the absorption coefficient is 4×10^4 cm^{-1}. The energy absorbed per second is

$$\Phi_o(1 - e^{-\alpha W}) = 10^{-2}[1 - \exp(-4 \times 10^4 \times 0.25 \times 10^{-4})]$$

$$= 0.0063 \text{ J/s} = 6.3 \text{ mW} .$$

The portion of each photon's energy that is converted to heat is

$$\frac{h\nu - E_g}{h\nu} = \frac{3 - 1.12}{3} = 62\% .$$

Therefore, the amount of energy dissipated per second to the lattice is

$$62\% \times 6.3 = 3.9 \text{ mW} .$$

Since the recombination radiation accounts for 2.4 mW (i.e., 6.3 mW− 3.9 mW) at 1.12 eV/photon, the number of photons per second from recombination is

$$\frac{2.4 \times 10^{-3}}{1.6 \times 10^{-19} \times 1.12} = 1.3 \times 10^{16} \text{ photons/s} .$$

7.2 LIGHT-EMITTING DIODE

Light-emitting diodes (LEDs) are *p–n* junctions that can emit spontaneous radiation in ultraviolet, visible, or infrared regions.[3] The visible LED has a multitude of applications as an information link between electronic instruments and their users. The infrared LEDs are useful in opto-isolators and for optical-fiber communication.

7.2.1 Visible LED

Figure 6 shows the relative eye response as a function of wavelength (or the corresponding photon energy). The maximum sensitivity of the eye is at 0.555 μm. The eye response falls to nearly zero at the extremes of the visible spectrum at about 0.4 and 0.7 μm. For normal vision at the peak response of the eye, 1 watt of radiant energy is equivalent to 683 lumen.

Since the eye is only sensitive to light with a photon energy $h\nu$ equal to or greater than 1.8 eV ($\lesssim$ 0.7 μm), semiconductors of interest must have an energy bandgap larger than this limit. Figure 6 also shows the bandgaps of various semiconductors. Among all the semiconductors shown, the most important one for visible LEDs is the alloy GaAs$_{1-y}$P$_y$ III–V compound system. An alloy III–V compound is formed when more than one Group III element are distributed randomly on Group III lattice sites (e.g., gallium sites) or more than one Group V element are distributed randomly on Group V lattice sites (e.g, arsenic sites). The notation used is A$_x$B$_{1-x}$C or AC$_{1-y}$D$_y$ for *ternary* (three elements) compounds and A$_x$B$_{1-x}$C$_y$D$_{1-y}$ for *quaternary* (four elements) compounds where A and B are the Group III elements, C and D are the Group V elements, and x and y are the mole fractions, that is, the ratios of the number of atoms of a given species to the total number of Group III or Group V atoms in the alloy compound.

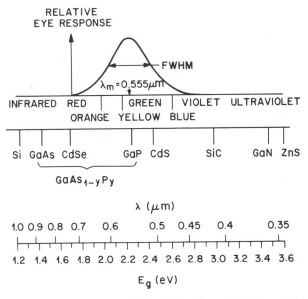

Fig. 6 Semiconductors of interest as visible LEDs. Figure includes relative response of the human eye.

Figure 7a shows the energy gap for GaAs$_{1-y}$P$_y$ as a function of the mole fraction y. For $0 < y < 0.45$, the bandgap is direct and increases from $E_g = 1.424$ eV at $y = 0$ to $E_g = 1.977$ eV at $y = 0.45$. For $y > 0.45$, the bandgap is indirect. Figure 7b shows the corresponding energy–momentum plots for selected alloy compositions.[4] As indicated, the conduction band has two minima. The one along $\bar{p} = 0$ is the direct minimum, and the one along $\bar{p} = \bar{p}_{max}$ is the indirect minimum. Electrons in the direct minimum of the conduction band and holes at the top of the valence band have equal momenta ($\bar{p} = 0$); electrons in the indirect minimum of the conduction band and holes at the top of the valence band have different momenta. The radiative transition mechanisms are found predominantly in direct-bandgap semiconductors, such as gallium arsenide and GaAs$_{1-y}$P$_y$ ($y < 0.45$), since the momentum is conserved. The photon energy is approximately equal to the bandgap energy of the semiconductor.

However, for GaAs$_{1-y}$P$_y$ with y greater than 0.45 and gallium phosphide, which are indirect-bandgap semiconductors, the probability for radiative transitions is very small, since lattice interactions or other scattering agents must participate in the process to conserve momentum. Therefore, for indirect-bandgap semiconductors, special recombination centers are incorporated to enhance the radiative processes. An efficient radiative recombination center in GaAs$_{1-y}$P$_y$ can be formed by incorporating nitrogen into the crystal lattice. When nitrogen is introduced, it replaces phosphorous atoms in the lattice sites. The outer electronic structure of nitrogen is similar to that of phosphorous (both are Group V elements in the periodic table), but the electronic core

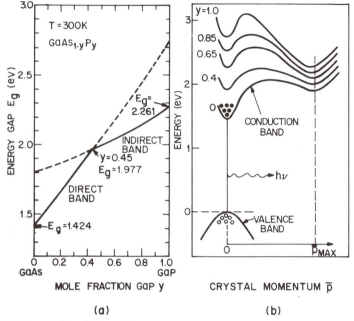

Fig. 7 (a) Compositional dependence of the direct- and indirect-energy bandgap[1] for GaAs$_{1-y}$P$_y$. (b) The alloy compositions shown correspond to red ($y = 0.4$), orange (0.65), yellow (0.85), and green light (1.0).[4]

structures of these atoms are quite different. This difference results in the creation of an electron trap level close to the bottom of the conduction band. A recombination center is thus produced and it is called an *isoelectronic center*. This recombination center can greatly enhance the probability of radiative transition in indirect-bandgap semiconductors.

Figure 8 shows the *quantum efficiency* (i.e., number of photons generated per electron–hole pair) versus alloy composition for GaAs$_{1-y}$P$_y$ with and without the isoelectronic impurity nitrogen.[5] The efficiency without nitrogen drops sharply in the composition range of $0.4 < y < 0.5$ because of the change over of the bandgap from direct to indirect at $y = 0.45$. The efficiency with nitrogen is considerably higher for $y > 0.5$ but nevertheless decreases steadily with an increasing y because of the increasing separation between the direct and indirect bandgap (Fig. 7b).

The basic LED structures are the flat-diode configurations shown[6] in Fig. 9. Generally, direct–bandgap LEDs (which emit red light) are fabricated on gallium arsenide substrates (Fig. 9a), and indirect–bandgap LEDs (orange, yellow, and green light) are fabricated on gallium phosphide substrates (Fig. 9b). A graded-alloy GaAs$_{1-y}$P$_y$ layer is grown epitaxially to minimize the nonradiative centers at the interface that result from lattice mismatch.

Three loss mechanisms reduce the quantity of emitted photons: (1) absorption within the LED material, (2) reflection loss when light passes from a semi-

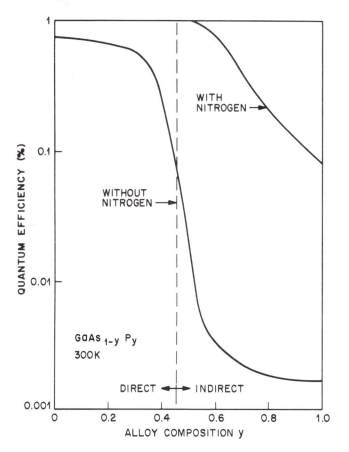

Fig. 8 Quantum efficiency versus alloy composition with and without isoelectronic impurity nitrogen.[5]

conductor to air due to differences in refractive index, and (3) total internal reflection of light at angles greater than the critical angle θ_c (Fig. 9a) defined by Snell's law:

$$\sin \theta_c = \frac{\bar{n}_1}{\bar{n}_2} \tag{10}$$

where the light passes from a medium with a refraction index of $\bar{n}_2$ (e.g., GaAs with $\bar{n}_2 = 3.66$ at $\lambda \simeq 0.8 \ \mu m$) to a medium of $\bar{n}_1$ (e.g., air with $\bar{n}_1 = 1$). For gallium arsenide, the critical angle is about 16 °; and for gallium phosphide with $\bar{n}_2 = 3.45$ at $\lambda \simeq 0.8 \ \mu m$, the critical angle is about 17 °.

The forward current–voltage behavior of a LED is similar to that of the GaAs p–n junction shown in Fig. 17 of Chapter 3. At low forward voltages, the diode current is dominated by the nonradiative recombination current due mainly to surface recombination near the perimeter of the LED chip. At

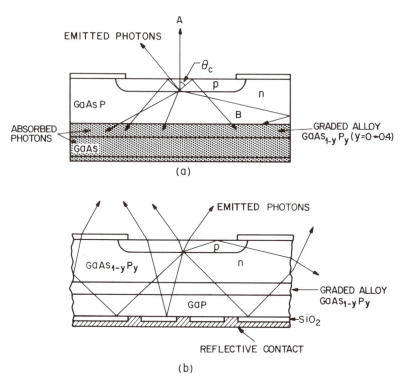

Fig. 9 Effects of (a) opaque substrate and (b) transparent substrate on photon emitted at the *p–n* junction.[6]

higher forward voltages, the diode current is dominated by the radiative diffusion current. At even higher voltages, the diode current will be limited by the series resistance. The total diode current can be written as

$$I = I_d \exp\left[\frac{q(V - IR_s)}{kT}\right] + I_r \exp\left[\frac{q(V - IR_s)}{2kT}\right] \qquad (11)$$

where R_s is the device series resistance and I_d and I_r are the saturation currents due to diffusion and recombination, respectively. To increase the power output of the LED, we must reduce I_r and R_s.

The emission spectra of LEDs are similar to the eye response curve shown in Fig. 6. The spectral width is given by the full width at half maximum intensity (FWHM). The spectral width generally varies as λ_m^2, where λ_m is the wavelength at the maximum intensity.[7] Thus, the FWHM becomes larger as the wavelength is increased from visible to infrared. For example, at $\lambda_m = 0.55\ \mu\text{m}$ (green color), FWHM is about 200 Å, but at 1.3 μm (infrared), FWHM is over 1200 Å.

Visible LEDs can be used as indicator lamps and displays. Figure 10 shows the diagrams of two LED lamps.[3] An LED lamp contains an LED chip and a

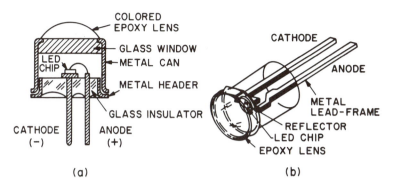

Fig. 10 Diagrams of two LED lamps.[3]

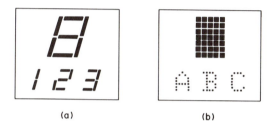

Fig. 11 LED display formats for numeric and alphanumeric: (a) 7-segment (numeric); (b) 5×7 array (alphanumeric).[3]

plastic lens which is usually colored to serve as an optical filter and to enhance contrast. The lamp in Fig. 10a uses a conventional diode header. Figure 10b shows a package that is suited for a transparent semiconductor, such as gallium phosphide, which emits light through all five facets (four sides and the top) of the LED chip.

Figure 11 shows the basic formats for LED displays. The seven segments (Fig. 11a) display numbers from 0 to 9. The 5×7 matrix array (Fig. 11b) displays alphanumerics (A to Z and 0 to 9). The displays can be made by monolithic processes similar to those used to make silicon integrated circuits (to be described in Chapters 8 through 12) or by using an individual LED chip mounted on a reflector to form a bar segment.

7.2.2 Infrared LEDs

Infrared LEDs include gallium arsenide LEDs which emit light near $0.9\,\mu m$, and many III–V compounds, such as the quaternary $Ga_x In_{1-x} As_y P_{1-y}$ LEDs, which emit light from 1.1 to $1.6\,\mu m$ (refer to Section 7.3 on quaternary compounds).

An important application of infrared LEDs is in opto-isolators, where an input or control signal is decoupled from the output. Figure 12 shows an opto-isolator having an infrared LED as the light source and a photodiode as the detector. When an input signal is applied to the LED, light is generated

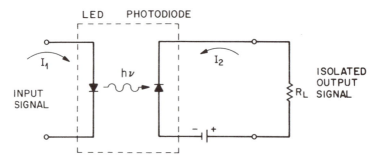

Fig. 12   An opto-isolator in which an input signal is decoupled from the output signal.

and subsequently detected by the photodiode. The light is then converted back to an electrical signal as a current that flows through a load resistor. Opto-isolators transmit signals at the speed of light and are electrically isolated because there is no electrical feedback from the output to the input.

Another important application of infrared LEDs is for transmission of an optical signal through an optical fiber as in a communication system. An optical fiber is a wave guide at optical frequencies. The fiber is usually drawn from a preform of glass to a diameter of about 100 μm. It is flexible and can guide optical signals over distances of many kilometers to a receiver, similar to the way a coaxial cable transmits electrical signals.

Two types of optical fibers are shown in Fig. 13. One type of the fiber has a cladding layer of relatively pure fused silica (SiO_2) surrounding a core of doped glass (e.g., germanium doped glass) that has a higher refractive index than the cladding layer.[8] This type of fiber is called a *step index fiber*. The light is transmitted along the length of the fiber by internal reflection at the step in the refractive index. The critical angle for internal reflection is about 86 ° for $\bar{n}_1 = 1.457$ (cladding layer) and $\bar{n}_2 = 1.460$ (core) as calculated from Eq. 10. Note that different rays will propagate with different path lengths (Fig. 13a). A light pulse reaching the end of a step index fiber will result in a pulse spread. In a *graded-index fiber* (Fig. 13b), the index decreases from the core center by a parabolic law. Now, rays traversing toward the cladding have a high velocity (due to lower index of refraction) than rays along the center of the core. The pulse spread is significantly reduced. As the light is transmitted along the optical fiber, the light signal will be attenuated. However, due to the transparency of ultrapure silica used for the fiber material in the wavelength region from 0.8 to 1.6 μm, the attenuation is quite low and is proportional to λ^{-4}. Typical attentuations are about 3 dB/km at a wavelength of 0.8 μm, 0.6 dB/km at 1.3 μm, and 0.2 dB/km at 1.55 μm.

A simple point-to-point optical-fiber communication system is shown in Fig. 14, where the electrical input signals are converted to optical signals using an optical source (LED or laser). The optical signals are coupled into the fiber and transmitted to the photodetector where they are converted back to electrical signals.

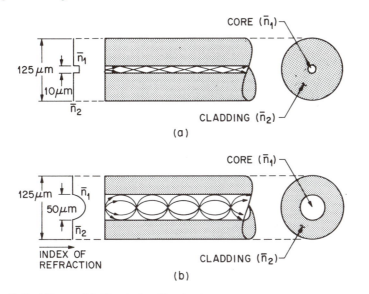

Fig. 13 Optical fibers. (*a*) Step index fiber having a core with slightly larger refractive index. (*b*) Graded-index fiber having a parabolic grading of the refractive index in the core.[8]

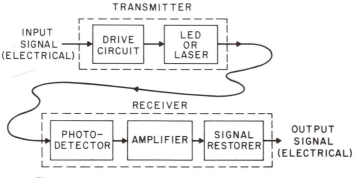

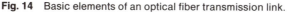

Fig. 14 Basic elements of an optical fiber transmission link.

One basic device configuration of infrared LEDs used for optical-fiber communication is the surface-emitting LED, shown[9] in the insert of Fig. 15. The light is emitted from the central surface area and coupled into the optical fiber. The semiconductor substrate through which the emission must be collected is made very thin, 10 to 15 μm, to minimize absorption and to allow the end of the fiber to be very close to the emitting surface. The use of heterojunctions (e.g., GaAs–Al$_x$Ga$_{1-x}$As) can increase the efficiency that results from the confinement of the carrier by the layers of the higher-bandgap semiconductor (e.g. Al$_x$Ga$_{1-x}$As) surrounding the radiative-recombination region (e.g., GaAs). We shall consider the carrier confinement in Section 7.3. The hetero-

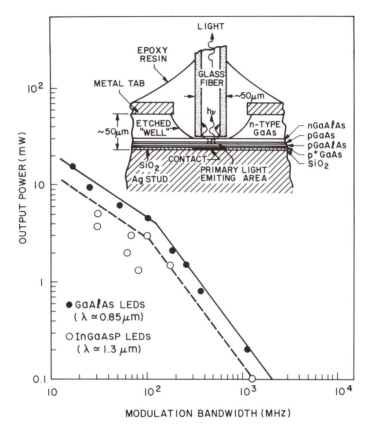

Fig. 15 Output power versus modulation bandwidth of LEDs.[7] Insert shows a small-area, high-radiance, double-heterostructure surface emitter LED with a fiber attached.[9]

junction can also serve as an optical window to the emitted radiation, because the higher-bandgap confining layers do not absorb radiation from the lower-bandgap emitting region.

The electrical input signal (e.g., an applied voltage) is generally modulated at high frequencies. This in turn gives rise to a direct modulation of the injected current in an LED. Parasitic elements such as the depletion layer capacitance and series resistance can cause a delay of carrier injection into the junction, and consequently a delay in the light output. The ultimate limit on how fast one can vary the light output depends on the carrier lifetime which is determined by various recombination processes such as the surface recombination discussed in Chapter 2. If the current is modulated at an angular frequency ω, the light output $P(\omega)$ is given by

$$P(\omega) = \frac{P(0)}{\sqrt{1 + (\omega\tau)^2}} \tag{12}$$

where $P(0)$ is the light output at $\omega = 0$, and τ is the carrier lifetime. The *modulation bandwidth* Δf is defined as the frequency at which the light output is reduced to $1/\sqrt{2}$ that at $\omega = 0$, that is

$$\Delta f \equiv \frac{\Delta \omega}{2\pi} = \frac{1}{2\pi\tau}. \tag{13}$$

Figure 15 shows the measured output power as a function of the modulation bandwidth for state-of-the-art LEDs.[7] It is apparent that the output power decreases with increasing Δf. The solid line and the dotted line in the figure are contours of best reported results of $Ga_x Al_{1-x} As$ and $In_x Ga_{1-x} As_y P_{1-y}$ LEDs, respectively. The output power of $Ga_x Al_{1-x} As$ LEDs is approximately 50% higher than that of $In_x Ga_{1-x} As_y P_{1-y}$ LEDs at all bandwidths. This can be explained by the fact that the photon energy at $0.85\mu m$ (i.e., 1.46 eV) is larger by a factor of 1.53 than that at $1.3\mu m$ (0.95 eV), and the output power of $Ga_x Al_{1-x} As$ LEDs is increased by the same factor.

7.3 SEMICONDUCTOR LASER

Semiconductor lasers are similar to the solid-state ruby laser and helium–neon gas laser in that the emitted radiation is highly monochromatic and produces a highly directional beam of light. However, the semiconductor laser differs from other lasers in that it is small (on the order of 0.1 mm long) and is easily modulated at high frequencies simply by modulating the biasing current. Because of these unique properties, the semiconductor laser is one of the most important light sources for optical-fiber communication. It is also used in video recording, optical reading, and high-speed laser printing. In addition, semiconductor lasers have significant applications in many areas of basic research and technology, such as high-resolution gas spectroscopy and atmospheric pollution monitoring.

7.3.1 Semiconductor Materials

All lasing semiconductors have direct bandgaps. This is expected, because the crystal momentum is conserved, and hence the radiative-transition probability in a direct-bandgap semiconductor is high. At present, the laser emission wavelengths cover the range from 0.3 to over $30\,\mu m$. Gallium arsenide was the first material to emit laser radiation, and its related III–V compound alloys are the most extensively studied and developed.

The two most important III–V compound alloy systems are $Al_x Ga_{1-x} As_y Sb_{1-y}$ and $Ga_x In_{1-x} As_y P_{1-y}$ solid solutions. Figure 16 shows the bandgaps plotted against the lattice constant for the III–V binary semiconductors and their intermediate ternary and quaternary compounds.[1] To achieve heterostructures with negligible interface traps, the lattices between the two semiconductors must be matched closely. If we use GaAs ($a = 5.6533$ Å) as the substrate, the ternary compound $Al_x Ga_{1-x} As$ can have a lattice mismatch less than 0.1%. Similarly, with InP ($a = 5.8686$ Å) as

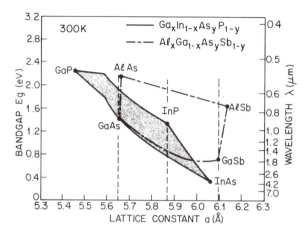

Fig. 16   Energy bandgap and lattice constant for two III–V compound solid alloy systems.[1]

the substrate, the quaternary compound $Ga_x In_{1-x} As_y P_{1-y}$ also can have a nearly perfect lattice match, as indicated by the center vertical line in Fig. 16.

Figure 17a shows the bandgap of ternary $Al_x Ga_{1-x} As$ as a function of aluminum composition.[1] The alloy has a direct bandgap up to $x = 0.45$, then becomes an indirect-bandgap semiconductor. Figure 17b shows the compositional dependence of the refractive index. For example, for $x = 0.3$, the bandgap of $Al_{0.3}Ga_{0.7}As$ is 1.789 eV, which is 0.365 eV larger than that of GaAs; its refractive index is 3.385, which is 6% smaller than that of GaAs. These properties are important for continuous operation of semiconductor lasers at and above room temperatures.

7.3.2   Laser Structures

Figure 18 shows three laser structures.[10] The first structure, Fig. 18a, is a basic *p–n* junction laser. This is called a *homojunction laser* because it has the same semiconductor material (e.g., GaAs) on both sides of the junction. A pair of parallel planes (or facets) are cleaved or polished perpendicular to the <110>-axis. Under appropriate biasing conditions, laser light will be emitted from these planes (only the front emission is shown in Fig. 18). The two remaining sides of the diode are roughened to eliminate lasing in the directions other than the main ones. This structure is called a *Fabry–Perot cavity*, with a typical cavity length L of about 300 μm. The Fabry–Perot cavity configuration is extensively used for modern semiconductor lasers.

Figure 18b shows a *double-heterostructure* (DH) laser, in which a thin layer of semiconductor (e.g., GaAs) is sandwiched between layers of a different semiconductor (e.g., $Al_x Ga_{1-x} As$). This laser can be fabricated using epitaxial crystal growth techniques (refer to Chapter 8). We show in the subsequent section that a DH laser requires much less current to operate than a homojunction laser with identical device geometry.[11]

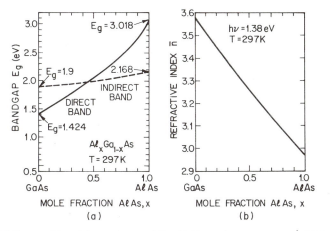

Fig. 17 (a) Compositional dependence of the $Al_xGa_{1-x}As$ energy gap.[1] (b) Compositional dependence of the refractive index at 1.38 eV.

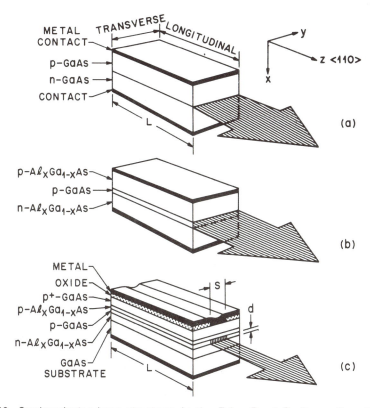

Fig. 18 Semiconductor laser structures in the Fabry–Perot–Cavity configuration. (a) Homojunction laser. (b) Double heterojunction (DH) laser. (c) Stripe geometry DH laser.[10, 11]

The laser structures shown in Figs. 18a and 18b are broad-area lasers, because the entire area along the junction plane can emit radiation. Figure 18c shows a DH laser having a stripe geometry. The oxide layer isolates all but the stripe contact, consequently the lasing area is restricted to a narrow region under the contact. The stripe widths S are typically 5 to 30 μm. The advantages of the stripe geometry are reduced operating current, elimination of multiple-emission areas along the junction, and improved reliability that is the result of removing most of the junction perimeter.

7.3.3 Laser Operation

Population Inversion As discussed in Section 7.1.1, to enhance the stimulated emission for laser operation we need population inversion. To achieve population inversion in a semiconductor laser, we shall consider a p–n junction formed between degenerate semiconductors, that is, one in which the doping levels on both sides of the junction are high enough so that the Fermi levels are below the valence band edge on the p-side and above the conduction band edge on the n-side. Figure 19a shows the band diagram of such a device at thermal equilibrium. When we apply a forward bias to the diode (Fig. 19b),

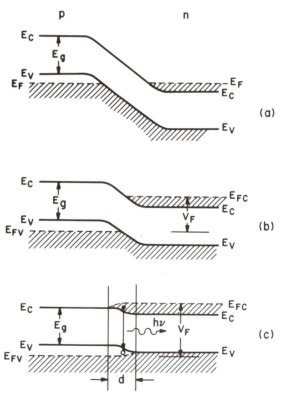

Fig. 19 Energy band diagrams of a degenerate p–n junction (a) at thermal equilibrium, (b) under forward bias, and (c) under high-injection condition.

electrons are injected from the n-side and holes are injected from the p-side into the transition region. When a sufficiently large bias is applied (Fig. 19c), high injection occurs, that is, large concentrations of electrons and holes are injected into the transition region. As a result, the region d (Fig. 19c) contains a large concentration of electrons in the conduction band and a large concentration of holes in the valence band; this is the condition of population inversion.

For band-to-band transition, the minimum energy required is the bandgap energy E_g. Therefore, from Fig. 19c, we can write the condition necessary for population inversion: $(E_{FC} - E_{FV}) > E_g$.

Carrier and Optical Confinement Figure 20 shows schematic representations of the band diagram under a forward-bias condition, the refractive-index profile, and the optical-field distribution of light generated at the junction of a homojunction laser (Fig. 20a) and a DH laser (Fig. 20b). As can be seen in the DH laser, the carriers are confined on both sides of the active region by the heterojunction barriers, while in the homojunction laser the carriers can move away from the active region where radiative recombination occurs.

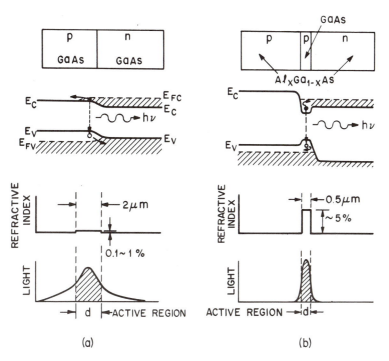

(a)   (b)

Fig. 20 Comparison of some characteristics of (a) homojunction laser and (b) double-heterojunction (DH) laser. Second from the top row shows energy band diagrams under forward bias. The refractive index change for a homojunction laser is less than 1%. The refractive index change for the DH laser is about 5%. The confinement of light is shown in the bottom row.[11]

The optical field is also confined within the active region by the abrupt reduction of the refractive index outside the active region. The optical confinement can be explained by Fig. 21, which shows a three-layer dielectric wave guide with refractive indices $\bar{n}_1$, $\bar{n}_2$, and $\bar{n}_3$, where an active layer is sandwiched between two confining layers (Fig. 21a). Under the condition $\bar{n}_2 > \bar{n}_1 \gtrsim \bar{n}_3$, the ray angle θ_{12} at the layer 1/layer 2 interface in Fig. 21b exceeds the critical angle given by Eq. 10. A similar situation occurs for θ_{23} at the layer 2/layer 3 interface. Therefore, when the refractive index in the active layer is larger than the index of its surrounding layers, the propagation of the electromagnetic radiation is guided (confined) in a direction parallel to the layer interfaces. We can define a *confinement factor* Γ, which is the ratio of the light intensity within the active layer to the sum of light intensity both within and outside the active layer. The confinement factor is given as

$$\Gamma \simeq 1 - \exp\left(-C\,\Delta\bar{n}d\right) \tag{14}$$

where C is a constant, $\Delta\bar{n}$ is the difference in the refractive index, and d is the thickness of the active layer. It is clear that the larger the $\Delta\bar{n}$ and d are, the higher the Γ will be.

Threshold Current Density   One of the most important parameters for laser operation is the threshold current density J_{th}, that is, the minimum current density required for lasing to occur. Figure 22 compares J_{th} versus operating temperature for a homojunction laser and a DH laser. Note that as the temperature increases, J_{th} for the DH laser increases much more slowly than J_{th} for the homojunction laser. Because of the low values of J_{th} for DH lasers at 300 K, DH lasers can be operated continuously at room temperature.

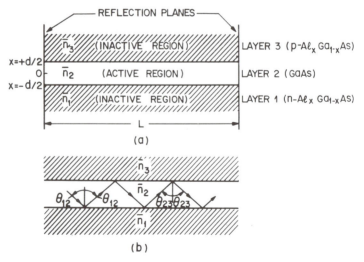

Fig. 21   (*a*) Representation of a three-layer dielectric wave guide. (*b*) Ray trajectories of the guided wave.

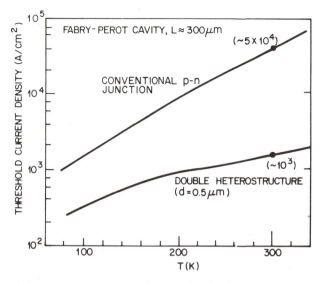

Fig. 22 Threshold current density versus temperature for the two laser structures shown[11] in Fig. 20.

This characteristic has led to the increased use of semiconductor lasers, especially in optical-fiber communication systems.

In a semiconductor laser, the gain g, that is, the incremental optical energy flux per unit length, depends on the current density. The gain g can be expressed as a function of a nominal current density J_{nom}, which is defined for unity quantum efficiency (i.e., number of carriers generated per photon, $\eta = 1$) as the current density required to excite a 1-μm-thick active layer uniformly. The actual current density is then given by

$$J \; (\text{A/cm}^2) = \frac{J_{nom}d}{\eta} \tag{15}$$

where d is the thickness of the active layer in μm. Figure 23 shows the calculated gain for a typical gallium arsenide DH laser. The gain increases linearly with J_{nom} for $50 \leqslant g \leqslant 400$ cm^{-1}. The linear dashed line can be written as[12]

$$g = (g_o/J_o)(J_{nom} - J_o) \tag{16}$$

where $g_o/J_o = 5 \times 10^{-2}$ cm-μm/A and $J_o = 4.5 \times 10^3$ A/cm^2-μm.

As discussed previously, at low currents there is spontaneous emission in all directions. As the current increases, the gain increases (Fig. 23) until the threshold for lasing is reached, that is, until the gain satisfies the condition that a light wave makes a complete traversal of the cavity without attenuation:

$$R \; \exp\left[(\Gamma g - \alpha)L\right] = 1 \tag{17}$$

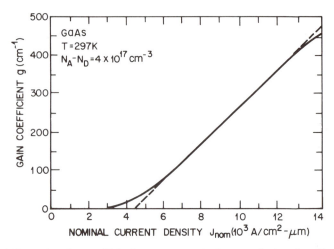

Fig. 23 Variation of gain coefficient versus nominal current density. Dashed line represents a linear dependence.[12]

or

$$\Gamma g \text{ (threshold gain)} = \alpha + \frac{1}{L} \ln \left[\frac{1}{R} \right] \qquad (18)$$

where Γ is the confinement factor, α is the loss per unit length from absorption and other scattering mechanism, L is the length of the cavity (shown in Figs. 18 and 21), and R is the reflectance of the ends of the cavity (assuming that R for both ends is equal). Equations 15, 16, and 18 may be combined to give the threshold current density as

$$J_{th} \text{ (A/cm}^2\text{)} = \frac{J_o d}{\eta} + J_o \frac{d}{g_o \eta \Gamma} \left[\alpha + \frac{1}{L} \ln \left[\frac{1}{R} \right] \right]. \qquad (19)$$

To reduce J_{th}, we can increase η, Γ, L, and R and reduce d and α.

7.3.4 Laser Characteristics

Figure 24 compares the calculated J_{th} from Eq. 19 to experimental results from $Al_x Ga_{1-x} As$–GaAs DH lasers, where x is the aluminum composition shown in Fig. 17. The threshold current density decreases with decreasing d, reaches a minimum, and then increases. The increase of J_{th} at very narrow active layer thickness is caused by poor optical confinement.[1]

Figure 25 shows the temperature dependence of J_{th} for a cw (continuous wave) stripe geometry $Al_x Ga_{1-x} As$–GaAs DH laser.[13] Figure 25a shows cw light outputs versus injection current at various temperatures between 25 and 115 °C. Note the excellent linearity in the light–current characteristics. The threshold current at a given temperature is the extrapolated value for zero output power. Figure 25b shows a plot of threshold currents as a function of

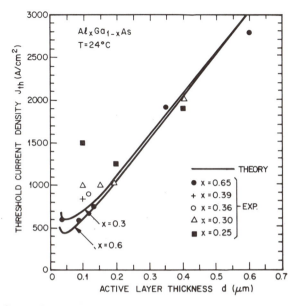

Fig. 24 Comparison of experimental and calculated threshold current density.[1]

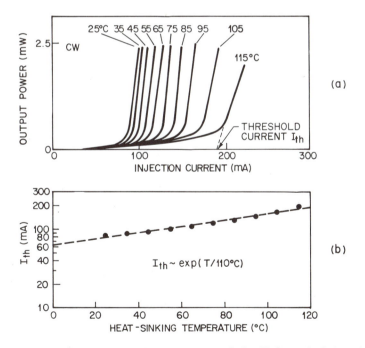

Fig. 25 (a) Light output versus diode current for a GaAs–Al$_x$Ga$_{1-x}$As heterostructure laser. (b) Temperature dependence of cw-current threshold.[13]

temperature. The threshold current increases exponentially with temperature as

$$I_{th} \sim \exp \left[\frac{T}{T_0} \right] \qquad (20)$$

where T is the temperature in °C and T_0 is 110 °C for this laser.

For optical fiber communications, the optical source must be able to be modulated at high frequencies. Unlike LEDs whose output power decreases with increasing modulation bandwidth, the output power of typical GaAs or GaInAsP laser remains at a constant level (e.g., 10 mW per facet) well into GHz range.

Figure 26 shows the emission spectra of a typical laser. At low currents, the spontaneous emission has broad spectral distribution with a full width of half-maximum intensity of 100 to 500 Å. As the current approaches the threshold, the spectral distribution becomes narrower. Above the threshold current, the laser may approach near-perfect monochromatic emission with a spectral width in the order of 1 to 10 Å.

Figure 27 shows a high-resolution emission spectrum for a stripe geometry InP–$Ga_x In_{1-x} As_y P_{1-y}$ DH laser.[14] The stripe is formed by proton bombardment of the area adjacent to the stripe to produce high-resistivity regions. The lasing area is restricted to the center region which is not bombarded (insert in Fig. 27). At a current above the threshold, many emission lines exist that are approximately evenly spaced with a separation of $\Delta\lambda \simeq 7.5$ Å. These emission lines belong to the longitudinal modes that will now be derived.

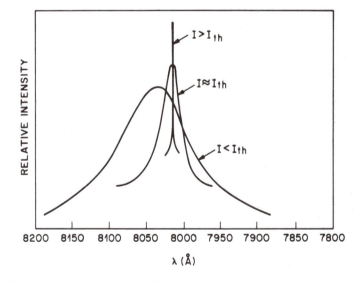

Fig. 26 Emission spectra of a laser below, just at, and above threshold, showing narrowing of the spectral distribution of the emission when lasing is initiated.

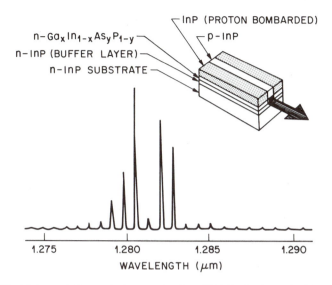

Fig. 27 High-resolution emission spectra for a DH InP-Ga$_x$In$_{1-x}$As$_y$P$_{1-y}$ laser.[14]

In a Fabry–Perot cavity (Figs. 18 and 21a), if an integral number of half-wavelengths fit between the two end planes, reinforced and coherent light will be reflected back and forth within the cavity. Therefore, for stimulated emission, the length L of the cavity must satisfy the condition

$$m \left[\frac{\lambda}{2\bar{n}} \right] = L \tag{21}$$

or

$$m\lambda = 2\bar{n}L \tag{21a}$$

where m is an integral number, and $\bar{n}$ is the refractive index in the semiconductor corresponding to the wavelength λ ($\bar{n}$ is generally a function of λ). The separation $\Delta\lambda$ between the allowed modes in the longitudinal direction is the difference in the wavelengths corresponding to m and $m + 1$. Differentiating Eq. 21a with respect to λ, we obtain

$$\Delta\lambda = \frac{\lambda^2 \, \Delta m}{2\bar{n}L \, [1 - (\lambda/\bar{n})(d\bar{n}/d\lambda)]} . \tag{22}$$

Because of these longitudinal modes, the stripe geometry laser is not a spectrally pure light source. The laser can emit light over a range of about 50 Å, as shown in Fig. 27. For optical-fiber communication systems, an ideal light source is one that has a single frequency. This is because light pulses of different frequencies travel through optical fiber at different speeds thus causing pulse spread.

The basic strip geometry laser has been modified in various ways to achieve single-frequency operation. One approach is the cleaved-coupled-cavity (C^3) laser shown[15] in Fig. 28a. The C^3 laser consists of two standard Fabry–Perot cavity laser diodes which are self-aligned and very closely coupled to form a two-cavity resonator. The active stripe from each laser is precisely aligned with respect to the other on a straight line, and they are separated from each other by about 5 μm. Because the laser light has to travel through an additional (modulator) cavity, the only radiation that is reinforced is at a wavelength that resonates both in the laser's cavity and also in the added (modulator) cavity. All other wavelengths are suppressed. The single-frequency lasers will allow an optical-fiber communication system to carry large amounts of information for very long distances.

The laser structures described previously use cavity facets that are formed by cleaving or polishing to obtain the feedback necessary for lasing. Feedback can also be obtained by a periodic variation of the refractive index within the wave guide, which can be produced by corrugating the interface between two dielectric layers. Figure 28b shows an example.[16] The periodic variation $\bar{n}$ can give rise to constructive interference. Lasers that utilize these corrugated structures are called *distributed-feedback* (DFB) lasers. Because of the small temperature dependence of the refractive index, the lasing wavelength of the DFB laser (Fig. 28b) has a very small temperature coefficient ($\sim$0.5 Å/°C); while the temperature coefficient for a corresponding Fabry–Perot laser is substantially larger ($\sim$3 Å/°C) because it follows the temperature dependence of the bandgap. The DFB lasers are particularly useful as optical sources in integrated optics which uses miniature optical wave guide components and circuits made by planar technology on rigid substrates.

7.4 PHOTODETECTOR

Photodetectors are semiconductor devices that can convert optical signals into electrical signals. The operation of a photodetector involves three steps: (1) carrier generation by incident light, (2) carrier transport and/or multiplication by whatever current gain mechanism may be present, and (3) interaction of current with the external circuit to provide the output signal.

Photodetectors have a broad range of applications including infrared sensors in opto-isolators and detectors for optical-fiber communications. For these applications, the photodetectors must have high sensitivity at the operating wavelengths, high response speed, and low noise. In addition, the photodetector should be compact, use low biasing voltages or currents, and be reliable under the required operating conditions.

7.4.1 Photoconductor

A photoconductor consists simply of a slab of semiconductor with ohmic contacts at both ends of the slab (Fig. 29). When incident light falls on the surface of the photoconductor, electron–hole pairs are generated either by

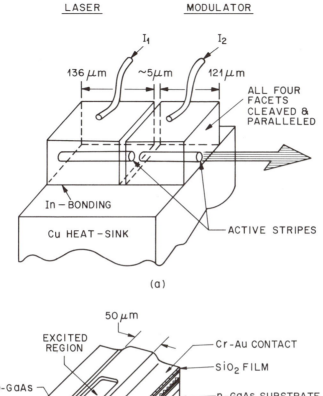

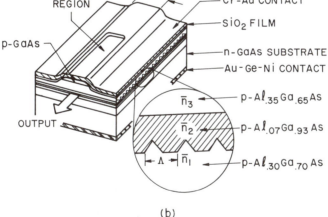

Fig. 28 (a) Schematic diagram of a cleaved-coupled-cavity, 1.3-μm-wavelength $Ga_x In_{1-x} As_y P_{1-y}$ laser.[15] (b) Schematic diagram of a distributed-feedback laser ($\lambda \cong 0.9$ μm).[16]

band-to-band transition (intrinsic) or by transitions involving forbidden-gap energy levels (extrinsic), resulting in an increase in conductivity.

For the intrinsic photoconductor, the conductivity is given by

$$\sigma = q(\mu_n n + \mu_p p) \tag{23}$$

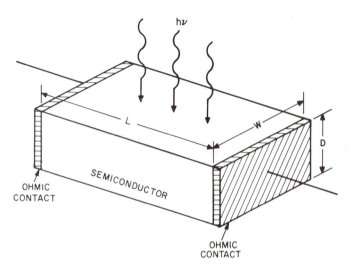

Fig. 29 Schematic diagram of a photoconductor that consists of a slab of semiconductor and two ohmic contacts at the ends.

and the increase in conductivity under illumination is mainly due to the increase in the number of carriers. The long-wavelength cutoff for this case is given by Eq. 9. For the extrinsic case, photoexcitation may occur between the band edge and an energy level in the energy gap. In this case, the long-wavelength cutoff is determined by the depth of the forbidden-gap energy level.

Consider the operation of a photoconductor under illumination. At time zero, the number of carriers generated in a unit volume by a given photon flux is n_o. At a later time t, the number of carriers $n(t)$ in the same volume decays by recombination as

$$n = n_o \exp \left[\frac{-t}{\tau} \right] \tag{24}$$

where τ is the carrier lifetime. In other words, the recombination rate is $1/\tau$. If we assume a steady flow of photon flux impinging uniformly on the surface of a photoconductor (Fig. 29) with area $A = WL$, the total number of photons arriving at the surface is $(P_{opt}/h\nu)$ per unit time, where P_{opt} is the incident optical power and $h\nu$ is the photon energy.

At steady state, the carrier generation rate must be equal to the recombination rate. If the detector thickness D is much larger than the light penetration depth $1/\alpha$, the total steady-state carrier generation rate per unit volume is

$$G = \frac{n}{\tau} = \frac{\eta(P_{opt}/h\nu)}{WLD} \tag{25}$$

where η is the quantum efficiency and n is the number of carriers per unit volume (carrier density). The photocurrent flowing between the electrodes is

$$I_p = (\sigma \mathscr{E})WD = (q\mu_n n \mathscr{E})WD = (qnv_d)WD \tag{26}$$

where $\mathscr{E}$ is the electric field inside the photoconductor and v_d is the carrier drift velocity. Substituting n in Eq. 25 into Eq. 26 gives

$$I_p = q \left[\eta \frac{P_{opt}}{h\nu} \right] \left[\frac{\mu_n \tau \mathscr{E}}{L} \right]. \tag{27}$$

If we define the primary photocurrent as

$$I_{ph} \equiv q \left[\eta \frac{P_{opt}}{h\nu} \right] \tag{28}$$

the photocurrent gain from Eq. 26 is

$$\text{Gain} \equiv \frac{I_p}{I_{ph}} = \frac{\mu_n \tau \mathscr{E}}{L} = \frac{\tau}{t_r} \tag{29}$$

where $t_r \equiv L/v_d$ is the carrier transit time. The gain depends upon the ratio of carrier lifetime to the transit time.

For a long-lifetime sample with short electrode spacing, the gain can be substantially greater than unity. Gains as high as 10^6 can be obtained from some photoconductors. The response time of a photoconductor is determined by the transmit time t_r. To achieve short transit time, small electrode spacing and a high electric field must be used. The response times of photoconductors cover a wide range from 10^{-3} to 10^{-10} seconds. They are extensively used for infrared detection especially for wavelengths greater than a few microns.

7.4.2 Photodiode

A photodiode is basically a p–n junction operated under reverse bias. When an optical signal impinges on the photodiode, the depletion region serves to separate photogenerated electron–hole pairs, and an electric current will flow in the external circuit. For high-frequency operation, the depletion region must be kept thin to reduce the transit time. On the other hand, to increase the quantum efficiency, the depletion layer must be sufficiently thick to allow a large fraction of the incident light to be absorbed. Thus, there is a trade-off between the response speed and quantum efficiency.

Quantum Efficiency The quantum efficiency as mentioned previously is the number of electron–hole pairs generated for each incident photon:

$$\eta = \left[\frac{I_p}{q} \right] \left[\frac{P_{opt}}{h\nu} \right]^{-1} \tag{30}$$

where I_p is the photogenerated current from the absorption of incident optical power P_{opt} at a wavelength λ (corresponding to a photon energy $h\nu$). One of the key factors that determines η is the absorption coefficient α (Fig. 5). Since α is a strong function of the wavelength, the wavelength range in which appreciable photocurrent can be generated is limited. The long-wavelength cutoff λ_c is established by the bandgap, Eq. 9, and is, for example, about 1.8 μm for germanium and 1.1 μm for silicon. For wavelengths longer than λ_c, the values of α are too small to give appreciable band-to-band absorption. The short-wavelength cutoff of the photoresponse comes about because for short wavelengths the values of α are very large ($\sim 10^5$ cm^{-1}), and hence the radiation is mostly absorbed very near the surface where recombination time is short. Therefore, the photocarriers can recombine before they can be collected in the p–n junction.

Figure 30 shows typical plots of quantum efficiency versus wavelength for some high-speed photodiodes.[17, 18] Note that in the ultraviolet and visible region, metal–semiconductor photodiodes show good quantum efficiencies. In the near-infrared region, silicon photodiodes (with an antireflection coating) can reach 100% quantum efficiency near the 0.8- to 0.9-μm region. In the 1.0- to 1.6-μm region, germanium photodiodes and Group III–V photodiodes (e.g., GaInAs) have shown high quantum efficiencies. For even longer wavelengths, photodiodes are cooled (e.g., to 77 K) for high-efficiency operation.

Response Speed   The response speed is limited by three factors: (1) diffusion of carriers, (2) drift time in the depletion region, and (3) capacitance of the depletion region. Carriers generated outside the depletion region must diffuse to the junction, resulting in considerable time delay. To minimize the diffusion effect, the junction should be formed very close to the surface. The greatest amount of light will be absorbed when the depletion region is sufficiently wide. However, the depletion layer must not be too wide, or transit time effects will limit the frequency response. It also should not be too thin, or excessive capacitance C will result in a large RC time constant, where R is the load resistance. The optimal compromise is the width at which the depletion layer transit time is approximately one half the modulation period. For exam-

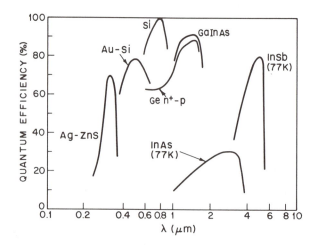

Fig. 30   Quantum efficiency versus wavelength for various photodetectors.[17, 18]

ple, for a modulation frequency of 2 GHz, the optimal depletion layer thickness in silicon (with a saturation velocity of 10^7 cm/s) is about 25 μm.

p–i–n Photodiode The p–i–n photodiode is one of the most common photodetectors because the depletion region thickness (the intrinsic layer) can be tailored to optimize the quantum efficiency and frequency response. Figure 31a shows a cross section of a p–i–n photodiode which has an antireflection coating to increase quantum efficiency.

Figures 31b and 31c show the energy band diagram of the p–i–n diode under reverse-bias condition and its optical absorption characteristics. Light absorption in the semiconductor produces electron–hole pairs. Pairs produced in the depletion region or within a diffusion length of it will eventually be separated by the electric field as shown in Fig. 31b, whereby a current flows in the external circuit as carriers drift across the depletion layer.

Metal–Semiconductor Photodiode The construction of a high-speed metal–semiconductor photodiode is shown in Fig. 32. To avoid large reflection and absorption losses when the diode is illuminated through the metal contact, the metal film must be very thin ($\sim$100 Å) and an antireflection coating must be used. Metal–semiconductor photodiodes are particularly use-

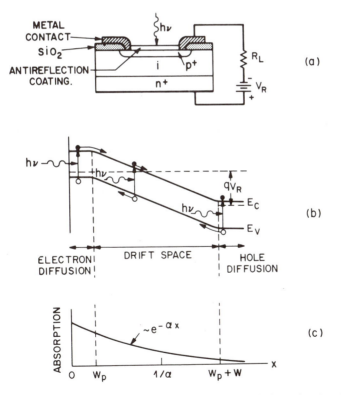

Fig. 31 Operation of a p–i–n photodiode. (a) Cross-sectional view of p–i–n diode. (b) Energy band diagram under reverse bias. (c) Carrier absorption characteristics.

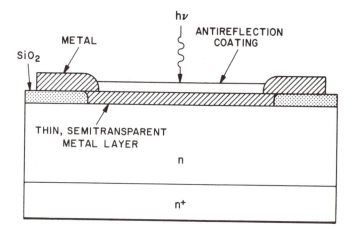

Fig. 32 Metal–semiconductor photodiode.

ful in the ultraviolet- and visible-light regions. In these regions the absorption coefficients α in most of the common semiconductors are very high, of the order of 10^4 cm^{-1} or more, which corresponds to an effective absorption length $1/\alpha$ of $1.0\,\mu$m or less. It is possible to choose a metal and an antireflection coating so that a large fraction of the incident radiation will be absorbed near the surface of the semiconductor. For example, for a gold–silicon photodetector having 100 Å gold and 500 Å zinc sulfide as the antireflection coating, more than 95% of the incident light with $\lambda = 6328$ Å (helium–neon laser wavelength, red light) will be transmitted into the silicon substrate.

Heterojunction Photodiode Another photodiode structure is a heterojunction device formed by depositing a large-bandgap semiconductor epitaxially on a smaller-bandgap semiconductor. One advantage of a heterojunction photodiode is that the quantum efficiency does not depend critically on the distance of the junction from the surface, because here the large-bandgap material can be used as a window for the transmission of optical power. In addition, the heterojunction can provide unique material combinations so that the quantum efficiency and response speed can be optimized for a given optical-signal wavelength.

To obtain a heterojunction with low leakage current, the lattice constants of the two semiconductors must be closely matched. Ternary III–V compounds $Al_xGa_{1-x}As$ epitaxially grown on a gallium arsenide substrate can form heterojunctions with perfectly matched lattices. These heterojunction photodetectors are important for photonic devices operated in the wavelength range from 0.65 to 0.85 μm. At longer wavelengths (1 to 1.6 μm), ternary compounds such as $Ga_{0.47}In_{0.53}As$ (with $E_g = 0.73$ eV) and quaternary compounds such as $Ga_{.27}In_{.73}As_{.63}P_{.37}$ (with $E_g = 0.95$ eV) can be used. These compounds have a nearly perfect lattice match to an indium phosphide substrate. The insert in Fig. 33 shows the back-illuminated mesa structure of a p-GaInAs/ν-

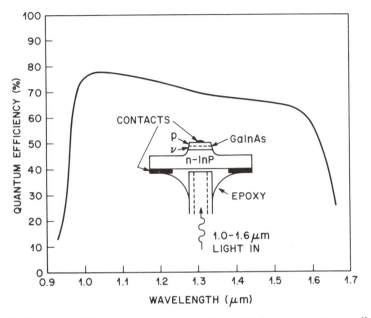

Fig. 33 Quantum efficiency versus wavelength for a GaInAs p–i–n photodiode.[19]

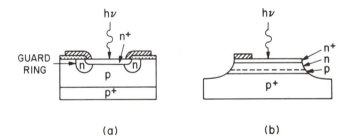

(a)   (b)

Fig. 34 Device configurations of two avalanche photodiodes. (a) Guard ring structure. (b) Mesa structure.

GaInAs/n^+-InP photodiode.[19] The quantum efficiency is greater than 55% over the wavelength range from 0.96 to 1.6 μm.

7.4.3 Avalanche Photodiode

An avalanche photodiode (APD) is operated under a reverse-bias voltage which is sufficient to enable avalanche multiplication to take place. The multiplication results in internal current gain, and the device can respond to light modulated at frequencies as high as microwave frequencies.

Figure 34 shows two APD configurations. The guard ring structure (Fig. 34a) has a low impurity gradient at the n–p guard ring junction and a sufficiently large radius of curvature so that the central n^+–p abrupt junction will break down before the guard ring does. A mesa structure (Fig. 34b) is a simpler structure which has a low surface field across the junction and uniform avalanche breakdown can occur inside the device.

One important consideration in the design of an APD is the need to minimize avalanche noise. The avalanche noise comes about from the random nature of the avalanche multiplication process in which every electron–hole pair generated at a given distance in the depletion region does not experience the same multiplication. The avalanche noise depends on the ratio of the ionization coefficients α_p/α_n; the smaller the ratio, the smaller is the avalanche noise. This is because when $\alpha_p = \alpha_n$, each incident photocarrier results in three carriers in the multiplicating region: the primary carrier and its secondary hole and electron. A fluctuation that changes the number of carriers by one represents a large percentage change, and the noise will be large. On the other hand, if one of the ionization coefficients approaches zero (e.g., $\alpha_p \to 0$), each incident photocarrier can result in a large number of carriers in the multiplication region. In this case, a fluctuation of one carrier is a relatively insignificant perturbation. To minimize the avalanche noise, we should use semiconductors with a large difference in α_n and α_p. The noise factor is given by

$$ F = M \left[\frac{\alpha_p}{\alpha_n} \right] + \left[2 - \frac{1}{M} \right] \left[1 - \frac{\alpha_p}{\alpha_n} \right] \tag{31} $$

where M is the multiplication factor (refer to Section 3.6.2). We can see from Eq. 31 that when $\alpha_p = \alpha_n$, the noise factor has a maximum value of M; while for $\alpha_p/\alpha_n = 0$ and for a large M, the minimum noise factor is 2.

Figure 35a shows a silicon APD having an $n^+-p-\pi-p^+$ doping profile.[20] The cross-sectional view is similar to that of Fig. 34a. Figure 35b shows the field distribution, which has a narrow avalanche region and a long drift region. The quantum efficiency is near 100% at a wavelength of about 0.8 μm for a device having a $SiO_2-Si_3N_4$ antireflection coating (Fig. 35c). Because the ratio α_p/α_n is about 0.04, the noise factor obtained from Eq. 31 is 2.3 for $M = 10$.

Many heterojunction APDs are made using the configurations shown in Fig. 34b, where a III–V binary semiconductor is used as the substrate. An example is shown in Fig. 36 for an AlGaAs–GaAs structure.[21] The top AlGaAs layer serves as a window for the transmission of incident light in the range from 0.5 to 0.9 μm. The quantum efficiency is about 70% at 0.53 μm and can be increased to 95% with silicon nitride antireflection coating. The ratio of α_p/α_n is 0.83 due to comparable ionization coefficients in gallium arsenide. This ratio results in the large noise factor of 8.6 for $M = 10$.

Figure 37 shows the minimum optical power needed for detecting of optical signals by a photodetector (i.e., detector sensitivity) as a function of the bit rate at which the information arrives at the surface of the photodetector.[22] As can be seen from the figure in the 0.85-μm spectral region, silicon APDs have excellent sensitivity and are therefore used in most receivers. In the 1.3-μm spectral region $In_xGa_{1-x}As$ $p-i-n$ diodes and $In_xGa_{1-x}As$ APDs have better sensitivities than germanium APDs.

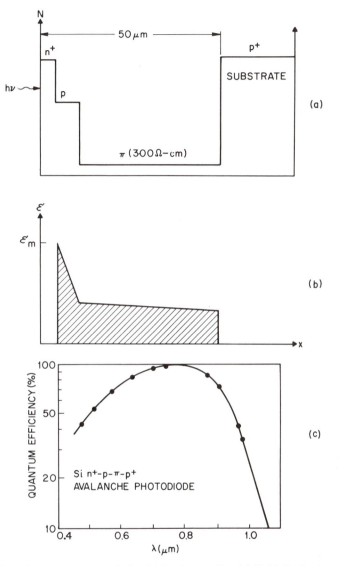

Fig. 35 Silicon avalanche photodiode. (*a*) Doping profile. (*b*) Field distribution. (*c*) Quantum efficiency.[20]

7.5 SOLAR CELL

Solar cells are useful for both space and terrestrial applications. Solar cells furnish the long-duration power supply for satellites. The solar cell is an important candidate for an alternative terrestrial energy source because it can convert sunlight directly to electricity with good conversion efficiency, can provide nearly permanent power at low operating cost, and is virtually nonpolluting.[23]

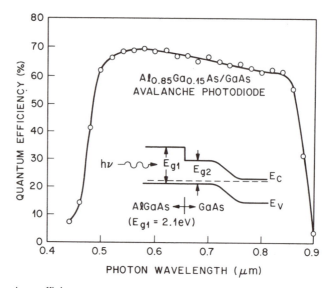

Fig. 36 Quantum efficiency versus wavelength for an AlGaAs/GaAs heterojunction avalanche photodiode. The energy band diagram is shown in the insert.[21]

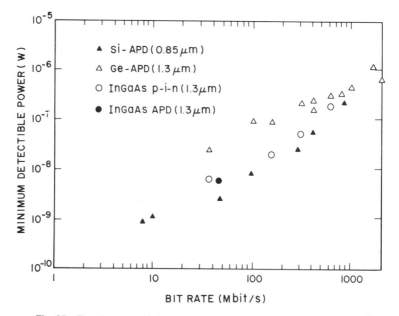

Fig. 37 Receiver sensitivity versus bit rate for various photodetectors.[22]

7.5.1 Solar Radiation

The radiative energy output from the sun derives from a nuclear fusion reaction. In every second, about 6×10^{11} kg hydrogen is converted to helium, with a net mass loss of about 4×10^{3} kg, which is converted through the Einstein relation ($E = mc^2$) to 4×10^{20} J. This energy is emitted primarily as electromagnetic radiation in the ultraviolet to infrared region (0.2 to 3 μm). The total mass of the sun is now about 2×10^{30} kg, and a reasonably stable life with a nearly constant radiative-energy output of over 10 billion (10^{10}) years is projected.

The intensity of solar radiation in free space at the average distance of the Earth from the sun is defined as the solar constant and has a value of 1353 W/m^2. The degree to which the atmosphere affects the sunlight received at the Earth's surface is defined by the *air mass*. Figure 38 shows two curves related to solar spectral irradiance (power per unit area per unit wavelength).[24] The upper curve, which represents the solar spectrum outside the Earth's atmosphere, is the air mass zero condition (AM0). The AM0 spectrum is the relevant one for satellite and space vehicle applications. The air mass one (AM1) spectrum represents the sunlight at the Earth's surface when the sun is overhead, at which point the incident power is about 925 W/m^2. The difference between AM0 and AM1 is caused by the atmospheric attenuation of sunlight, mainly due to ultraviolet absorption in the ozone, to infrared absorption in the water vapor, and to scattering by airborne dust and aerosols.

7.5.2 *p–n* Junction Solar Cell

A schematic representation of a *p–n* junction solar cell is shown in Fig. 39. It consists of a shallow *p–n* junction formed on the surface, a front ohmic-

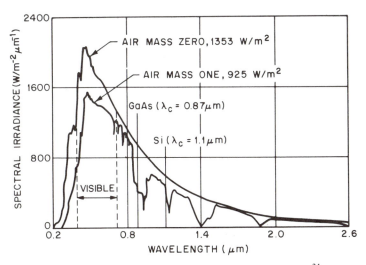

Fig. 38 Two curves related to solar spectral irradiance.[24]

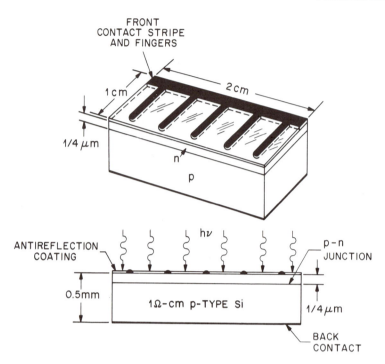

Fig. 39 Schematic representation of a silicon p–n junction solar cell.

contact stripe and fingers, a back ohmic contact that covers the entire back surface, and an antireflection coating on the front surface.

   When the cell is exposed to the solar spectrum, a photon that has an energy
less than the bandgap E_g makes no contribution to the cell output. A photon
that has an energy greater than E_g contributes an energy E_g to the cell output.
Energy greater than E_g is wasted as heat. To derive the conversion efficiency,
we shall consider the energy band diagram of a p–n junction under solar radiation, shown in Fig. 40a. The equivalent circuit is shown in Fig. 40b, where a
constant-current source is in parallel with the junction. The source I_L results
from the excitation of excess carriers by solar radiation, I_s is the diode saturation current as derived in Chapter 3, and R_L is the load resistance.

   The ideal I–V characteristics of such a device are given by

$$I = I_s (e^{qV/kT} - 1) - I_L \tag{32}$$

and

$$J_s = \frac{I_s}{A} = qN_CN_V \left[\frac{1}{N_A} \sqrt{\frac{D_n}{\tau_n}} + \frac{1}{N_D} \sqrt{\frac{D_p}{\tau_p}} \right] e^{-E_g/kT} \tag{32a}$$

where A is the device area. A plot of Eq. 32 is given in Fig. 41a for
$I_L = 100$ mA, $I_s = 1$ nA, cell area $A = 4$ cm^2, and $T = 300$ K. The curve

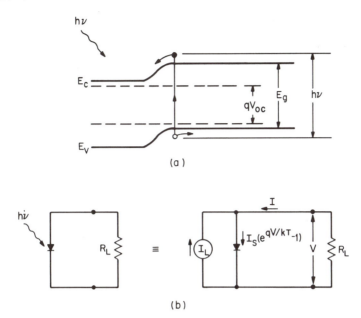

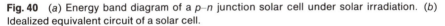

Fig. 40 (a) Energy band diagram of a p–n junction solar cell under solar irradiation. (b) Idealized equivalent circuit of a solar cell.

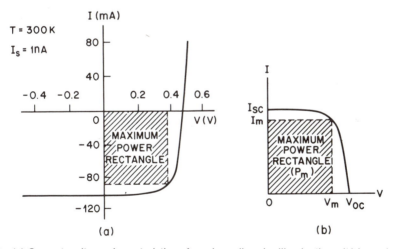

Fig. 41 (a) Current–voltage characteristics of a solar cell under illumination. (b) Inversion of (a) about the voltage axis.

passes through the fourth quadrant, and therefore power can be extracted from the device. The I–V curve is more generally represented by Fig. 41b, which is an inversion of Fig. 41a about the voltage axis. By choosing a proper load, close to 80% of the product $I_{sc}V_{co}$ can be extracted, where I_{sc} is the short-circuit current equal to I_L and V_{oc} is the open-circuit voltage of the cell;

the shaded area in the figure is the maximum-power rectangle. Also defined in Fig. 41b are the quantities I_m and V_m that correspond to the current and voltage, respectively, for the maximum power output $P_m(I_m V_m)$.

From Eq. 32 we obtain for the open-circuit voltage ($I = 0$)

$$V_{oc} = \frac{kT}{q} \ln \left[\frac{I_L}{I_s} + 1 \right] \simeq \frac{kT}{q} \ln \left[\frac{I_L}{I_s} \right]. \tag{33}$$

Hence, for a given I_L, V_{oc} increases logarithmically with decreasing saturation current I_s. The output power is given by

$$P = IV = I_s V (e^{qV/kT} - 1) - I_L V. \tag{34}$$

The condition for maximum power is obtained when $dP/dV = 0$, or

$$V_m = \frac{kT}{q} \ln \left[\frac{1 + (I_L/I_s)}{1 + (qV_m/kT)} \right] \simeq V_{oc} - \frac{kT}{q} \ln \left[1 + \frac{qV_m}{kT} \right] \tag{35a}$$

$$I_m = I_s \left[\frac{qV_m}{kT} \right] e^{qV_m/kT} \simeq I_L \left[1 - \frac{1}{qV_m/kT} \right]. \tag{35b}$$

The maximum output power P_m is then

$$P_m = I_m V_m \simeq I_L \left[V_{oc} - \frac{kT}{q} \ln \left[1 + \frac{qV_m}{kT} \right] - \frac{kT}{q} \right]. \tag{36}$$

7.5.3 Conversion Efficiency

The power conversion efficiency of a solar cell is given by

$$\eta = \frac{I_m V_m}{P_{in}} = \frac{I_L \left[V_{oc} - \frac{kT}{q} \ln \left[1 + \frac{qV_m}{kT} \right] - \frac{kT}{q} \right]}{P_{in}}$$
$$= \frac{FF \cdot I_L V_{oc}}{P_{in}} \tag{37}$$

where P_{in} is the incident power and FF is the fill factor defined as

$$FF \equiv \frac{I_m V_m}{I_L V_{oc}}. \tag{38}$$

To maximize the efficiency, we should maximize all three items in the numerator of Eq. 37.

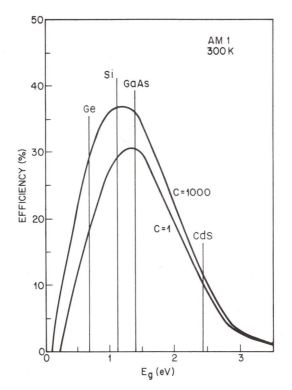

Fig. 42 Ideal solar-cell efficiency at 300 K for 1 sun and for a 1000-sun concentration.[25]

The ideal solar-cell efficiency at 300 K is shown in Fig. 42 as a function of energy bandgap.[25] The ideal efficiency is obtained from the ideal $I-V$ characteristic defined by Eq. 32. For a given semiconductor, the saturation current density J_s is obtained from Eq. 32a. For a given air mass condition (e.g., AM1), the short-circuit current I_L is the product of q and the number of the available photons with $h\nu \geqslant E_g$ in the solar spectrum. The input power P_{in} is the integration of all the photons in the solar spectrum (Fig. 38).

The curve marked $C = 1$ in Fig. 42 is under one-sun AM1 condition. Note that the efficiency has a broad maximum and does not depend critically on E_g. Therefore, semiconductors with bandgaps between 1 and 2 eV can all be considered solar cell materials. Figure 42 also shows the ideal efficiency at an optical concentration of 1000 suns (i.e., 925 kW/m²). Details on optical concentration will be considered in Section 7.5.5.

Many factors degrade the ideal efficiency. One of the major factors is the series resistance R_s from the ohmic loss in the front surface. The equivalent circuit is shown in the insert of Fig. 43. If the diode current is given by Eq. 32, the $I-V$ characteristics are found to be

$$\ln\left[\frac{I + I_L}{I_s} + 1\right] = \frac{q}{kT}(V - IR_s). \tag{39}$$

Plots of this equation are shown in Fig. 43, with $R_s = 0$ and 5 Ω and where the other parameter I_s, I_L, and T are the same as those in Fig. 41. It can be seen that a series resistance of only 5 Ω reduces the available power to less than 30% of the maximum power with $R_s = 0$. The output current and output power are

$$I = I_s \left\{ \exp \left[\frac{q(V - IR_s)}{kT} \right] - 1 \right\} - I_L \tag{40}$$

$$P = I \left[\frac{kT}{q} \ln \left[\frac{I + I_L}{I_s} + 1 \right] + IR_s \right]. \tag{41}$$

The series resistance depends on the junction depth, the impurity concentrations of p-type and n-type regions, and the arrangement of the front surface ohmic contacts. For a typical silicon solar cell with the geometry shown in Fig. 39, the series resistance is about 0.7 Ω for n^+-p cells and 0.4 Ω for p^+-n cells. The difference in resistance is mainly the result of the lower resistivity in n-type substrates.

Another factor is the recombination current in the depletion region. For single-level centers, the recombination current can be expressed as

$$I_{rec} = I_s' \left[\exp \left[\frac{qV}{2kT} \right] - 1 \right] \tag{42}$$

with

$$\frac{I_s'}{A} = \frac{q n_i W}{\sqrt{\tau_p \tau_n}}$$

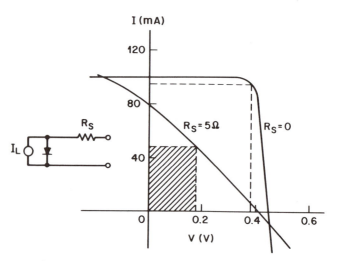

Fig. 43 Current–voltage characteristics for solar cells that have series resistances. Insert shows the equivalent circuit.

where I_s' in the saturation current. The energy conversion equation can be put into a closed form yielding equations similar to Eqs. 33 through 36, with the exception that I_s is replaced by I_s' and the exponential factor is divided by 2. The efficiency for the recombination current case is found to be much less than the ideal current case due to the degradation of both V_{oc} and the fill factor. For silicon solar cells at 300 K, the recombination current can cause 25% reduction in efficiency.

Amorphous silicon (α-Si) is also a material for solar cells. Layers a few microns thick are deposited by radio-frequency glow discharge decomposition of silane onto metal or glass substrates. The optical-absorption characteristics of α-Si, which has an effective bandgap of 1.5 eV, is shown in Fig. 5. Although the efficiency of α-Si solar cells ($\sim 10\%$) is lower than that of single-crystal silicon solar cells, their production costs are considerably lower. Therefore, the α-Si solar cell is one of the major candidates for large-scale use of solar energy.

7.5.4   Heterojunction and Interface Solar Cells

A heterojunction solar cell can have a similar energy band diagram as that shown in the insert of Fig. 36. The wide-gap semiconductor will act as a window, admitting photons of energy less than E_{g1}. Those photons with energies between E_{g1} and E_{g2} will create carriers in the p–n junction (with bandgap E_{g2}). If the absorption coefficient is high in the lower-gap semiconductor, the carriers are generated in the depletion region or close to it, so that the collection efficiency is high.

Figure 44 shows two interface solar cells. The Schottky barrier solar cell, Fig. 44a, must have a very thin metal to allow a substantial amount of light to

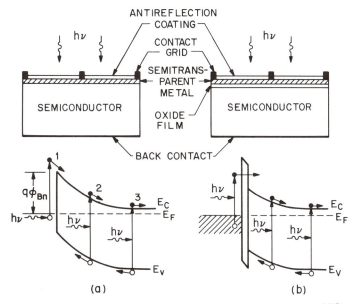

Fig. 44   (a) Schottky barrier solar cell. (b) Metal–insulator–semiconductor (MIS) solar cell.

reach the semiconductor. As indicated in Fig. 44a, there are three photo-current components: (1) light with energy $h\nu > q\phi_{Bn}$ (the barrier height), which can be absorbed in the metal and excite electrons over the barrier into the semiconductor (1 in Fig. 44a); (2) short-wavelength light ($h\nu > E_g$) entering the semiconductor, which is absorbed mainly in the depletion region (2 in Fig. 44a); and (3) long-wavelength light ($h\nu \simeq E_g$), which is absorbed in the neutral region (3 in Fig. 44a).

The advantages of Schottky barrier solar cells include: low-temperature processing (because high-temperature diffusion is not required), adaptability to polycrystalline and thin-film solar cells, and high current output (because of the presence of a depletion region at the semiconductor surface).

When we form a thin insulating layer between the metal and the semiconductor, we have an MIS (metal–insulator–semiconductor) solar cell shown in Fig. 44b. The saturation current density is similar to that for the Schottky barrier cells but has an additional tunneling term:

$$J_s = A^*T^2 \exp \left[\frac{-q\phi_{Bn}}{kT} \right] \exp(-a\delta) \qquad (43)$$

where A^* equals 110 A/K²-cm² for n-type silicon, a is a constant, and δ is the insulating layer thickness. From Eqs. 33 and 43 we obtain

$$V_{oc} = \phi_{Bn} + \frac{kTa\delta}{q} + \frac{kT}{q} \ln \left[\frac{J_L}{A^*T^2} \right]. \qquad (44)$$

Therefore, the V_{oc} of an MIS solar cell will be larger than that of a Schottky barrier solar cell. V_{oc} increases with increasing δ. However, as δ increases, the short-circuit current decreases, causing a degradation of the conversion efficiency. An optimum oxide thickness for a metal–SiO₂–Si system is about 20 Å. At AM1, efficiencies of up to 18% have been obtained.

7.5.5 Optical Concentration

Sunlight can be focused by using mirrors and lenses. Optical concentration offers an attractive and flexible approach to reducing high cell costs by substituting a concentrator area for much of the cell area. It also offers other advantages, such as increased efficiency as shown in the $C = 1000$ curve in Fig. 42.

Figure 45a shows a standard planoconvex lens, and Fig. 45b shows an equivalent Fresnel lens. These lenses can focus the sunlight onto solar cells. Under high sunlight concentrations, the carrier density approaches that of the substrate doping and a high-injection condition prevails. Figure 46 shows the measured results of a silicon solar cell mounted in a concentrated system.[26] Note that device performances improve as the concentration increases from one sun toward 1000 suns. The short-circuit current density increases linearly

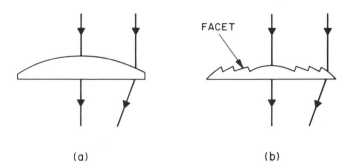

(a)   (b)

Fig. 45 Lenses used for solar concentration. (a) Standard planoconvex lens. (b) Fresnel lens.

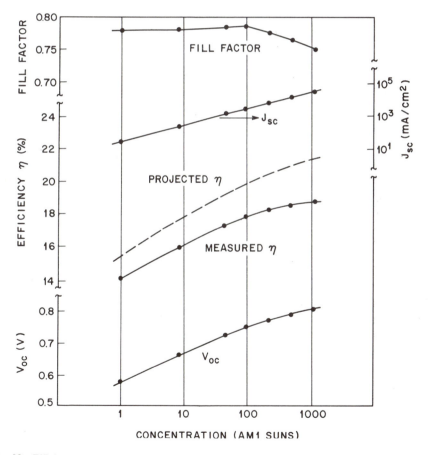

CONCENTRATION (AM1 SUNS)

Fig. 46 Efficiency, open-circuit voltage, short-circuit current, and fill factor versus solar concentration.[26]

with concentration. The open-circuit voltage increases at a rate of 0.1 V per decade, while the fill factor varies slightly. The efficiency, which is the product of the foregoing three factors divided by the input power, increases at a rate of about 2% per decade. With a proper antireflection coating, we project an efficiency of 22% at 1000 suns. Therefore, one cell operated under 1000-sun concentration can produce the same power output as 1300 cells under one sun. Potentially, the optical concentration approach can replace expensive solar cells with less expensive concentrator materials and a related tracking and heat removal system to minimize the overall system cost.

REFERENCES

1   H. C. Casey, Jr., and M. B. Panish, *Heterostructure Lasers*, Academic, New York, 1978.

2   H. Melchior, "Demodulation and Photodetection Techniques," in F. T. Arecchi and E. O. Schulz-Dubois, Eds., *Laser Handbook*, Vol. 1, North-Holland, Amsterdam, 1972, pp. 725–835.

3   A. A. Bergh and P. J. Dean, *Light Emitting Diodes*, Clarendon, Oxford, 1976.

4   M. G. Craford, "Recent Developments in LED Technology," *IEEE Trans. Electron Devices*, **ED-24**, 935 (1977).

5   W. O. Groves, A. H. Herzog, and M. G. Craford, "The Effect of Nitrogen Doping on GaAsP Electroluminescent Diodes," *Appl. Phys. Lett.*, **19**, 184 (1971).

6   S. Gage, D. Evans, M. Hodapp, and H. Sorenson, *Optoelectronic Application Manual*, McGraw-Hill, New York, 1977.

7   R. H. Saul, T. P. Lee, and C. A. Burrus, "Light-Emitting Diode Device Design," in R. K. Willardon and A. C. Bear, Eds., *Semiconductors and Semimetals*, Academic, New York, 1984.

8   S. E. Miller and A. G. Chynoweth, Eds., *Optical Fiber Communications*, Academic, New York, 1979.

9   C. A. Burrus and B. I. Miller, "Small-Area, Double Heterostructure AlGaAs Electroluminescent Diode Source for Optical-Fiber Transmission Lines," *Opt. Comm.*, **4**, 307 (1971).

10 T. E. Bell, "Single-Frequency Semiconductor Lasers," *IEEE Spectrum*, **20**, No. 12, 38 (1983).

11 M. B. Panish, I. Hayashi, and S. Sumski, "Double-Heterostructure Injection Lasers with Room Temperature Threshold As Low As 2300 A/cm^2," *Appl. Phys. Lett.*, **16**, 326 (1970).

12 F. Stern, "Calculated Spectral Dependence of Gain in Excited GaAs," *J. Appl. Phys.*, **47**, 5382 (1976).

13 W. T. Tsang, R. A. Logan, and J. P. Van der Ziel, "Low-Current-Threshold Stripe-Buried-Heterostructure Laser with Self-Aligned Current Injection Stripes," *Appl. Phys. Lett.*, **34**, 644 (1979).

14 A. G. Foyt, "1.0–1.6 μm Source and Detectors for Fiber Optics Applications," IEEE Device Research Conf., Boulder, Colo., June 1979.

15 W. T. Tsang, N. A. Olsson, and R. A. Logan, "High Speed Direct Single-Frequency Modulation with Large Tuning Rate and Frequency Excursion in Cleaved-Coupled-Cavity Semiconductor Lasers," *Appl. Phys. Lett.*, **42**, 650 (1983).

16 K. Aiki, M. Nakamura, and J. Umeda, "Lasing Characteristics of Distributed-Feedback GaAs-GaAlAs Diode Lasers with Separate Optical and Carrier Confinement," *IEEE J. Quantum Electron.*, **QE-12**, 597 (1976).

17 S. M. Sze, *Physics of Semiconductor Devices*, 2nd ed., Wiley, New York, 1981, Chapters 12–14.

18 S. R. Forrest, "Photodiodes for Long-Wavelength Communication Systems," *Laser Focus*, **18**, 81 (1982).

19 T. P. Lee, C. A. Burrus, and A. G. Dentai, "InGaAs/InP *p–i–n* Photodiodes for Lightwave Communications at 0.95 to 1.65 μm Wavelengths," *IEEE J. Quantum Electron.*, **QE-17**, 232 (1981).

20 H. Melchior, A. R. Hartman, D. P. Schinke, and T. E. Seidel, "Planar Epitaxial Silicon Avalanche Photodetector," *Bell Syst. Tech. J.*, **57**, 1791 (1978).

21 H. D. Law, K. Nakano, and L. R. Tomasetta, "III–V Alloy Heterostructure High Speed Avalanche Photodiodes," *IEEE J. Quantum Electron.*, **QE-15**, 549 (1979).

22 T. P. Lee, "Photodetectors" in J. C. Daly, ed., *Fiber Optics*, CRC Press, Inc., Boca Raton, 1984.

23 D. M. Chapin, C. S. Fuller, and G. L. Pearson, "A New Silicon *p–n* Junction Photocell for Converting Solar Radiation into Electrical Power," *J. Appl. Phys.*, **25**, 676 (1954).

24 M. P. Thekaekara, "Data on Incident Solar Energy," Suppl. Proc. 20th Ann. Meet. Inst. Environ. Sci., 1974, p. 21.

25 C. H. Henry, "Limiting Efficiency of Ideal Single and Multiple Energy Gap terrestrial Solar Cells," *J. Appl. Phys.*, **51**, 4494 (1980).

26 R. I. Frank, J. L. Goodrich, and R. Kaplow, "A Novel Silicon High-Intensity Photovoltaic Cell," Conf. Rec. 14th IEEE Photovoltaic Conf. IEEE, New York, 1980, p. 1350.

PROBLEMS

1 A gallium arsenide sample is illuminated with a light having a wavelength of 0.6 μm. The incident power is 15 mW. If one third of the incident power is reflected and another third exits from the other end of the sample, what is the thickness of the sample? Find the thermal energy dissipated to the lattice per second.

2 The efficiency for electrical-to-optical conversion in a LED is given by $4\bar{n}_1\bar{n}_2(1 - \cos\theta_c)/(\bar{n}_1 + \bar{n}_2)^2$, where $\bar{n}_1$ and $\bar{n}_2$ are the refractive index of air and the semiconductor, respectively, and θ_c is the critical angle. Find the efficiency of a $Al_{0.3}Ga_{0.7}As$ LED operated at 0.898 μm.

3 For high-temperature laser operation, it is important to have a low temperature coefficient of the threshold current $\xi \equiv (dI_{th}/dT)/dI_{th}$. What is

the coefficient ξ for the laser shown in Fig. 25? If $T_0 = 50$ °C, is this laser better or worse for high-temperature operation?

4   Derive Eq. 22 for the separation $\Delta\lambda$ between the allowed modes in the longitudinal direction. For a GaAs laser diode operated at $\lambda = 0.89$ μm, with $\bar{n} = 3.58$, $L = 300$ μm, and $d\bar{n}/d\lambda = 2.5$ μm^{-1}, find $\Delta\lambda$.

5   A photoconductor with dimensions $L = 6$ mm, $W = 2$ mm, and $D = 1$ mm (Fig. 29) is placed under uniform radiation. The absorption of the light increases the current by 2.83 mA. A voltage of 10 V is applied across the device. When the radiation is suddenly cut off, the current falls, initially at a rate 23.6 A/s. Let $\mu_n = 3600$ and $\mu_p = 1700$ cm^2/V-s. Find (a) the equilibrium density of electron–hole pairs generated under radiation, (b) the minority carrier lifetime, and (c) the excess density of electrons and holes remaining 1 ms after the radiation is cut off.

6   A silicon n^+–p–π–p^+ avalanche photodiode operated at 0.8 μm has a p-layer of 3 μm and a π-layer 9 μm thick. The biasing voltage must be high enough to cause avalanche breakdown in the p-region and velocity saturation in the π-region. Find the minimum required biasing voltage and the corresponding doping concentration of the p-region. Estimate the transit time of the device.

7   A p–n junction photodiode can be operated under photovoltaic conditions similar to a solar cell (Fig. 40). State three major differences between a photodiode and a solar cell.

8   Consider a silicon p–n junction solar cell of area 2 cm^2. If the dopings of the solar cell are $N_A = 1.7 \times 10^{16}$ cm^{-3} and $N_D = 5 \times 10^{19}$ cm^{-3}, and given $\tau_n = 10$ μs, $\tau_p = 0.5$ μs, $D_n = 9.3$ cm^2/s, $D_p = 2.5$ cm^2/s and $I_L = 95$ mA, (a) calculate and plot the I–V characteristics of the solar cell, (b) calculate the open-circuit voltage, and (c) determine the maximum output power of the solar cell, all at room temperature.

9   For the solar cell shown in Fig. 43, find the relative maximum power output for an R_s of 0 and 5 Ω.

10   For the solar cell operated under solar-concentration conditions (Fig. 46 with the measured η), how many such solar cells operated under one-sun conditions are needed to produce the same power output as one cell operated under 10-sun, 100-sun, or 1000-sun concentration?

8

Crystal Growth and Epitaxy

As discussed previously in Chapter 1, the two most important semiconductors for discrete devices and integrated circuits are silicon and gallium arsenide. In this chapter we describe the common techniques for growing single crystals of these two semiconductors. The basic process flow from starting materials to polished wafers is shown in Fig. 1. The starting materials (e.g., silicon dioxide for a silicon wafer) are chemically processed to form a high-purity polycrystalline semiconductor from which single crystals are grown. The single-crystal ingots are shaped to define the diameter of the material and sawed into wafers. These wafers are etched and polished to provide smooth, specular surfaces on which devices will be made.

A technology closely related to crystal growth involves the growth of single–crystal semiconductor layers upon a single-crystal semiconductor substrate. This is called *epitaxy*, from the Greek words epi (meaning "on") and taxis (meaning "arrangement"). The epitaxial process offers an important means of controlling the doping profiles so that device and circuit performances can be optimized. For example, a semiconductor layer with a relatively low doping concentration can be grown epitaxially upon a substrate

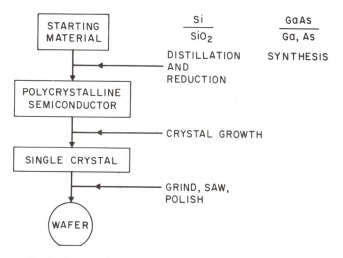

Fig. 1 Process flow from starting material to polished wafer.

which contains the same type of dopant in a much higher concentration (e.g., n-type silicon on an n^+-silicon substrate). In this way the series resistance associated with the substrate can be substantially reduced. Many novel device structures, especially for microwave and photonic devices, can be made by epitaxial processes. Later in this chapter we consider some important epitaxial growth techniques.

8.1 CRYSTAL GROWTH FROM THE MELT

There are basically two techniques for crystal growth from the melt (i.e., material in liquid form): the Czochralski technique and the Bridgman technique. A substantial percentage ($\sim 90\%$) of the silicon crystals for the semiconductor industry are prepared by the Czochralski technique; virtually all the silicon used for fabricating integrated circuits is prepared by this technique. Most gallium arsenide, on the other hand, is grown by the Bridgman technique. However, the Czochralski technique is becoming more popular for the growth of large-diameter gallium arsenide.

8.1.1 Starting Materials

The starting material for silicon is a relatively pure form of sand (SiO_2) called quartzite. This is placed in a furnace with various forms of carbon (coal, coke, and wood chips). While a number of reactions take place in the furnace, the overall reaction is

$$SiC(solid) + SiO_2(solid) \longrightarrow Si(solid) + SiO(gas) + CO(gas) . \quad (1)$$

This process produces metallurgical-grade silicon with a purity of about 98%. Next, the silicon is pulverized and treated with hydrogen chloride (HCl) to form trichlorosilane ($SiHCl_3$):

$$Si(solid) + 3HCl(gas) \xrightarrow{300°C} SiHCl_3(gas) + H_2(gas) . \quad (2)$$

The trichlorosilane is a liquid at room temperature (boiling point 32°C). Fractional distillation of the liquid removes the unwanted impurities. The purified $SiHCl_3$ is then used in a hydrogen reduction reaction to prepare the electronic-grade silicon (EGS):

$$SiHCl_3(gas) + H_2(gas) \longrightarrow Si(solid) + 3HCl(gas) . \quad (3)$$

This reaction takes place in a reactor containing a resistance-heated silicon rod, which serves as the nucleation point for the deposition of silicon. The EGS, a polycrystalline material of high purity, is the raw material used to prepare device quality, single-crystal silicon. Pure EGS generally has impurity concentrations in the parts-per-billion range.[1, 2]

The starting materials for gallium arsenide are the elemental, chemically pure gallium and arsenic which are used for the synthesis of polycrystalline gallium arsenide. Because gallium arsenide is a combination of two materials, its behavior is quite different from that of a single material such as silicon. The behavior of a combination can be described by a *phase diagram*. A phase is a state (e.g., solid, liquid, or gaseous) in which a material may exist. A phase diagram shows the relationship between two components (e.g., gallium and arsenic) as a function of temperature.

Figure 2 shows the phase diagram of the gallium–arsenic system. The abscissa represents various compositions of the two components in terms of atomic percent (lower scale) or weight percent (upper scale).[3, 4] Consider a melt that is initially of composition x (e.g., 85 atomic percent arsenic shown in Fig. 2). When the temperature is lowered, its composition will remain fixed until the *liquidus* line is reached. At the point (T_1, x), material of 50 atomic percent arsenic (i.e., gallium arsenide) will begin to solidify.

Unlike silicon, which has a relatively low vapor pressure at its melting point ($\sim 10^{-6}$ atm at 1420°C), both gallium and arsenic have much higher vapor pressures at the melting point of gallium arsenide (1238°C). In its vapor phase, arsenic has As_2 and As_4 as its major species. Figure 3 shows the vapor pressures of gallium and arsenic along the liquidus curve.[5] Also shown for comparison is the vapor pressure of silicon. The vapor pressure curves for gallium arsenide are double-valued. The dashed curves are for arsenic-rich gallium arsenide melt (right side of liquidus line in Fig. 2), and the solid curves are for gallium-rich gallium arsenide melt (left side of liquidus line in Fig. 2). Because there is a larger amount of arsenic in an arsenic-rich melt than in a gallium-rich melt, more arsenic (As_2 and As_4) will be vaporized from the arsenic-rich melt, thus resulting in a higher vapor pressure. A similar argument can explain the higher vapor pressure of gallium in a gallium-rich melt. Note that long before the melting point is reached, the surface layers of liquid gallium arsenide may decompose into gallium and arsenic. Since the vapor

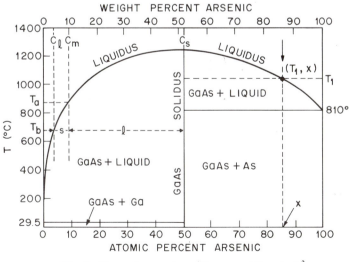

Fig. 2 Phase diagram for gallium–arsenic system.[3]

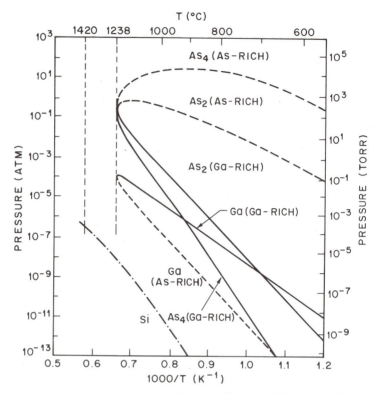

Fig. 3 Partial pressures of gallium and arsenic over gallium arsenide as a function of temperature.[5] Also shown is the partial pressure of silicon.

pressure of gallium and arsenic are quite different, there is a preferential loss of the more volatile arsenic species, and the liquid becomes gallium rich.

To synthesize gallium arsenide, an evacuated, sealed quartz tube system with a two-temperature furnace is commonly used. The high-purity arsenic is placed in a graphite boat and heated to 610 to 620°C, while the high-purity gallium is placed in another graphite boat and heated to slightly above the gallium arsenide melting temperature (1240 to 1260°C). Under these conditions, an overpressure of arsenic is established (1) to cause the transport of arsenic vapor to the gallium melt, converting it into gallium arsenide, and (2) to prevent decomposition of the gallium arsenide while it is being formed in the furnace. When the melt cools, a high-purity polycrystalline gallium arsenide results. This serves as the raw material to grow single-crystal gallium arsenide.[4]

8.1.2 The Czochralski Technique

The Czochralski technique for silicon crystal growth uses an apparatus called a *puller*, as shown in Fig. 4. The puller has three main components: (1) a furnace, which includes a fused-silica (SiO_2) crucible, a gra-

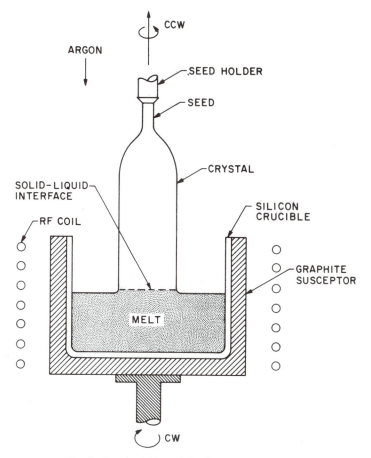

Fig. 4 Czochralski crystal puller.

phite susceptor, a rotation mechanism (clockwise as shown), a heating element, and a power supply; (2) a crystal-pulling mechanism, which includes a seed holder and a rotation mechanism(counter-clockwise); and (3) an ambient control, which includes a gas source (such as argon), a flow control, and a exhaust system. In addition, the puller has an overall microprocessor-based control system to control process parameters such as temperature, crystal diameter, pull rate, and rotation speeds, as well as to permit programmed process steps. Also, various sensors and feedback loops allow the control system to respond automatically, thereby reducing operator intervention.

In the crystal-growing process, polycrystalline silicon is placed in the crucible and the furnace is heated above the melting temperature of silicon. A suitably oriented seed crystal (e.g., <111>) is suspended over the crucible in a seed holder. The seed is inserted into the melt. Part of it melts, but the tip of the remaining seed crystal still touches the liquid surface. It is then slowly withdrawn. Progressive freezing at the solid–liquid interface yields a large, single crystal. A typical pull rate is a few millimeters per minute.

For Czochralski growth of gallium arsenide, the basic puller is identical to that for silicon. However, to prevent decomposition of the melt during crystal growth, a liquid encapsulation method is employed. The liquid encapsulant is a molten boron trioxide (B_2O_3) layer about 1 cm thick. Molten boron trioxide is inert to gallium arsenide at the growth temperature. The layer adheres to the gallium arsenide surface and serves as a cap to cover the melt. This cap prevents decomposition of the gallium arsenide as long as the pressure on its surface is higher than 1 atm (760 Torr). Since boron trioxide can dissolve silicon dioxide, the fused-silica crucible is replaced with a graphite crucible.

8.1.3 Distribution of Dopant

In crystal growth, a known amount of dopant is added to the melt to obtain the desired doping concentration in the grown crystal. For silicon, boron and phosphorus are the most common dopants for p- and n-type materials, respectively. For gallium arsenide, cadmium and zinc are commonly used for p-type material, while selenium, silicon, and tellurium are used for n-type material; chromium is used for semi-insulating material.

As a crystal is pulled from the melt, the doping concentration incorporated into the crystal (solid) is usually different from the doping concentration of the melt (liquid) at the interface. The ratio of these two concentrations is defined as the *equilibrium segregation coefficient* k_0:

$$k_0 \equiv \frac{C_s}{C_l} \tag{4}$$

where C_s and C_l are respectively the equilibrium concentrations of the dopant in the solid and liquid near the interface. Table 1 lists values of k_0 for the commonly used dopants for silicon and gallium arsenide. Note that most values are below 1, which means that during growth the dopants are rejected into the melt. Consequently, the melt becomes progressively enriched with the dopant as the crystal grows.

Consider a crystal being grown from a melt having an initial weight M_0 with an initial doping concentration C_0 in the melt (i.e., the weight of the dopant per 1 gram melt). At a given point of growth when a crystal of weight M has been grown, the amount of dopant remaining in the melt (by weight) is S. For an incremental amount of the crystal with weight dM, the corresponding reduction of the dopant ($-dS$) from the melt is $C_s\,dM$, where C_s is the doping concentration in the crystal (by weight):

$$-dS = C_s\,dM . \tag{5}$$

Now, the remaining weight of the melt is $M_0 - M$, and the doping concentration in the liquid (by weight), C_l, is given by

$$C_l = \frac{S}{M_0 - M} . \tag{6}$$

Table 1 Equlibrium Segregation Coefficients for Dopants in Si and GaAs

Si			GaAs		
Dopant	k_0	Type	Dopant	k_0	Type
As	0.3	n	S	0.5	n
Bi	7×10^{-4}	n	Se	0.1	n
C	0.07	n	Sn	0.08	n
Li	10^{-2}	n	Te	0.064	n
O	0.5	n	C	1	n/p
P	0.35	n	Ge	0.018	n/p
Sb	0.023	n	Si	2	n/p
Te	2×10^{-4}	n	Be	3	p
Al	2.8×10^{-3}	p	Mg	0.1	p
Ga	8×10^{-3}	p	Zn	0.42	p
B	0.8	p	Cr	5.7×10^{-4}	Semi-insulating
Au	2.5×10^{-5}	Deep lying	Fe	3×10^{-3}	Semi-insulating

Combining Eqs. 5 and 6 and substituting $C_s/C_l = k_0$ yields

$$\frac{dS}{S} = -k_0 \left[\frac{dM}{M_0 - M} \right]. \tag{7}$$

Given the initial weight of the dopant, $C_0 M_0$, we can integrate Eq. 7:

$$\int_{C_0 M_0}^{S} \frac{dS}{S} = k_0 \int_0^M \frac{-dM}{M_0 - M}. \tag{8}$$

Solving Eq. 8 and combining with Eq. 6 gives

$$C_s = k_0 C_0 \left[1 - \frac{M}{M_0} \right]^{k_0 - 1} \tag{9}$$

Figure 5 illustrates the doping distribution as a function of the fraction solidified (M/M_0) for several segregation coefficients.[2, 6] As crystal growth progresses, the composition initially at $k_0 C_0$ will increase continually for $k_0 < 1$ and decrease continually for $k_0 > 1$. When $k_0 \simeq 1$, a uniform impurity distribution can be obtained.

Problem

A silicon ingot, which should contain 10^{16} boron atoms/cm^3, is to be grown by the Czochralski technique. What concentration of boron atoms should be in the melt to give the required concentration in the ingot? If the initial load of silicon in the crucible is 60 kg, how many grams boron (atomic weight 10.8) should be added?

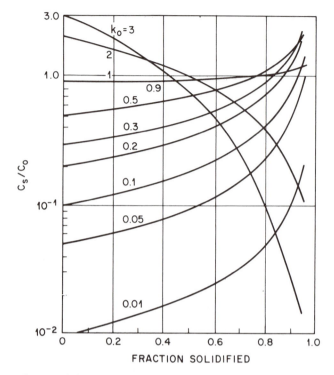

Fig. 5 Curves for growth from the melt, showing the doping concentration in a solid as a function of the fraction solidified.[6]

Solution

The segregation coefficient k_0 for boron is 0.8. We assume that $C_s = k_0 C_l$ throughout the growth. Thus, the initial concentration of boron in the melt should be

$$\frac{10^{16}}{0.8} = 1.25 \times 10^{16} \text{ boron atoms/cm}^3 .$$

Since the amount of boron concentration is so small, the volume of melt can be calculated from the weight of silicon. The density of molten silicon is 2.53 g/cm^3. Therefore, the volume of 60 kg of silicon is

$$\frac{60 \times 10^3}{2.53} = 2.37 \times 10^4 \text{ cm}^3$$

The total number of boron atoms in the melt is

$$1.25 \times 10^{16} \text{ atoms/cm}^3 \times 2.37 \times 10^4 \text{ cm}^3 = 2.96 \times 10^{20} \text{ boron atoms}$$

so that

$$\frac{2.96 \times 10^{20} \text{ atoms} \times 10.8 \text{ g/mole}}{6.02 \times 10^{23} \text{ atoms/mole}} = 5.31 \times 10^{-3} \text{ g boron}$$

$$= 5.31 \text{ mg boron} .$$

Note the small amount of boron needed to dope such a large load of silicon.

8.1.4 Effective Segregation Coefficient

While the crystal is growing, dopants are constantly being rejected into the melt (for $k_0 < 1$). If the rejection rate is higher than the rate of which the dopant can be transported away by diffusion or stirring, then a concentration gradient will develop at the interface, as illustrated in Fig. 6. The segregation coefficient (given in Section 8.1.3) is $k_0 = C_s / C_l(0)$. We can define an effective segregation coefficient k_e, which is the ratio of C_s and the impurity concentration far away from the interface:

$$k_e \equiv \frac{C_s}{C_l} . \tag{10}$$

Consider a small, virtually stagnant layer of melt with width δ in which the only flow is that required to replace the crystal being withdrawn from the melt. Outside this stagnant layer, the doping concentration has a constant value C_l. Inside the layer, the doping concentration can be described by the continuity equation (Eq. 81) derived in Chapter 2. At steady state, the only significant terms are the second and third terms on the right-hand side (we replace n_p by C and $\mu_n \mathscr{E}$ by v):

$$0 = v \frac{dC}{dx} + D \frac{d^2C}{dx^2} \tag{11}$$

where D is the dopant diffusion coefficient in the melt, v is the crystal growth velocity, and C is the doping concentration in the melt.

The solution of Eq. 11 is

$$C = A_1 e^{-vx/D} + A_2 \tag{12}$$

where A_1 and A_2 are constants to be determined by the boundary conditions. The first boundary condition is that $C = C_l(0)$ at $x = 0$. The second boundary condition is the conservation of the total number of dopants; that is, the sum of the dopant fluxes at the interface must be zero. By considering

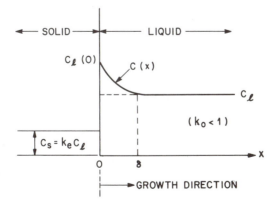

Fig. 6 Doping distribution near the solid–melt interface.

diffusion of dopant atoms in the melt (neglecting diffusion in the solid), we have

$$D \left[\frac{dC}{dx} \right]_{x=0} + [C_l(0) - C_s] v = 0 . \tag{13}$$

Substituting these boundary conditions into Eq. 12 and noting that $C = C_l$ at $x = \delta$ gives

$$e^{-v\delta/D} = \frac{C_l - C_s}{C_l(0) - C_s} . \tag{14}$$

Therefore,

$$k_e \equiv \frac{C_s}{C_l} = \frac{k_0}{k_0 + (1 - k_0)e^{-v\delta/D}} . \tag{15}$$

The doping distribution in the crystal is given by the same expression as in Eq. 9, except that k_0 is replaced by k_e. Values of k_e are larger than those of k_0 and can approach 1 for large values of the growth parameter $v\delta/D$. Uniform doping distribution ($k_e \to 1$) can be obtained by employing a high pull rate and a low rotation speed (since δ is inversely proportional to the rotation speed).

8.1.5 The Bridgman Technique

Figure 7 shows a Bridgman system in which a two-zone furnace is used for growing single-crystal gallium arsenide. The left-hand zone is held at a temperature ($\sim 610°C$) to maintain the required overpressure of arsenic, while the right-hand zone is held just above the melting point of gallium arsenide ($\sim 1240°C$). The sealed tube is made of quartz and the boat is made of graphite. In operation, the boat is loaded with a charge of polycrystalline gallium arsenide, with the arsenic kept at another end of the tube.

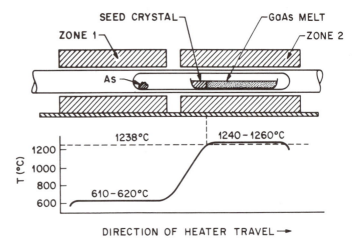

Fig. 7   Bridgman technique for gallium arsenide with temperature profile of the furnace.

As the furnace is moved toward the right, the melt cools at one end. Usually there is a seed placed at the left end of the boat to establish a specific crystal orientation. The gradual freezing (solidification) of the melt allows a single crystal to propagate at the liquid–solid interface. Eventually, a single crystal of gallium arsenide is grown. The impurity distribution can be described essentially by Eqs. 9 and 15, where the growth rate is given by the traversing speed of the furnace.

8.2 THE FLOAT ZONE PROCESS

The float zone process can be used to grow silicon that has lower contaminations than that normally obtained from the Czochralski technique. A schematic setup of the float zone process is shown in Fig. 8a. A high-purity polycrystalline rod with a seed crystal at the bottom is held in a vertical position and rotated. The rod is enclosed in a quartz envelope within which an inert atmosphere (argon) is maintained. During the operation, a small zone (a few centimeters in length) of the crystal is kept molten by a radio-frequency heater, which is moved from the seed upward so that this *floating zone* traverses the length of the rod. The molten silicon is retained by surface tension between the melting and growing solid-silicon faces. As the floating zone moves upward, a single-crystal silicon freezes at the zone's retreating end and grows as an extension of the seed crystal. Materials with higher resistivities can be obtained from float zone process than from the Czochralski process, because it can be used to purify the crystal more easily. Furthermore, since no crucible is used in the float zone process, there is no contamination from the crucible (as with Czochralski growth). At the present time, float zone crystals

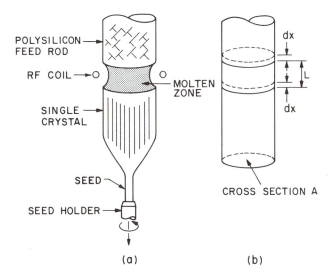

Fig. 8 Float zone process. (*a*) Schematic setup. (*b*) Simple model for doping evaluation.

are used mainly for high-power, high-voltage devices, where high-resistivity materials are required.

To evaluate the doping distribution of a float zone process, consider a simplified model as shown in Fig. 8b. The initial, uniform doping concentration in the rod is C_0 (by weight). Let L be the length of the molten zone at a distance x along the rod, A the cross-sectional area of the rod, ρ_d the specific density of silicon, and S the amount of dopant present in the molten zone. As the zone traverses a distance dx, the amount of dopant added to it at its advancing end is $C_0\rho_d A\, dx$, while the amount of dopant removed from it at the retreating end is $k_e(S\,dx/L)$, where k_e is the effective segregation coefficient. Thus,

$$dS = C_0\rho_d A dx - \frac{k_e S}{L}\,dx = \left[C_0\rho_d A - \frac{k_e S}{L} \right] dx \qquad (16)$$

so that

$$\int_0^x dx = \int_{S_0}^{S} \frac{dS}{C_0\rho_d A - (k_e S/L)} \qquad (16a)$$

where $S_0 = C_0\rho_d AL$ is the amount of dopant in the zone when it was first formed at the front end of the rod. From Eq. 16a we obtain

$$\exp\left[\frac{k_e x}{L} \right] = \frac{C_0\rho_d A - (k_e S_0/L)}{C_0\rho_d A - (k_e S/L)} \qquad (17)$$

or

$$S = \frac{C_0 A\rho_d L}{k_e} [1 - (1 - k_e)^{-k_e x/L}] . \qquad (17a)$$

Since C_s (the doping concentration in the crystal at the retreating end) is given by $C_s = k_e(S/A\rho_d L)$, then

$$C_s = C_0[1 - (1 - k_e)^{-k_e x/L}] . \qquad (18)$$

Figure 9 shows the doping concentration versus the solidified zone length for various values of k_e.

These two crystal growth techniques can also be used to remove impurities. A comparison of Fig. 9 with Fig. 5 shows that a single pass in the float zone process does not produce as much purification as a single Czochralski growth. For example, for $k_0 = k_e = 0.1$, C_s/C_0 is smaller over most of the solidified ingot by the Czochralski growth. However, multiple passes can be performed on a rod much more easily than a crystal can be grown, cropped off the end region, and regrown from the melt. Figure 10 shows the impurity distribution for an element with $k_e = 0.1$ after a number of successive passes of the zone along the length of the rod.[6] Note that there is a substantial reduction of impurity concentration in the rod after each pass. Therefore, the float zone process is ideally suited for crystal purification.

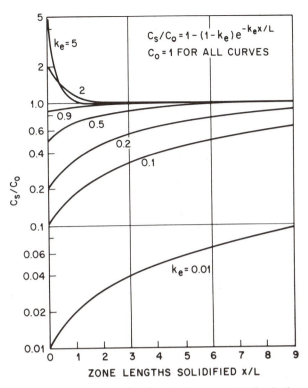

Fig. 9 Curves for float zone process, showing doping concentration in the solid as a function of zone lengths solidified.[6]

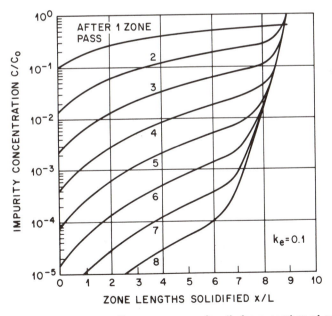

Fig. 10 Relative impurity concentration versus zone length for a number of passes. L denotes zone length.[6]

If it is desirable to dope the rod rather than to purify it, consider the case in which all the dopants are introduced in the first zone ($S_0 = C_1 A \rho_d L$) and the initial concentration C_0 is negligibly small. Equation 17 gives

$$S_0 = S \exp\left[\frac{k_e x}{L}\right]. \qquad (19)$$

Since $C_s = k_e(S / A \rho_d L)$, we obtained from Eq. 19

$$C_s = k_e C_1 e^{-k_e x /L}. \qquad (20)$$

Therefore, if $k_e x /L$ is small, C_s will remain nearly constant with distance, except at the end that is last to solidify.

For certain switching devices (e.g., high-voltage thyristors, discussed in Chapter 4) large chip areas are used, frequently an entire wafer for a single device. This size imposes stringent requirements on the uniformity of the starting material. To obtain homogeneous distribution of dopants, we use a float zone silicon slice that has an average doping concentration well below the required amount. The slice is then irradiated with thermal neutrons. This process, called *neutron irradiation*, gives rise to fractional transmutation of silicon into phosphorus and dopes the silicon *n*-type:

$$\mathrm{Si}_{14}^{30} + \text{neutron} \longrightarrow \mathrm{Si}_{14}^{31} + \gamma\ \text{ray} \xrightarrow{2.62\ \text{hr}} \mathrm{P}_{15}^{31} + \beta-\text{ray}. \quad (21)$$

The half-life of the intermediate element Si_{14}^{31} is 2.62 h. Because the penetration depth of neutrons in silicon is about 100 cm, doping is very uniform throughout the slice. Figure 11 compares the lateral resistivity distributions in conventionally doped silicon and in silicon doped by neutron irradiation.[7] Note that the resistivity variations for the neutron-irradiated silicon are much smaller than that for the conventionally doped silicon.

8.3 WAFER SHAPING AND MATERIAL CHARACTERIZATION

8.3.1 Wafer Shaping

After a crystal is grown, the first shaping operation is to remove the seed and the other end of the ingot, which is last to solidify.[1] The next operation is to grind the surface so that the diameter of the material is defined. After that, one or more flat regions are ground along the length of the ingot. These regions, or *flats*, mark the specific crystal orientation of the ingot and the conductivity type of the material. The largest flat, the *primary flat*, allows a mechanical locator in automatic processing equipment to position the wafer and to orient the devices relative to the crystal in a specific manner. Other smaller flats, called *secondary flats*, are ground to identify the orientation and conductivity type of the crystal, as shown in Fig. 12.

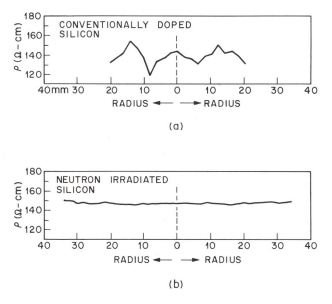

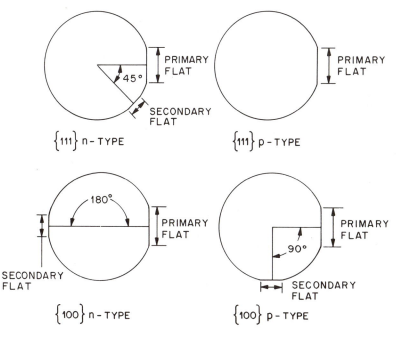

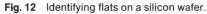

Fig. 11 Typical lateral resistivity distribution in conventionally doped silicon (a) and in silicon doped by neutron irradiation (b).[7]

Fig. 12 Identifying flats on a silicon wafer.

Once these operations have been done, the ingot is ready to be sliced by diamond saw into wafers. Slicing determines four wafer parameters: surface orientation (e.g., $<111>$ or $<100>$), thickness (e.g., 0.5 to 0.7 mm depending on wafer diameter), taper (i.e., wafer thickness variations from one end to another), and bow (i.e., surface curvature of the wafer, measured from the center of the wafer to its edge).

After slicing, both sides of the wafer are lapped using a mixture of Al_2O_3 and glycerine to produce a typical flatness uniformity within 2 μm. The lapping operation usually leaves the surface and edges of the wafer damaged and contaminated. The damaged and contaminated regions can be removed by chemical etching (refer to Section 11.3). The final step of wafer shaping is polishing. Its purpose is to provide a smooth, specular surface where device features can be defined by lithographic processes (see Chapter 11). Table 2 shows the specifications for 100-, 125-, and 150-mm diameter polished silicon wafers from the Semiconductor Equipment and Materials Institute (SEMI).

Gallium arsenide is a softer and more fragile material than silicon. Although the basic shaping operation of gallium arsenide is essentially the same as that for silicon, greater care must be exercised in gallium arsenide wafer preparation. The state of gallium arsenide technology is relatively primitive compared with that of silicon. However, the technology of Group III–V compounds has advanced partly because of the advances in silicon technology.

8.3.2 Crystal Characterization

Crystal Defects A real crystal (such as a silicon wafer) differs from the ideal crystal in important ways. It is finite; thus, surface atoms are incompletely bonded. Furthermore, it has defects, which strongly influence the electrical, mechanical, and optical properties of the semiconductor. There are four categories of defects: (1) point defects, (2) line defects, (3) area defects, and (4) volume defects.

Figure 13 shows several forms of point defects.[1] Any foreign atom incorporated into the lattice at either a substitutional site (i.e., at a regular lattice site) or interstitial site (i.e., between regular lattice sites) is a point defect. A missing atom in the lattice creates a vacancy, also considered a point defect. A

Table 2 Specification for Polished Monocrystalline Silicon Slices

Parameter	100 mm	125 mm	150 mm
Diameter (mm)	100 ± 1	125 ± 1	150 ± 1
Thickness (mm)	0.5–0.55	0.6–0.65	0.65–0.7
Primary flat length (mm)	30–35	40–45	55–60
Secondary flat length (mm)	16–20	25–30	35–40
Bow (μm)	60	70	60
Total thickness variation (μm)	50	65	50
Surface orientation	$(100) \pm 1°$	Same	Same
	$(111) \pm 1°$	Same	Same

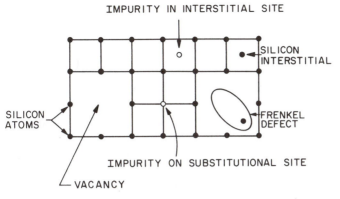

Fig. 13 Types of point defects in a simple lattice.[1]

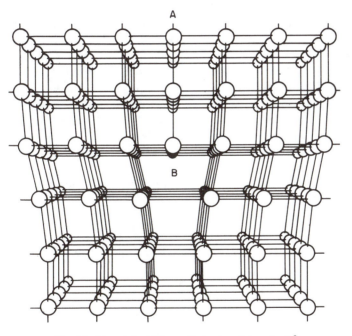

Fig. 14 Line defect (also called edge dislocation).[8]

host atom that is situated between regular lattice sites and adjacent to a vacancy is called a *Frenkel defect.* Point defects are particularly important subjects in the kinetics of diffusion and oxidation processes. These topics are considered in subsequent chapters.

The next class of defects is the *line defect,* also called the edge dislocation. Figure 14 shows a line defect in a cubic crystal.[8] The defect is an extra plane of atoms AB in the lattice. Line defects in devices are undesirable because they act as precipitation sites for metallic impurities which may degrade device performance.

Area defects represent a large area discontinuity in the lattice. Typical defects are twins and grain boundaries. Twinning represents a change in the crystal orientation across a plane. A grain boundary is a transition between crystals having no particular orientational relationship to one another. Such defects appear during crystal growth. Crystals having these defects are not usable for integrated-circuit manufacture and are discarded.

Precipitates of impurities or dopant atoms make up the fourth class of defects. These defects arise because of the inherent solubility of the impurity in the host; and there is a specific concentration of impurity that the host lattice can accept in a solid solution of itself and the impurity. Figure 15 shows solubility versus temperature for a variety of elements in silicon.[9] The solubility of most impurities decreases with decreasing temperature. Thus, at a given temperature, if an impurity is introduced to the maximum concentration allowed by its solubility and the crystal is then cooled to a lower temperature, the crystal can only achieve an equilibrium state by precipitating the impurity atoms in excess of the solubility level. However, the volume mismatch between the host lattice and the precipitates results in dislocations.

Material Properties   Table 3 compares present silicon characteristic and the requirements for very-large-scale integration (VLSI). Note that although the current capabilities can meet most of the wafer specifications listed in Table 2, there are many improvements needed to satisfy the stringent requirements of VLSI technology.[10]

Table 3   Comparison of Silicon Material Characteristics and Requirements for VLSI[10]

Property[†]	Characteristics		Requirements for VLSI
	Czochralski	Float zone	
Resistivity (phosphorus) n-type (ohm-cm)	1–50	1–300 and up	5–50 and up
Resistivity (antimony) n-type (ohm-cm)	0.005–10	—	0.001–0.02
Resistivity (boron) p-type (ohm-cm)	0.005–50	1–300	5–50 and up
Resistivity gradient (four-point probe) (%)	5–10	20	< 1
Minority carrier lifetime (μs)	30–300	50–500	300–1000
Oxygen (ppma)	5–25	Not detected	Uniform and controlled
Carbon (ppma)	1–5	0.1–1	< 0.1
Dislocation (before processing)(per cm^2)	⩽ 500	⩽ 500	⩽ 1
Diameter (mm)	Up to 200	Up to 100	Up to 150
Slice bow (μm)	⩽ 25	⩽ 25	< 5
Slice taper (μm)	⩽ 15	⩽ 15	< 5
Surface flatness (μm)	⩽ 5	⩽ 5	< 1
Heavy-metal impurities (ppba)	⩽ 1	⩽ 0.01	< 0.001

[†] ppma = Parts per million atoms; ppba = parts per billion atoms.

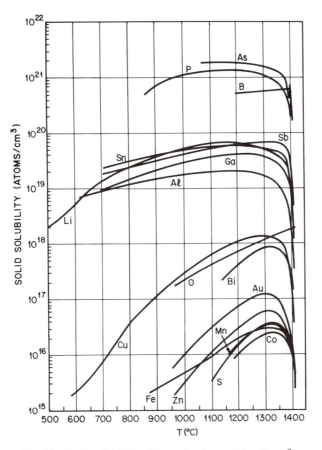

Fig. 15   Solid solubilities of impurity elements in silicon.[9]

The oxygen and carbon concentrations are substantially higher in Czochral-ski crystals than in float zone crystals due to the dissolution of the silica cruci-ble (for oxygen) and transport to the melt from the graphite susceptor (carbon) during crystal growth. Typical carbon concentrations range from 10^{16} to about 10^{17} atoms/cm³ (carbon in silicon occupies substitutional lattice sites). The presence of carbon is undesirable because it aids the formation of defects. Typical oxygen concentrations range from low 10^{17} to 10^{18} atoms/cm³. Oxy-gen, however, has both deleterious and beneficial effects. It can act as a donor, distorting the resistivity of the crystal caused by intentional doping. On the other hand, oxygen in an interstitial lattice site can increase the yield strength of silicon. In addition, the precipitates of oxygen due to the solubility effect, can be used for *gettering*. Gettering is a general term meaning a process that removes harmful impurities or defects from the region in a wafer where dev-ices are fabricated. When the wafer is subjected to high-temperature treat-ment (e.g., 1050°C in N_2), oxygen evaporates from the surface. This lowers the oxygen content near the surface, so that precipitation of oxygen does not occur there. The treatment creates a defect-free (or *denuded*) zone for device fabrication as shown in the insert of Fig. 16. Additional thermal cycles can be used to promote the formation of oxygen precipitates in the interior of the wafer for gettering of impurities. The depth of the defect-free zone depends on the time and temperature of the thermal cycle and on the diffusivity of oxygen in silicon. Measured results for the denuded zone are shown[1] in Fig. 16.

It is possible to obtain Czochralski crystals of silicon that are virtually free of dislocations. In practice 500 dislocations/cm^2 or less can be obtained from both Czochralski and float zone crystals.

Commercial melt-grown materials of gallium arsenide are heavily contaminated by the crucible. However, for photonic applications, most requirements call for heavily doped materials (between 10^{17} and 10^{18} cm^{-3}). For integrated circuits or for discrete MESFET devices, gallium arsenide can be doped with chromium to give a starting resistivity of 10^9 Ω-cm. Oxygen is an undesirable impurity in gallium arsenide because it can form an impurity complex and undesirably increase the resistivity. Oxygen contamination can be minimized by using graphite crucibles for melt growth. The dislocation content for Czochralski-grown GaAs gallium arsenide crystals is about two orders of magnitude higher than that for silicon. For Bridgman GaAs crystals, the dislocation density is about an order of magnitude lower than that for Czochralski-grown GaAs crystals.

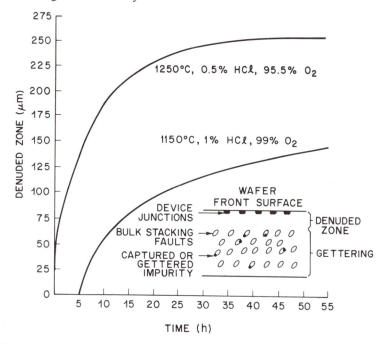

Fig. 16 Denuded zone width for two sets of processing conditions. Insert shows a schematic of the denuded zone and gettering sites in a wafer cross section.[1]

8.4 VAPOR-PHASE EPITAXY

In an epitaxial process, the substrate wafer acts as a seed crystal. Epitaxial processes are differentiated from the melt growth processes (described in Section 8.1) in that the epitaxial layer can be grown at a temperature substantially below the melting point (typically 30 to 50% lower). Among various epitaxial processes, vapor phase epitaxy (VPE) is by far the most important for silicon devices. VPE is also important for gallium arsenide, but other epitaxial processes (e.g., molecular-beam epitaxy) can provide certain advantages not obtainable from VPE.

Figure 17 shows three common susceptors for epitaxial growth. Note that the geometric shape of the susceptor provides the name for the reactor: horizontal, pancake, and barrel susceptors—all made from graphite blocks. Susceptors in the epitaxial reactors are analogous to crucibles in the crystal growing furnaces. Not only do they mechanically support the wafer, but in induction-heated reactors they also serve as the source of thermal energy for the reaction.

Four silicon sources have been used for vapor phase epitaxial growth. They are silicon tetrachloride ($SiCl_4$), dichlorosiliane (SiH_2Cl_2), trichlorosilane ($SiHCl_3$), and silane (SiH_4). Silicon tetrachloride has been the most studied and has the widest industrial use. The typical reaction temperature is 1200°C. Other silicon sources are used because of lower reaction temperatures. The

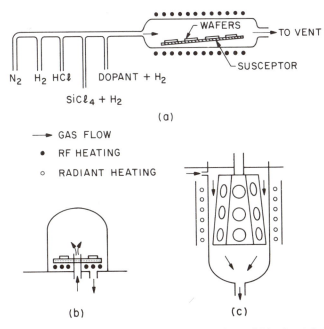

Fig. 17 Three common susceptors for vapor phase epitaxy: (a) horizontal, (b) pancake, and (c) barrel susceptor.

substitution of a hydrogen atom for each chlorine atom from silicon tetrachloride permits about a 50°C reduction in the reaction temperature. The overall reaction of silicon tetrachloride that results in the growth of silicon layers is

$$SiCl_4(gas) + 2H_2(gas) \rightleftharpoons Si(solid) + 4HCl(gas) . \qquad (22)$$

An additional competing reaction is taking place along with that given in Eq. 22:

$$SiCl_4(gas) + Si(solid) \rightleftharpoons 2SiCl_2(gas) . \qquad (23)$$

As a result, if the silicon tetrachloride concentration is too high, etching rather than growth of silicon will take place. Figure 18 shows the effect of the concentration of silicon tetrachloride in the gas on the reaction, where the *mole fraction* is defined as the ratio of the number of molecules of a given species to the total number of molecules.[11] Note that initially the growth rate increases linearly with increasing concentration of silicon tetrachloride. As the concentration of silicon tetrachloride is increased, a maximum growth rate is reached. Beyond that, the growth rate starts to decrease, and eventually etching of the silicon will occur. Silicon is usually grown in the low concentration region, as indicated in Fig. 18.

The reaction of Eq. 22 is reversible, that is, it can take place in either direction. If the carrier gas entering the reactor contains hydrochloric acid, removal or etching will take place. Actually, this etching operation is used for in-situ cleaning of the silicon wafer prior to epitaxial growth.

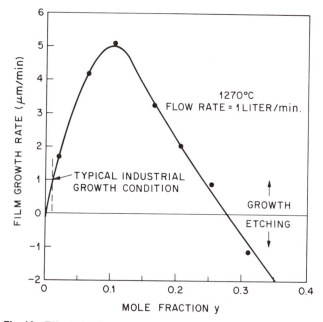

Fig. 18 Effect of SiCl_4 concentration on silicon epitaxial growth.[11]

The dopant is introduced at the same time as the silicon tetrachloride during epitaxial growth (Fig. 17a). Gaseous diborane (B_2H_6) is used as the p-type dopant, while phosphine (PH_3) and arsine (AsH_3) are used as n-type dopants. Gas mixtures are ordinarily used with hydrogen as the diluent to allow reasonable control of flow rates for the desired doping concentration. The dopant chemistry for arsine is illustrated in Fig. 19, which shows arsine being adsorbed on the surface, decomposing, and being incorporated into the growing layer. Figure 19 also shows the growth mechanisms at the surface, which are based on the surface adsorption of host atoms (silicon) as well as the dopant atom (e.g., arsenic) and the movement of these atoms toward the ledge sites.[12] To give these adsorbed atoms sufficient mobility for finding their proper positions within the crystal lattice, epitaxial growth needs relatively high temperatures.

For gallium arsenide vapor phase epitaxy, the basic setup is similar to that shown in Fig. 17a. Since gallium arsenide decomposes into gallium and arsenic upon evaporation, its direct transport in the vapor phase is not possible. One approach is the use of As_4 for the arsenic component and gallium chloride ($GaCl_3$) for the gallium component. The overall reaction leading to epitaxial growth of gallium arsenide is

$$As_4 + 4GaCl_3 + 6H_2 \longrightarrow 4GaAs + 12HCl . \qquad (24)$$

The As_4 is generated by thermal decomposition of arsine (AsH_3):

$$4AsH_3 \longrightarrow As_4 + 6H_2 \qquad (24a)$$

and the gallium chloride is generated by the reaction

$$6HCl + 2Ga \longrightarrow 2GaCl_3 + 3H_2 . \qquad (24b)$$

The reactants are introduced into a reactor with a carrier gas (e.g., H_2). The gallium arsenide wafers are typically held within the 650 to 850°C temperature range. There must be sufficient arsenic overpressure to prevent thermal decomposition of the substrate and the growing layer.

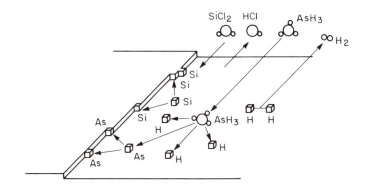

Fig. 19 Schematic representation of arsenic doping and growth processes.[12]

Another approach is the MOCVD (metalorganic chemical-vapor deposition) which uses metalorganic compounds such as trimethylgallium $Ga(CH_3)_3$. One can use trimethylgallium for the gallium component and arsine (AsH_4) for the arsenic component. Both chemicals can be transported in vapor form into the reactor. The overall reaction is

$$AsH_3 + Ga(CH_3)_3 \longrightarrow GaAs + 3CH_4 . \qquad (25)$$

During epitaxy the doping of gallium arsenide is done by the introduction of dopants in vapor form. The hydrides of sulfur and selenium or tetramethyltin are used for n-type doping; diethylzinc or diethylcadmium is used for p-type doping; and chromyl chloride is used to dope chromium onto gallium arsenide to form semi-insulating layers. As with silicon epitaxial growth, an in-situ etching is done to remove any contaminants prior to the growth.

8.4.1 Kinetics of Growth

We shall now study the kinetics of epitaxial growth using a simple model as shown in Fig. 20. The concentration of the reactant species in the gas stream (e.g., $SiCl_4$) is C_g in the bulk of the gas (far away from the gas–semiconductor interface) but becomes C_s at the interface. The term *flux* is defined as the number of molecules crossing a unit area in a unit time. The flux of the reactant species from the bulk of the gas to the interface is indicated as F_1, while the flux corresponding to the reactant species consumed in the epitaxial reactions is indicated as F_2.

We assume that the flux F_1 can be expressed as

$$F_1 = h_g(C_g - C_s) \qquad (26)$$

where h_g is the vapor phase mass transfer coefficient (h_g has the dimensions of velocity in cm/s). The flux consumed by the chemical reaction taking place at the surface of the growing layer can be expressed as

$$F_2 = k_s C_s \qquad (27)$$

where k_s is the surface reaction rate constant (also in units of cm/s). These linear approximations are analogous to Ohm's law: they describe a flux which is proportional to its driving force. The driving force for F_1 is the concentration difference, and that for F_2 is the concentration of the reactant species.

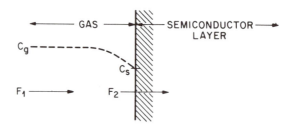

Fig. 20   Model of the epitaxial-growth process.

In the steady state, $F_1 = F_2 = F$. Given this condition, we can obtain the surface concentration of the reactant species:

$$C_s = \frac{C_g}{1 + (k_s/h_g)} . \tag{28}$$

The growth rate v of the semiconductor layer is given by the steady-state flux divided by the number of semiconductor atoms incorporated into a unit volume of the layer (C_a), or from Eqs. 27 and 28:

$$v = \frac{F}{C_1} = \frac{k_s h_g}{k_s + h_g} \left[\frac{C_g}{C_a} \right]. \tag{29}$$

The value of C_a is 5×10^{22} atoms/cm^3 for silicion and 4.4×10^{22} atoms/cm^3 for gallium arsenide. Since $C_g = yC_t$, where y is the mole fraction of the reactant species and C_t is the total number of molecules per cm^3 in the gas, we obtain the following expression for the growth rate:

$$v = \frac{k_s h_g}{k_s + h_g} \left[\frac{C_t}{C_a} \right] y . \tag{29a}$$

Equation 29a states that the growth rate is proportional to the mole fraction y of the reactant species. This is indeed the case for small values of y, as illustrated in Fig. 18. The growth rate at a given mole fraction is determined by the smaller of either h_g or k_s. If k_s is much smaller than h_g, the growth rate is determined by how fast the surface reaction can take place. This is referred to as surface reaction-controlled. On the other hand, if k_s is much larger than h_g, the growth rate is determined by how fast the reactant species can be transported to the wafer surface. This is referred to as mass transfer-controlled. Therefore, in these two limiting cases, we have

$$v \simeq k_s \left[\frac{C_t}{C_a} \right] y \qquad \text{surface reaction-controlled} \tag{30a}$$

or

$$v \simeq h_g \left[\frac{C_t}{C_a} \right] y \qquad \text{mass transfer-controlled} . \tag{30b}$$

Figure 21 shows the temperature dependence of growth rate for various silicon sources.[13] At low temperatures (region A), the growth rate follows an exponential law: $v \sim \exp(-E_a/kT)$. The activation energy E_a is about 1.5 eV. At higher temperature (region B), the growth rate is essentially independent of temperature. Since chemical reactions generally follow an exponential temperature dependence while the mass transfer process is independent of temperature, region A is characterized as surface reaction-controlled, and region B is characterized as mass transfer-controlled.

To obtain a high-quality epitaxial layer, growth temperatures must be relatively high. Furthermore, epitaxial growth should be done at a temperature

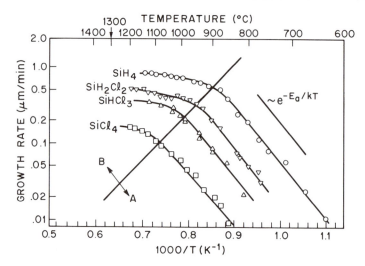

Fig. 21   Temperature dependence of the growth rate for various silicon sources.[13]

where the growth rate is insensitive to minor temperature variations. As a result, most vapor phase epitaxial growth takes place in the mass transfer region. In the next subsection, we shall derive the vapor phase mass transfer coefficient h_g.

8.4.2   Physics of Vapor Phase Mass Transfer

The reactant species is delivered to the semiconductor substrates by transport in the gas stream, with a finite velocity. At the substrate, this velocity is zero because of friction. It is also expected that there is a region, next to the substrates, over which the gas velocity is extremely low. This region represents a boundary layer through which the reactant species must diffuse in order to reach the semiconductor surface. Figure 22a shows schematically the character of this boundary layer in a horizontal reactor. Above the boundary layer, the gas stream is assumed to follow a laminar flow pattern (i.e., parallel flow).

Figure 22b shows an expanded view of the boundary layer on top of a flat plate of length L. Far away from the plate, the gas flows in a laminar flow pattern with a uniform velocity v. Right next to the plate, the velocity of the gas is zero. The frictional force per unit area along the x direction, acting on a gas flow element next to the plate, is given by

$$F(\text{friction}) = \mu \frac{\partial v}{\partial y} \tag{31}$$

where μ is the viscosity of the gas. Consider the crosshatched area in Fig. 22b, which represents a volume element whose size is unity in the direction normal to the paper. The principal force acting on this element is $F(\text{friction})$. The acceleration of this element is

$$a \equiv \frac{dv}{dt} = \frac{dv}{dx}\frac{dx}{dt} = \frac{dv}{dx} v \; . \tag{32}$$

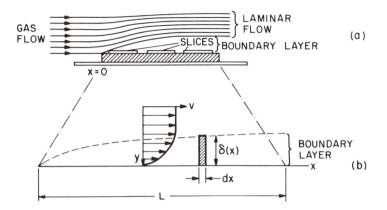

Fig. 22 (a) Development of a boundary layer in gas flow over a flat plate. (b) Expanded view of the boundary layer.

The mass of this element is $\rho_d \delta(x)\, dx$, where ρ_d is the density of the gas. From Newton's second law, $F = ma$, we obtain from Eqs. 31 and 32

$$F = \mu \frac{dv}{dy} = \rho_d \delta(x) v \frac{dv}{dx}. \tag{33}$$

Replacing the differentials by the respective differences gives

$$\mu \frac{v}{\delta(v)} = \rho_d \delta(x)\, v \frac{v}{x} \tag{34}$$

or

$$\delta(x) \simeq \sqrt{\frac{\mu x}{\rho_d v}} \tag{34a}$$

where $\delta(x)$ is the boundary layer thickness, which depends on the viscosity, density, and velocity of the gas; note that $\delta(x)$ increases with the square root of the distance. The average boundary layer thickness $\bar{\delta}$ over the whole plate is given by

$$\bar{\delta} = \frac{1}{L} \int_0^L \delta(x)\, dx = \frac{2}{3} \sqrt{\frac{\mu L}{\rho_d v}}. \tag{35}$$

We can now use $\bar{\delta}$ to obtain an expression for the mass transfer coefficient. As mentioned previously, the reactant species must diffuse through the boundary layer to reach the semiconductor surface. The flux of the diffusing impurities can be expressed as

$$F_1 = D_g \frac{dC}{dy} \approx \frac{D_g}{\bar{\delta}} (C_g - C_s) \tag{36}$$

where D_g is the diffusivity of the reactant species in the gas. Comparing

Eq. 36 with Eq. 26 gives

$$h_g = \frac{D_g}{\delta} = \frac{3}{2} D_g \sqrt{\frac{\rho_d v}{\mu L}}. \tag{37}$$

Problem

Calculate h_g for a silicon epitaxial process at 1200°C. The diffusivity of gas is 25 cm²/s, $\rho_d = 1.5 \times 10^{-5}$ g/cm³, $\mu = 2 \times 10^{-4}$ poise (or g/cm–s), $v = 10$ cm/s, and $L = 50$ cm.

Solution

$$h_g = \left[\frac{3}{2}\right] (25) \sqrt{\frac{1.5 \times 10^{-5} \times 10}{2 \times 10^{-4} \times 50}} = 4.5 \text{ cm/s}.$$

This is in reasonable agreement with typical experimental results. (Some properties of the gases are given in Section 8.6.1.)

8.4.3 Silicon on Insulators

Silicon devices have problems with inherent parasitic circuit elements due to junction capacitances. This is unlike gallium arsenide, which can take advantage of its semi-insulating substrates (such as a chromium-doped substrate) to minimize these capacitances. One way to circumvent the problem is to fabricate silicon devices on an insulating substrate. The initial approach was to grow silicon epitaxially on a substrate of sapphire (called silicon-on-sapphire, or SOS). A more recent approach is the silicon-on-insulator (SOI) process.

For the SOI process, silicon film is deposited onto an amorphous substrate (such as SiO_2) using reactors similar to the ones described previously. How-

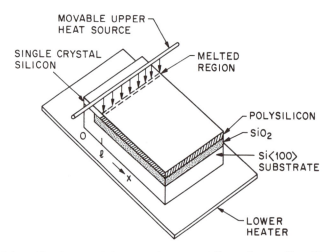

Fig. 23 Schematic of one technique used to recrystallize polycrystalline silicon on silicon dioxide.[14]

ever, only amorphous or polycrystalline silicon is formed on the substrate. Recrystallization is needed to obtain single-crystal silicon. Figure 23 shows one way to recrystallize the deposited silicon using a strip heat source.[14] Heat energy can be supplied from a heater, an incoherent lamp, a laser, or an electron beam. The silicon substrate is processed to give a flat surface with a pattern of exposed silicon area within the surrounding oxide. The surface is then coated by vapor phase deposition with the polycrystalline to a thickness of 0.5 to 1 μm. Next, the heat source (Fig. 23) melts the polysilicon through to the substrate. Now the substrate acts as a seed to the molten silicon; and as the heat source is moved laterally, single-crystal silicon is grown laterally over the oxide-covered regions. Capping the molten zone with silicon dioxide and silicon nitride (Si_3N_4) layers will improve the zone's thermal stability.

8.5 LIQUID-PHASE EPITAXY *

Liquid phase epitaxy (LPE) is the growth of epitaxial layers on crystalline substrates by direct precipitation from the liquid phase. This process is particularly useful for growing gallium arsenide and related III–V compounds. Liquid phase epitaxy is suited to grow thin epitaxial layers ($\gtrsim 0.2$ μm) because it has a slow growth rate. It is also useful to grow multilayered structures in which precise doping and composition controls are required.

Figure 24a shows the boat configuration for liquid phase epitaxy.[15] Here, one or more wells are machined in a high-purity graphite block which serves to hold the reactant solutions. A graphite slider holding the substrates is moved so as to locate them under the wells. The entire assembly is placed in a furnace (Fig. 24b), in a neutral carrier gas (H_2) ambient. During operation while the system is brought up to the required temperature, the substrate is fully covered by part of the graphite block. When the target temperature is reached, the substrate is moved under the first well and the furnace temperature is lowered at a given rate (e.g., 1°C/min). To terminate the layer growth, the wafer is moved out from under the solution. To grow additional layers, the substrate can be moved successively under other wells (with appropriate temperature changes).

In liquid phase epitaxy it is necessary that the material to be grown (e.g., GaAs) dissolves in a solvent, and the solution must melt at a temperature well below the melting point of the semiconductor substrate. Gallium is most commonly used as the solvent for the growth of gallium arsenide layers. For a given temperature, there is a specific amount of gallium arsenide that can saturate the melt resulting in an equilibrium solution. The specific amount is given by the liquidus line of the phase diagram (Fig. 2). When a melt is cooled below the liquidus line, solidification occurs. The equilibrium compositions can be obtained as follows. Consider a melt of initial composition C_m (weight percent scale in Fig. 2) that is cooled from T_a (on the liquidus line) to

---

* See footnote to Section 2.4.2, p. 48.

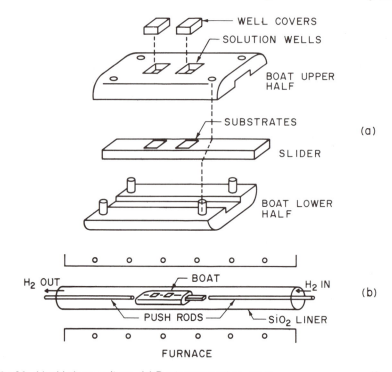

Fig. 24 Liquid phase epitaxy. (*a*) Boat construction. (*b*) Furnace arrangement.[15]

T_b. At T_b, M_l is the weight of the liquid, M_s the weight of the solid (i.e., GaAs), and C_l and C_s are the compositions of the liquid and the solid, respectively (weight percents of arsenic as indicated in Fig. 2). The weight of arsenic in the liquid and solid are $M_l C_l$ and $M_s C_s$, respectively. Because the total arsenic weight is $(M_l + M_s)C_m$, we have

$$M_l C_l + M_s C_s = (M_l + M_s)C_m \tag{38}$$

or

$$\frac{M_s}{M_l} = \frac{\text{weight of GaAs at } T_b}{\text{weight of liquid at } T_b} = \frac{C_m - C_l}{C_s - C_m} = \frac{s}{l} \tag{38a}$$

where s and l are the lengths of the two lines measured from C_m to the liquidus and solidus lines, respectively. As can be seen from Fig. 2, a small fraction ($\sim 10\%$ for the present case) of the melt is solidified.

One technique of LPE is the equilibrium-cooling process. In this process the solution is initially saturated with gallium arsenide at a given temperature T_a with a composition C_m. A substrate, also at T_a, is inserted into the solution, and the temperature of the system is lowered slowly. This drives the solution into a supersaturation state, from which it proceeds toward equilibrium (via precipitation), resulting in growth of gallium arsenide. Growth will continue until the substrate is removed from the solution.

A second technique is known as the step-cooling process. In this process a gallium solution is saturated with gallium arsenide at a temperature T_a. The temperature is lowered by a few degrees (between 5 and 20°C) to T_b so that the melt becomes supersaturated. The substrate, which is also at T_b, is now inserted into the melt, where it is held at T_b indefinitely. Initially, growth occurs because of the supersaturation in the melt. However, the solution becomes depleted of gallium arsenide as the growth continues. Therefore, the growth rate will decrease with increasing time.

The distribution of gallium arsenide in gallium solution during the epitaxy process is similar to the melt growth situation shown in Fig. 6. Diffusion of arsenic will take place through the boundary layer δ and results in the growth of the epitaxial layer. The magnitude of the solute concentration $C(0, t)$ depends on the growth process used. The one-dimensional continuity equation (as derived in Chapter 2) can be simplified to a diffusion equation:

$$\frac{\partial C}{\partial t} = D \frac{\partial^2 C}{\partial x^2} \tag{39}$$

where D is the diffusion coefficient of arsenic in gallium solution. The amount of solute transported through the boundary layer and deposited on the substrate is given by M per unit area:

$$M = \int_0^t D\left[\frac{\partial C}{\partial x}\right]_{x=0} dt . \tag{40}$$

The layer thickness x_0 is given by

$$x_0 = \frac{M}{C_s} \tag{41}$$

where C_s is the solute concentration in the grown layer.

For the equilibrium-cooling process, consider a cooling rate α, such that

$$C(0, t) = C_l - \alpha t . \tag{42}$$

A boundary condition is $C(x, 0) = C_l$, where C_l is the solute concentration at $x = \delta$. The solution of the diffusion equation (Eq. 39) subject to these boundary conditions is

$$C(x, t) = C_l - 4\alpha t \int_x^\infty \int_y^\infty \mathrm{erfc}\ \xi\ d\xi\ dy \tag{43}$$

where $\xi \equiv x/2\sqrt{Dt}$ and erfc is the error function complement.[†] From Eqs. 40 and 41, we have

$$x_0 = \frac{4}{3}\left[\frac{\alpha}{C_s}\right]\left[\frac{D}{\pi}\right]^{1/2} t^{3/2} . \tag{44}$$

[†] Some properties of the error function (erf) and error function complement (erfc) are summarized in Table 1 of Chapter 10.

The thickness of the epitaxial layer increases as $t^{3/2}$, and the growth rate dx_0/dt also increases with time as $t^{1/2}$.

For the step-cooling process, the boundary conditions (as shown in Fig. 6) are

$$C(x,\ 0) = C_l \tag{45a}$$

$$C(0,\ t) = C_l(0). \tag{45b}$$

For this situation, the solution of the diffusion equation (Eq. 39) is

$$C - C_l(0) = [C_l - C_l(0)]\ \mathrm{erf}\left(\frac{x}{2\sqrt{Dt}}\right) \tag{46}$$

where erf is the error function. Substituting in Eqs. 40 and 41 gives

$$x_0 = \frac{2[C_l - C_l(0)]}{C_s}\left[\frac{D}{\pi}\right]^{1/2}t^{1/2}. \tag{47}$$

The thickness of the epitaxial layer increases as $t^{1/2}$; however, the growth rate is proportional to $t^{-1/2}$, which decreases with increasing time.

Representative results for the above two processes are shown[16] in Fig. 25. Note that the time dependences of the epitaxial-layer thickness are in excellent agreement with Eq. 44 (for the equilibrium-cooling process) and with Eq. 47 (for the step-cooling process).

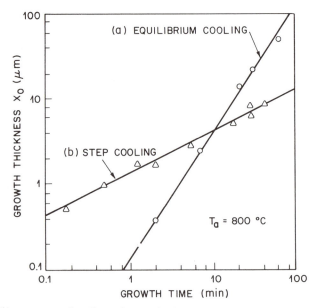

Fig. 25 Thickness as a function of growth time for GaAs layers grown by (a) the equilibrium-cooling process and (b) the step-cooling process.[16]

8.6 MOLECULAR-BEAM EPITAXY

Molecular-beam epitaxy (MBE) is an epitaxial process involving the reaction of one or more thermal beams of atoms or molecules with a crystalline surface under ultrahigh vacuum conditions ($\sim 10^{-10}$ Torr).[§] MBE can achieve precise control in both chemical compositions and doping profiles.[17, 18] Single-crystal multilayer structures with dimensions of the order of atomic layers can be made using MBE. Because MBE uses an evaporation method, we shall consider first the basic kinetic theory of gases in a vacuum system.

8.6.1 Kinetic Theory of Gases*

The ideal gas law states that

$$PV = RT = N_{AVO}kT \tag{48}$$

where P is the pressure, V is the volume of one mole of gas, R is the gas constant (1.98 cal/mole-K, or 82 atm-cm^3/mole-K), T is the absolute temperature in K, N_{AVO} the Avogadro constant (6.02×10^{23} molecules/mole), and k is the Boltzmann constant (1.38×10^{-23} J/K, or 1.37×10^{-22} atm-cm^2/K). Since real gases behave more and more like the ideal gas as the pressure is lowered, Eq. 48 is valid for most vacuum processes. We can use Eq. 48 to calculate the molecular concentration n (the number of molecules per unit volume):

$$n = \frac{N_{AVO}}{V} = \frac{P}{kT} \tag{49}$$

$$= 9.65 \times 10^{18} \frac{P}{T} \quad \text{molecules/cm}^3 \tag{49a}$$

where P is in Torr. The density ρ_d of a gas is given by the product of its molecular weight and its concentration:

$$\rho_d = \text{molecular weight} \times \left[\frac{P}{kT}\right]. \tag{50}$$

The densities of three gases—oxygen, nitrogen, and hydrogen—are shown in Fig. 26 as a function of temperature. Also shown are the viscosities of the same three gases. The diffusivities of gases at room temperature are on the order of 1 cm^2/s and increase with absolute temperature approximately as T^2. These results have been used to evaluate the mass transfer coefficient in vapor phase epitaxy.

The gas molecules are in constant motion and their velocities are temperature dependent. The distribution of velocities is described by the Maxwell–

[§] The international unit for pressure is the Pascal (Pa); and 1 Pa $= 1$ N/m^2. However, various other units have been used. The conversion of these units are as follows:

$$1 \text{ atm} = 760 \text{ mm Hg} = 760 \text{ Torr} = 1.013 \times 10^5 \text{ Pa}$$

* See footnote to Section 2.4.2, p. 48.

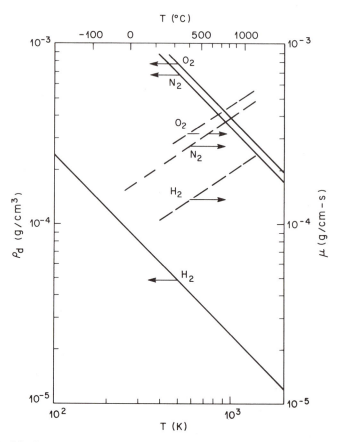

Fig. 26 Density and viscosity of H_2, N_2, and O_2 as a function of temperature.[11]

Boltzmann distribution law, which states that for a given speed v,

$$\frac{1}{n}\frac{dn}{dv} \equiv f_v = \frac{4}{\sqrt{\pi}}\left[\frac{m}{2kT}\right]^{3/2}v^2\exp\left[-\frac{mv^2}{2kT}\right] \tag{51}$$

where m is the mass of a molecule.[19] This equation states that if there are n molecules in the volume, there will be dn molecules having a speed between v and $v + dv$. The average speed is obtainable from Eq. 51:

$$v_{av} = \frac{\int_0^\infty vf_v\,dv}{\int_0^\infty f_v\,dv} = \frac{2}{\sqrt{\pi}}\sqrt{\frac{2kT}{m}}. \tag{52}$$

An important parameter for vacuum technology is the molecular *impingement rate*, that is, how many molecules impinge on a unit area per unit time. To obtain this parameter, first consider the distribution function f_{v_x} for the velocities of molecules in the x-direction. This function can be expressed by an equation similar to Eq. 51:

$$\frac{1}{n}\frac{dn_x}{dv_x} \equiv f_{v_x} = \left[\frac{m}{2\pi kT}\right]^{1/2} v_x^2 \exp\left[\frac{-mv_x^2}{2kT}\right]. \tag{53}$$

The molecular impingement rate ϕ is given by

$$\phi = \int_0^\infty v_x \, dn_x . \tag{54}$$

Substituting dn_x from Eq. 53 and integrating gives

$$\phi = n \sqrt{\frac{kT}{2\pi m}}. \tag{55}$$

The relationship between the impingement rate and the gas pressure is obtained by using Eq. 49:

$$\phi = P(2\pi mkT)^{-1/2} \tag{56}$$

$$= 3.51 \times 10^{22} \left[\frac{P}{\sqrt{MT}}\right] \tag{56a}$$

where P is the pressure in Torr and M is the molecular weight. Therefore, at 300 K and 10^{-6} Torr pressure, the impingement rate is 3.6×10^{14} molecules/cm²-s for oxygen ($M = 32$).

Problem

Find the time required to form a monolayer of oxygen at pressures of 1, 10^{-6}, and 10^{-10} Torr.

Solution

The effective molecular diameter of oxygen is 3.64 Å. Assuming close packing, the number N_s of molecules per unit area for oxygen is 8.6×10^{14} cm^{-2}. The time required to form a monolayer (assuming 100% sticking) is obtained from the impingement rate:

$$t = \frac{N_s}{\phi} = \frac{N_s \sqrt{2\pi mkT}}{P}.$$

Therefore,

$$t = 2.43 \times 10^{-6} \simeq 2 \ \mu s \quad \text{at 1 Torr}$$
$$= 2.43 \text{ s} \qquad \text{at } 10^{-6} \text{ Torr}$$
$$= 6 \text{ hr} \qquad \text{at } 10^{-10} \text{ Torr}$$

To avoid contamination of the epitaxial layer, it is of paramount importance to maintain ultrahigh-vacuum conditions ($\sim 10^{-10}$ Torr) for the MBE process.

Another important parameter is the mean free path. During their motion, the molecules suffer collisions among themselves. The average distance traversed by all the molecules between successive collisions with each other is defined as the mean free path. A molecule having a diameter d and a velocity v will move a distance $v\,\delta t$ in the time δt. The molecule suffers a collision with another molecule if its center is anywhere within the distance d of the center of another molecule. Therefore, it sweeps out (without collision) a cylinder of diameter $2d$. The volume of the cylinder is

$$\delta V = \frac{\pi}{4}(2d)^2 v\,\delta t \ . \tag{57}$$

Since there are n molecules/cm^3, the volume associated with one molecule is on the average $1/n$ cm^3. When the volume δV is equal to $1/n$, it must contain on the average one other molecule; thus, a collision has occurred. If $\tau = \delta t$ is the average time between collisions,

$$\frac{1}{n} = \pi d^2 v\,\tau \tag{58}$$

and the mean free path λ is then

$$\lambda = v\,\tau = \frac{1}{\pi n d^2} = \frac{kT}{\pi P d^2} \ . \tag{59}$$

A more rigorous derivation gives

$$\lambda = \frac{kT}{\sqrt{2}\pi P d^2} \tag{60}$$

and

$$\lambda = \frac{5 \times 10^{-3}}{P\ (\text{in Torr})} \quad \text{cm} \tag{61}$$

for air molecules (equivalent molecular diameter of 3.7 Å) at room temperature. Therefore, at a system pressure of 10^{-10} Torr, λ would be 500 km.

8.6.2 Molecular-Beam Epitaxy Process

A schematic of an MBE growth system is shown in Fig. 27 for gallium arsenide and related III–V compounds such as $Al_x Ga_{1-x} As$. Separate effusion ovens made of pyrolytic boron nitride are used for gallium, arsenic, and the dopants. All the effusion ovens are housed in an ultrahigh-vacuum chamber ($\sim 10^{-10}$ Torr). The temperature of each oven is adjusted to give the desired evaporation rate. The substrate holder rotates continuously to achieve uniform epitaxial layers (e.g., $\pm 1\%$ in doping variations and $\pm 0.5\%$ in thickness variations). The sample exchange load lock permits the maintenance of ultrahigh vacuum while changing substrates. To grow gallium arsenide, an overpressure of arsenic is maintained, since the sticking coefficient of gallium to gallium arsenide is unity while that for arsenic is zero, unless there is a previously deposited gallium layer. For a silicon MBE system, an electron gun is used to evaporate silicon. One or more effusion ovens are used for the dopants.

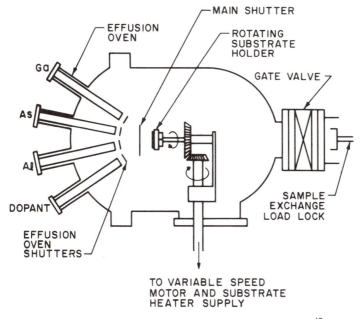

Fig. 27 Schematic of a molecular-beam epitaxial system.[17]

Problem

Assume an effusion oven geometry of area $A = 5$ cm^2 and a distance L between the top of the oven and the gallium arsenide substrate of 10 cm. Calculate the MBE growth rate for the effusion oven filled with gallium arsenide at 900°C.

Solution

Upon heating gallium arsenide, the volatile arsenide vaporizes first, leaving a gallium-rich solution. Therefore, only the pressures marked Ga-rich in Fig. 3 are of interest. The pressure at 900°C is 5.5×10^{-7} atm (4.2×10^{-4} Torr) for gallium and 1.1×10^{-5} atm (8.3×10^{-3} Torr) for arsenic (As$_2$). The arrival rate can be obtained from the impingement rate (Eq. 56a) by multiplying it by $A / \pi L^2$:

$$\text{Arrival rate} = 3.51 \times 10^{22} \left[\frac{P_{\text{Torr}}}{\sqrt{MT}} \right] \left[\frac{A}{\pi L^2} \right] \quad \text{molecules/cm}^2\text{-s .}$$

The molecular weight M is 69.72 for Ga and 74.92×2 for As$_2$. Substituting values of P, M, and T (1173 K) into the above equation gives

$$\text{Arrival rate} = 8.2 \times 10^{14}/\text{cm}^2\text{-s for Ga}$$
$$= 1.1 \times 10^{16}/\text{cm}^2\text{-s for As}_2$$

The growth rate of gallium arsenide is governed by the arrival rate of gallium. Since the surface density of gallium atoms is about 6×10^{14} cm^{-2}, and the average thickness of a monolayer is 2.8 Å, the growth rate is

$$\frac{8.2 \times 10^{14} \times 2.8}{6 \times 10^{14}} \approx 3.8 \text{ Å/s} = 230 \text{ Å/min} .$$

Note that the growth rate is relatively low compared to that of vapor phase epitaxy.

There are two ways to clean a surface in situ for MBE. High-temperature baking can decompose native oxide and remove other adsorbed species by evaporation or diffusion into the wafer. Another approach is to use a low-energy ion beam of an inert gas to sputter-clean the surface, followed by a low-temperature annealing to reorder the surface lattice structure.

MBE can use a wide variety of dopants (compared to vapor phase epitaxy), and the doping profile can be exactly controlled. However, the doping process is similar to the vapor phase growth process: a flux of evaporated dopant atoms arrives at a favorable lattice site and is incorporated along the growing interface. Fine control of the doping profile is achieved by adjusting the dopant flux relative to the flux of silicon atoms (for silicon epitaxial films) or gallium atoms (for gallium arsenide epitaxial films). It is also possible to dope the epitaxial film using a low-current, low-energy ion beam to implant the dopant (see Chapter 10).

The substrate temperatures for MBE range from 400 to 900°C; and the growth rates range from 0.001 to 0.3 μm/min. Because of the low-temperature process and low-growth rate, many unique doping profiles and alloy compositions not obtainable from conventional VPE and LPE can be produced in MBE. Many novel structures have been made using MBE. These include the *superlattice*, which is a periodic structure consisting of alternating ultrathin layers with its period less than the electron mean free path (e.g., GaAs/$Al_x Ga_{1-x}$As, with each layer 100 Å or less in thickness), and the heterojunction MESFET (as discussed in Chapter 5).

REFERENCES

1 For a recent review on crystal growth and epitaxy of silicon, see C. W. Pearce, "Crystal Growth and Wafer Preparation" and "Epitaxy," in S. M. Sze, Ed., *VLSI Technology*, McGraw-Hill, New York, 1983.

2 W. R. Runyan, *Silicon Semiconductor Technology*, McGraw-Hill, New York, 1965.

3 M. Hansen, *Constitution of Binary Alloys*, McGraw-Hill, New York, 1958.

4 For a recent review on crystal growth and epitaxy of silicon and gallium arsenide, see S. K. Ghandhi, *VLSI Fabrication Principles*, Wiley, New York, 1983.

5 J. R. Arthur, "Vapor Pressures and Phase Equilibria in the GaAs System," *J. Phys. Chem. Solids*, **28**, 2257 (1967).

6 W. G. Pfann, *Zone Melting*, 2nd ed., Wiley, New York, 1966.

7 E. W. Hass and M. S. Schnoller, "Phosphorus Doping of Silicon by Means of Neutron Irradiation," *IEEE Trans. Electron Devices*, **ED-23**, 803 (1976).

8   C. A. Wert and R. M. Thomson, *Physics of Solids*, McGraw-Hill, New York, 1964.

9   F. A. Trumbore, "Solid Solubilities of Impurity Elements in Germanium and Silicon," *Bell Syst. Tech. J.*, **39**, 205 (1960).

10 J. A. Keenar and G. B. Larrabee, "Characterization of Silicon Materials for VLSI," in N. G. Einspruch and C. B. Larrabee, Eds., *VLSI Electronics*, Vol. 6, Academic, New York, 1983.

11 A. S. Grove, *Physics and Technology of Semiconductor Devices*, Wiley, New York, 1967.

12 R. Reif, T. I. Kamins, and K. C. Saraswat, "A Model for Dopant Incorporation into Growing Silicon Epitaxial Films," *J. Electrochem. Soc.*, **126**, 644 and 653 (1979).

13 F. C. Eversteyn, "Chemical-Reaction Engineering in the Semiconductor Industry," *Philips Res. Rep.*, **29**, 45 (1974).

14 B. Y. Tsuar et al., "Improved Techniques for Growth of Large Area Single Crystal Si Sheets over SiO₂ Using Lateral Epitaxy by Seeded Solidification," *Appl. Phys. Lett.*, **39**, 561 (1981).

15 W. D. Johnston, *Solar Voltaic Cells*, Marcel Dekker, New York, 1980.

16 J. J. Hsieh, "Liquid-Phase Epitaxy," In S. P. Keller, Ed., *Handbook on Semiconductors*, Vol. 3, North-Holland, Amsterdam, 1980.

17 A. Y. Cho, "Growth of III–V Semiconductors by Molecular Beam Epitaxy and Their Properties," *Thin Solid Films*, **100**, 291 (1983).

18 L. L. Chang, "Molecular Beam Epitaxy," in S. P. Keller, Ed., *Handbook on Semiconductors*, Vol. 3, North-Holland, Amsterdam, 1980.

19 A. Roth, *Vacuum Technology*, North-Holland, Amsterdam, 1976.

PROBLEMS

1   Plot the doping distribution of arsenic at distances of 10, 20, 30, 40 and 45 cm from the seed in a 50-cm-long silicon ingot that has been pulled from a melt with an initial doping concentration of 10^{17} cm^{-3}.

2   A boron-doped ingot is required to have a resistivity of 10.0 Ω-cm when one half of the ingot is grown. Assuming that a 10-kg pure silicon charge is used, what is the amount of boron-doped silicon having a resistivity of 0.01 Ω-cm that must be added.

3   The seed crystal used in the Czochralski process is usually necked down to a small diameter (3 mm) as a means to initiate dislocation-free growth. If the critical yield strength of silicon is 2×10^6 g/cm^2, calculate the maximum length of a 125-mm-diameter silicon ingot that can be supported by such a seed.

4   We use the float zone process to purify a silicon ingot that contains a uniform gallium concentration of 5×10^{16} cm^{-3}. One pass is made with a molten zone 2 cm long. Over what distance is the resulting gallium concentration below 5×10^{15} cm^{-3}?

5   If $p^{+} - n$ abrupt-junction diodes are fabricated using the silicon materials shown in Fig. 11, find the ranges of breakdown voltages for the conventionally doped silicon and the neutron-irradiated silicon. (Note that the resistivity variations are $\pm$ 15% for the conventionally doped silicon and $\pm$ 1% for neutron doping.)

6   The equilibrium density of vacancy n_s (also called Schottky defect) is given by $N \exp\left(- E_s /kT\right)$, where N is the density of semiconductor atoms and E_s is the energy of formation. Calculate n_s in silicon at 27, 900, and 1200°C. (Assume $E_s = 2.3$ eV.)

7   Calculate the growth rate of a silicon epitaxy layer grown from a silicon tetrachloride source at 1200°C. The vapor phase mass transfer coefficient of the reactor is $h_g = 5$ cm/s, the surface reaction rate coefficient is $k_s = 10^7 \exp\left(- 1.9 \text{ eV}/kT\right)$ cm/s, and $C_g = 3 \times 10^{16} \text{ cm}^{-3}$. What will be the change in growth rate if the reaction temperature is increased by 1%?

8   Find the average molecular velocity of air at 300 K (the molecular weight for air is 29).

9   Find the number of atoms per unit area, N_s, needed to form a monolayer under close-packing condition (i.e., each atom is in contact with its six neighboring atoms), assuming the diameter d of the atom is 4.68 Å.

10   Assume an effusion oven geometry of $A = 5 \text{ cm}^2$ and $L = 12$ cm. (a) Calculate the arrival rate of gallium and the MBE growth rate for the effusion oven filled with gallium arsenide at 970°C. (b) To maintain the growth rate, what mole fraction of arsenic would be required in a liquid composition of gallium and arsenic? (c) For a tin effusion oven operated at 700°C under the same geometry, calculate the doping concentration (assuming tin atoms are fully incorporated in the gallium arsenide grown at the aforementioned rate. The molecular weight for tin is 118.69, and the pressure at 700°C for tin is 2×10^{-8} Torr).

9

Oxidation and Film Deposition

To fabricate discrete devices and integrated circuits we use many different kinds of thin film. We can classify thin films into four groups: thermal oxides, dielectric layers, polycrystalline silicon, and metal films. Figure 1 shows a schematic view of a conventional silicon *n*-channel MOSFET that uses all four groups of film. The first important thin film from the thermal oxide group is the *gate oxide* layer under which a conducting channel can be formed between the source and the drain. A related layer is the *field oxide*, which provides isolation from other device structures. Both gate and field oxides generally are grown by a thermal oxidation process because only thermal oxidation can provide the highest-quality oxides having the lowest interface trap densities.

Dielectric layers such as the deposited silicon dioxide and silicon nitride are used for insulation between conducting layers, for diffusion and ion implantation masks, for capping doped films to prevent the loss of dopants, and for passivation to protect devices from impurities, moisture, and scratches. Polycrystalline silicon, usually referred to as polysilicon, is used as gate electrode material in MOS devices, as a conductive material for multilevel metallization, and as a contact material for devices with shallow junctions. Metal films such as aluminum and silicides are used to form low-resistance interconnections, ohmic contacts to n^+-, p^+-, and polysilicon layers, and rectifying metal–semiconductor barriers.

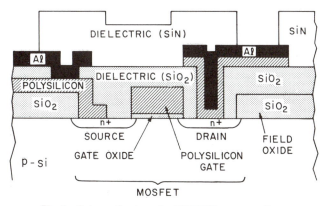

Fig. 1 Schematic view of a MOSFET cross section.

When a film is formed (e.g., by oxidation or chemical vapor deposition), device features generally are defined by lithographic and etching processes (refer to Chapter 11). Each film must both perform its intended function and be compatible with the overall processing sequence, that is, the film must withstand the required chemical treatment and thermal cycle while its structure remains stable. In this chapter we consider the formation and characteristics of these films.

9.1 THERMAL OXIDATION

Semiconductors can be oxidized by various methods. These include thermal oxidation, electrochemical anodization, and plasma reaction.[1] Among these methods thermal oxidation is by far the most important for silicon devices. It is the key process in modern silicon integrated-circuit technology. For gallium arsenide, however, thermal oxidation results in generally non-stoichiometric films containing oxides of gallium and arsenic as well as arsenic ions. The oxides provide poor electrical insulation and semiconductor surface protection; hence, these oxides are rarely used in gallium arsenide technology. Consequently, in this section we shall concentrate on thermal oxidation of silicon.[2]

The basic thermal oxidation setup is shown[3] in Fig. 2. The reactor consists of a resistance-heated furnace, a cylindrical fused-quartz tube containing the silicon wafers held vertically in a slotted quartz boat, and a source of either pure, dry oxygen or pure water vapor. The loading end of the furnace tube protrudes into a vertical flow hood where a filtered flow of air is maintained. Flow is directed as shown by the arrow in Fig. 2. The hood reduces dust and particulate matter in the air surrounding the wafers and minimizes contamination during wafer loading. The oxidation temperature is generally in the range of 900 to 1200°C, and the typical gas flow rate is about 1 cm/s. The oxidation

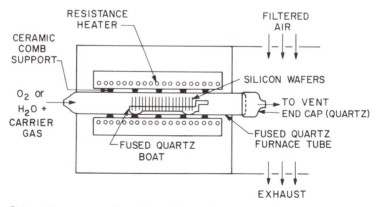

Fig. 2 Schematic cross section of a resistance-heated oxidation furnace. The silicon wafer loading area is shown in a laminar hood.[3]

system uses microprocessors to regulate the gas flow sequence, to control the automatic insertion and removal of silicon wafers, to ramp the temperature up (i.e., to increase the furnace temperature linearly from a low temperature to the oxidation temperature so that the wafers will not warp due to sudden temperature change), to maintain the oxidation temperature to within $\pm 1°C$, and to ramp the temperature down when oxidation is completed.

9.1.1 Kinetics of Growth

The following chemical reactions describe the thermal oxidation of silicon in oxygen or water vapor:

$$\text{Si(solid)} + \text{O}_2(\text{gas}) \longrightarrow \text{SiO}_2(\text{solid}) \tag{1}$$

$$\text{Si(solid)} + 2\text{H}_2\text{O(gas)} \longrightarrow \text{SiO}_2(\text{solid}) + 2\text{H}_2(\text{gas}) . \tag{2}$$

The silicon–silicon dioxide interface moves into the silicon during the oxidation process. This creates a fresh interface region, with surface contamination on the original silicon ending up on the oxide surface. The densities and molecular weights of silicon and silicon dioxide are used in the following example to show that growing an oxide of thickness x consumes a layer of silicon $0.44x$ thick (Fig. 3).

Problem
If a silicon oxide layer of thickness x is grown from thermal oxidation, what is the thickness of silicon being consumed?

Solution
The volume of 1 mole silicon is

$$\frac{\text{Molecular weight of Si}}{\text{Density of Si}} = \frac{28.09 \ \text{g/mole}}{2.33 \ \text{g/cm}^3} = 12.06 \ \text{cm}^3/\text{mole} .$$

The volume of 1 mole silicon dioxide is

$$\frac{\text{Molecular weight of SiO}_2}{\text{Density of SiO}_2} = \frac{60.08 \ \text{g/mole}}{2.21 \ \text{g/cm}^3} = 27.18 \ \text{cm}^3/\text{mole} .$$

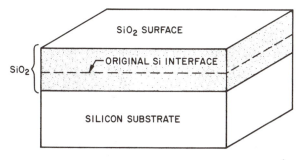

Fig. 3 Growth of silicon dioxide by thermal oxidation.[2]

Since 1 mole silicon is converted to 1 mole silicon dioxide,

$$\frac{\text{Thickness of Si} \times \text{area}}{\text{Thickness of SiO}_2 \times \text{area}} = \frac{\text{volume of 1 mole of Si}}{\text{volume of 1 mole of SiO}_2}$$

$$\frac{\text{Thickness of Si}}{\text{Thickness of SiO}_2} = \frac{12.06}{27.18} = 0.44$$

Thickness of silicon = 0.44(thickness of SiO$_2$).

That is, to grow 1000 Å of silicon dioxide, a layer of 440 Å of silicon is consumed.

The basic structural unit of thermally grown silicon dioxide is a silicon ion surrounded tetrahedrally by four oxygen ions, as illustrated[3] in Fig. 4a. The silicon-to-oxygen internuclear distance is 1.6 Å, and the oxygen-to-oxygen internuclear distance is 2.27 Å. These tetrahedra are joined together at their corners by oxygen bridges in a variety of ways to form the various phases or structures of silicon dioxide (also called silica). Silica has several crystalline structures (e.g., quartz) and an amorphous structure. When silicon is thermally oxidized, the silicon dioxide structure is amorphous. Typically amorphous silica has a density of 2.21 g/cm^3, as compared to 2.65 g/cm^3 for quartz.

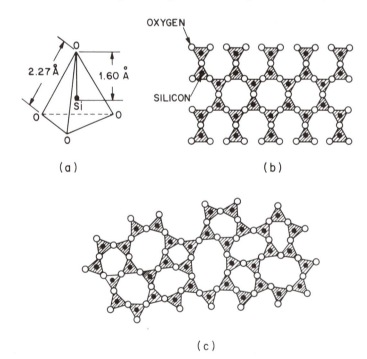

Fig. 4 (a) Basic structural unit of silicon dioxide. (b) Two dimensional representation of a quartz crystal lattice. (c) Two dimensional representation of the amorphous structure of silicon dioxide.[3]

The basic difference between the crystalline and amorphous structures is that the former is a periodic structure, extending over many molecules, while the latter has no periodic structure at all. Figure 4*b* is a two-dimensional schematic diagram of a quartz crystalline structure made up of rings with six silicon atoms. Figure 4*c* is a two-dimensional schematic diagram of an amorphous structure for comparison. In the amorphous structure there is still a tendency to form characteristic rings with six silicon atoms. Note that the amorphous structure in Fig. 4*c* is quite open because only 43% of the space is occupied by silicon dioxide molecules. The relatively open structure accounts for the lower density and allows a variety of impurities (such as sodium) to enter and diffuse readily through the silicon dioxide layer.

The kinetics of thermal oxidation of silicon can be studied based on a simple model illustrated[4] in Fig. 5. A silicon slice contacts the oxidizing species (oxygen or water vapor), resulting in a surface concentration of C_0 molecules/cm³ for this species. The gas phase mass transfer coefficient (described in Chapter 8) is very high, so the magnitude of C_0 equals the equilibrium bulk concentration of the species at the oxidation temperature. The equilibrium concentration generally is proportional to the partial pressure of the oxidant adjacent to the oxide surface. At 1000°C and at a pressure of 1 atm, the concentration C_0 is 5.2×10^{16} molecules/cm³ for dry oxygen and 3×10^{19} molecules/cm³ for water vapor.

The oxidizing species diffuses through the silicon dioxide layer, resulting in a concentration C_s at the surface of silicon. The flux F_1 can be written as

$$F_1 = D \frac{dC}{dx} \simeq \frac{D(C_0 - C_s)}{x} \tag{3}$$

where D is the diffusion coefficient of the oxidizing species and x is the thickness of the oxide layer already present.

At the silicon surface, the oxidizing species reacts chemically with silicon. Assuming the rate of reaction is proportional to the concentration of the

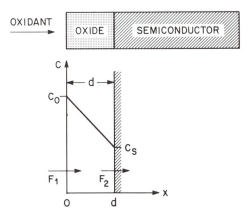

Fig. 5 Basic model for the thermal oxidation of silicon.[4]

species at the silicon surface, the flux F_2 is given by

$$F_2 = kC_s \tag{4}$$

where k is the surface reaction rate constant for oxidation. At the steady state, $F_1 = F_2 = F$. Combining Eqs. 3 and 4 gives

$$F = \frac{DC_0}{x + (D/k)} . \tag{5}$$

The reaction of the oxidizing species with silicon forms silicon dioxide. Let C_1 be the number of molecules of the oxidizing species in a unit volume of the oxide. There are 2.2×10^{22} silicon dioxide molecules/cm^3 in the oxide, and we add one oxygen molecule (O_2) to each silicon dioxide molecule while we add two water molecules (H_2O) to each silicon dioxide molecule. Therefore, C_1 for oxidation in dry oxygen is 2.2×10^{22} cm^{-3}, and for oxidation in water vapor it is twice this number. Thus, the growth rate of the oxide layer thickness is given by

$$\frac{dx}{dt} = \frac{F}{C_1} = \frac{DC_0/C_1}{x + (D/k)} . \tag{6}$$

We can solve this differential equation subject to the initial condition, $x(0) = d_o$, where d_o is the initial oxide thickness; d_o can also be regarded as the thickness of oxide layer grown in an earlier oxidation step. Solving Eq. 6 yields the general relationship for the oxidation of silicon

$$x^2 + \frac{2D}{k} x = \frac{2DC_0}{C_1} (t + \tau) \tag{7}$$

where $\tau \equiv (d_o^2 + 2Dd_o/k)C_1/2DC_0$, which represents a time coordinate shift to account for the initial oxide layer d_o.

The oxide thickness after an oxidizing time t is given by

$$x = \frac{D}{k} \left[\sqrt{1 + \frac{2C_0 k^2(t + \tau)}{DC_1}} - 1 \right] . \tag{8}$$

For small values of t, Eq. 8 reduces to

$$x \simeq \frac{C_0 k}{C_1} (t + \tau) \tag{9}$$

and for larger values of t, it reduces to

$$x \simeq \sqrt{\frac{2DC_0}{C_1} (t + \tau)} . \tag{10}$$

Thus, during the early stages of oxide growth, when surface reaction is the rate-limiting factor, the oxide thickness varies linearly with time. As the oxide layer becomes thicker, the oxidant must diffuse through the oxide layer to react at the silicon–silicon dioxide interface and the reaction becomes diffusion limited. The oxide growth then becomes proportional to the square root of the oxidizing time, which results in a parabolic growth rate.

Equation 7 is often written in a more compact form:

$$x^2 + Ax = B(t + \tau) \tag{11}$$

where $A \equiv 2D/k$, $B \equiv 2DC_0/C_1$ and $B/A \equiv kC_0/C_1$. Using this form, Eqs. 9 and 10 can be written as

$$x = \frac{B}{A}(t + \tau) \tag{12}$$

for the linear region and as

$$x^2 = B(t + \tau) \tag{13}$$

for the parabolic region. For this reason, the term B/A is referred to as the linear rate constant and B as the parabolic rate constant. Experimentally measured results agree with the predictions of this model over a wide range of oxidation conditions. For wet oxidation, the initial oxide thickness d_o is very small, or $\tau \simeq 0$. However, for dry oxidation, the extrapolated value of d_o at $t = 0$ is about 200 Å.

The temperature dependence of the linear rate constant B/A is shown in Fig. 6 for both dry and wet oxidation and for (111)- and (100)-oriented silicon wafers.[4] The linear rate constant varies as $\exp(-E_a/kT)$, where the activation energy E_a is about 2 eV for both dry and wet oxidations. This closely agrees with the energy required to break silicon–silicon bonds, that is, 1.83 eV/molecule. Under a given oxidation condition, the linear rate constant depends on crystal orientation. This is because the rate constant is related to the rate of incorporation of silicon atoms into the silicon dioxide network. This rate depends on the surface density of silicon atoms, making it orientation dependent. Because the density of silicon atoms on the (111)-plane is larger than that on the (100)-plane, the linear rate constant for (111)-silicon is larger.

Figure 7 shows the temperature dependence of the parabolic rate constant B, which can also be described by $\exp(-E_a/kT)$. The activation energy E_a is 1.24 eV for dry oxidation. The comparable activation energy for oxygen diffusion in fused silica is 1.18 eV. The corresponding value for wet oxidation, 0.71 eV, compares favorably with the value of 0.79 eV for the activation energy of diffusion of water in fused silica. The parabolic rate constant is independent of crystal orientation. This independence is expected, because it is a measure of the diffusion process of the oxidizing species through a random network layer of amorphous silica.

While oxides grown in dry oxygen have the best electrical properties, considerably more time is required to grow the same oxide thickness at a given temperature in dry oxygen than in water vapor. For relatively thin oxides such as the gate oxide in a MOSFET (typically $\lesssim$ 1000 Å), dry oxidation is

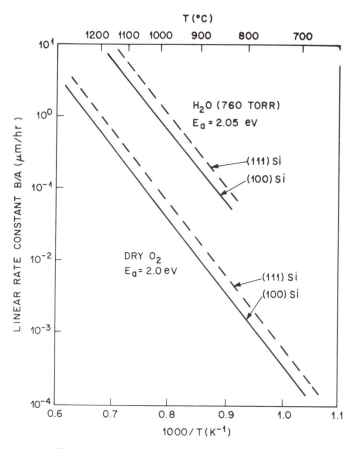

Fig. 6 Linear rate constant versus temperature.[4]

used. However, for thicker oxides such as field oxides ($\gtrsim$ 5000 Å) in MOS integrated circuits and for bipolar devices, oxidation in water vapor (or steam) is used to provide both adequate isolation and passivation. Figure 8 shows the experimental results of silicon dioxide thickness as a function of reaction time and temperature for two substrate orientations.[5] Under a given oxidation condition, the oxide thickness grown on a (111)-substrate is larger than that grown on a (100)-substrate because of the larger linear rate constant of the (111)-orientation. Note that for a given temperature and time, the oxide film obtained using wet oxidation is about 5 to 10 times thicker than that using dry oxidation.

9.1.2 Thin-Oxide Growth

We noted earlier that for dry oxidation, there is an apparently rapid oxidation that gives rise to an initial oxide thickness d_o of about 200 Å. Therefore, the simple model presented in Section 9.1.1 is not valid for dry oxidation with

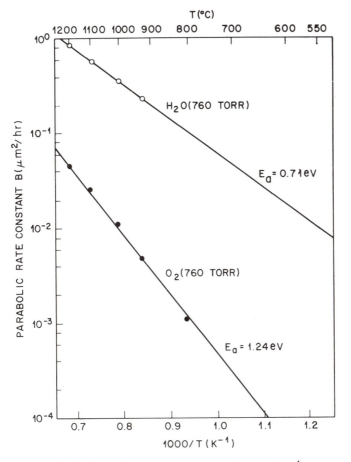

Fig. 7 Parabolic rate constant versus temperature.[4]

an oxide thickness $\lesssim 200$ Å. For very large-scale integration, the ability to grow thin ($100\sim300$ Å), uniform, high-quality reproducible gate oxides has become increasingly important. We shall briefly consider the growth mechanisms of such thin oxides.

In the early stage of growth in dry oxidation, there is a large compressive stress in the oxide layer. Such stress reduces the oxygen diffusion coefficient in the oxide. As the oxide becomes thicker, the stress will be reduced due to the viscous flow of silica and the diffusion coefficient will approach its stress-free value. Therefore, for thin oxides, the value of D/k may be sufficiently small that we can neglect the term Ax in Eq. 11 and obtain

$$x^2 - d_o^2 = Bt \qquad (14)$$

where d_o is equal to $\sqrt{2DC_0\tau/C_1}$, which is the initial oxide thickness when time is extrapolated to zero, and B is the parabolic rate constant defined previ-

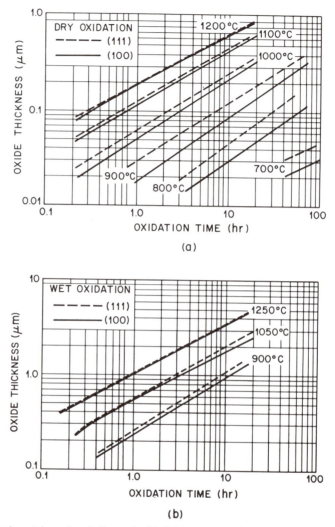

Fig. 8 Experimental results of silicon dioxide thickness as a function of reaction time and temperature for two substrate orientations. (a) Dry oxygen growth. (b) Steam growth.[5]

ously.[6] We therefore expect the initial growth in dry oxidation to follow a parabolic form.

Figure 9 shows a plot of oxide thickness as a function of oxidation time for different temperatures and oxygen partial pressure.[7] The solid lines are the parabolic oxidation equations. The value of d_o at time zero is found to be 27 Å. This initial oxide thickness is reasonable, since even after a cleaning in hydrofluoric acid the silicon surface remains covered with a few monolayers of oxide, and upon exposure to room temperature air the oxide will grow gradually to a thickness of about 30∼50 Å. The parabolic rate constant B is proportional to the product of D and C_0; C_0 in turn is proportional to the partial

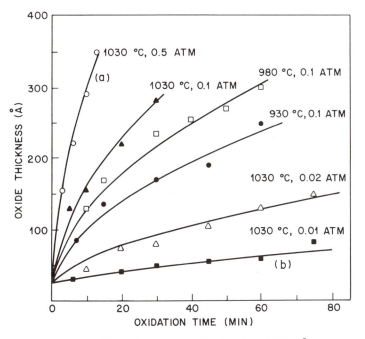

Fig. 9 Oxide thickness versus time for dry oxidation.[7]

pressure P of the oxidizing species in the gas phase. Therefore, the ratio B/P, which is proportional to the diffusion coefficient D, is expected to be smaller for smaller oxide thicknesses. This is indeed the case. For example, at 1030°C the value of B/P for thicker oxides is 1.5×10^{-2} $\mu m^2/hr$-atm (from Fig. 7). For thin oxides, B/P is 0.7×10^{-2} $\mu m^2/hr$-atm for curve a in Fig. 9 and becomes 0.22×10^{-2} $\mu m^2/hr$-atm for curve b.

9.1.3 High-Pressure Oxidation

Oxidation in high-pressure steam can produce substantial acceleration in the growth rate because the parabolic rate constant B is proportional to the partial pressure of the oxidizing species in the gas phase. High-pressure oxidation for growing thick oxides such as field oxide offers the advantages of low-temperature processing at growth rates comparable to high-temperature, 1-atm conditions. The advantages include minimized movement of previously diffused or implanted impurities and minimized lateral diffusion; these advantages are of particular importance as device dimensions decrease.

Figure 10 is a graph of oxide thickness as a function of steam pressure at two temperatures for 1-hr.[2] The substantial acceleration in the oxidation rate caused by the increased pressure can be seen in the graph. A linear–parabolic model similar to Eq. 11 is used to analyze the data in Fig. 10 as well as related data at different times and temperatures. From these analyses we observe a linear pressure dependence for both the linear and the parabolic rate constants.

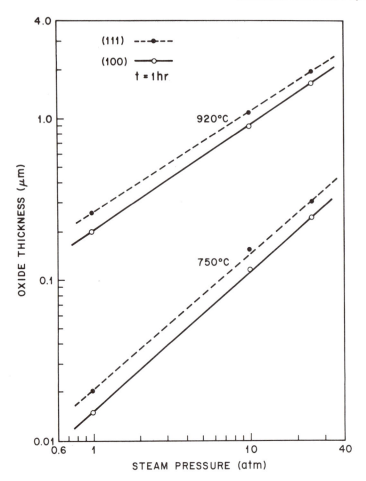

Fig. 10 Oxide thickness as a function of steam pressure and temperature.[2]

9.1.4 Effects of Impurities

Both linear and parabolic rate constants are sensitive to impurities either in the oxidizing gas (such as water and chlorine) or in the silicon substrate (such as boron and phosphorus). Water has already been discussed as an oxidant. Now, we shall treat it as an impurity in the oxygen gas for the dry-oxidation process. It has been found that adding 15 ppm water vapor to the oxygen gas significantly increases both oxidation rate constants. This is expected because both the linear and parabolic rate constants are significantly higher in water vapor than in oxygen as shown previously in Figs. 6 and 7. However, during dry oxidation, a precombustor and cold trap can be used to minimize the water vapor level to less than 1 ppm.

In modern integrated circuit processing, chlorine is introduced into the oxidation ambient to improve both the oxide quality and the Si–SiO$_2$ interface property. Chlorine, contained either in chlorine gas or in such compounds as anhydrous HCl and trichloroethylene, removes certain impurities at the Si–SiO$_2$ interface by changing them into volatile chlorides; for example, sodium ions are removed by the formation of volatile NaCl. The addition of chlorine also increases the dielectric breakdown strength and reduces the interface trap density.

Experimental results for dry O$_2$–HCl mixtures shows that adding HCl increases the oxidation rate. Figure 11 shows the rate constants as a function of HCl concentration.[8] Although the linear rate constant B/A initially increases when 1% HCl is added, there is no further increase with subsequent HCl additions. The parabolic rate constant B increases linearly with HCl additions above 1%. At 1000 and 1100°C, large increases in B are initially observed. Typical HCl additions to the dry oxidation process range from 1 to 5%. Care must be taken while handling and using halogens such as HCl, since they can corrode the metal parts of the oxidation reactor. In addition, high concentrations of halogens at high temperatures can pit the silicon surface.

When present at high concentration levels, the common dopant elements for silicon such as boron and phosphorus can enhance oxidation behavior. The dopant impurities will be redistributed at the growing silicon–silicon dioxide interface. The impurities will segregate either into the silicon or into the oxide, depending on the segregation coefficient. For boron, the dopant segregates into the oxide and remains there, weakening the silicon dioxide bond structure. This weakened structure enhances diffusion of the oxidizing species through the oxide and thus increases the oxidation rate. This in turn enhances the parabolic rate constant for boron-doped silicon.

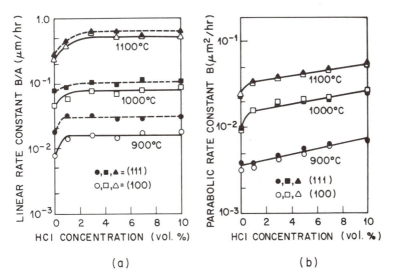

Fig. 11 Linear and parabolic rate constants versus % HCl concentration for (111)- and (100)-oriented silicon at different temperatures.[8]

For oxidation of phosphorus-doped silicon, we observe a concentration dependence only at lower temperatures ($\leqslant 1000°$ C) where the surface reaction is important. This dependence may result from the segregation of phosphorus into silicon. The oxidation rate constants for dry oxygen as a function of the phosphorus doping level are plotted[9] in Fig. 12. At high phosphorus concentrations, B/A becomes significantly larger, since the reaction rate is influenced by the phosphorus concentration at the silicon–silicon dioxide interface. B is relatively independent of concentration since the phosphorus concentration in the oxide is low and the diffusion constant of the oxidant in the oxide remains essentially the same. This type of enhanced oxidation may present a problem with devices containing $n^+–p$ or $p^+–n$ lateral junctions where the oxide growth rate of the heavily doped regions can be many times faster than the oxide growth rate of neighboring low-doping regions. These rate differences can create large steps in the oxide which in turn may cause breaks in the metal interconnections deposited over them.

9.2   DIELECTRIC DEPOSITION

Deposited dielectric films are used mainly for insulation and passivation of discrete devices and integrated circuits. There are three commonly used deposition methods: atmospheric-pressure chemical vapor deposition (CVD), low-pressure chemical vapor deposition (LPCVD), and plasma-assisted chemical vapor deposition (PCVD, or plasma deposition). PCVD is an energy-enhanced CVD method, because plasma energy is added to the thermal energy of a conventional CVD system. Considerations in selecting a deposi-

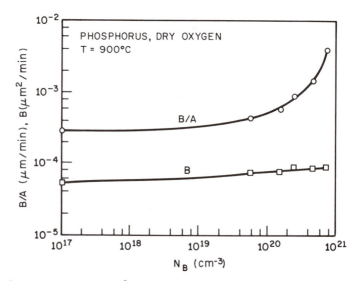

Fig. 12   Oxidation rate constants[9] for dry oxygen as a function of the phosphorus doping level at 900° C.

tion process are the substrate temperature, the deposition rate and film uniformity, the morphology, the electrical and mechanical properties, and the chemical composition of the dielectric films.

The reactor for atmospheric-pressure CVD is similar to the one shown in Fig. 2, except that different gases are used at the gas inlet. In a hot-wall, reduced-pressure reactor as shown in Fig. 13a, the quartz tube is heated by a three-zone furnace, and gas is introduced at one end and pumped out at the opposite end. The semiconductor wafers are held vertically in a slotted quartz boat.[10] The quartz tube wall is hot because it is adjacent to the furnace, in contrast to a cold-wall reactor such as the horizontal epitaxial reactor that uses rf heating. Typical reaction chamber process parameters are as follows: pressure varies from 30 to 250 Pa (0.25 to 2.0 Torr), gas flows at rates of 1 to 10 cm/s, and temperatures are normally 300 to 900°C. Among the benefits of this reactor are that it deposits films with excellent uniformity and that its large batch size allows processing of several hundred wafers each run. However, the deposition process is slow and the gases used may be toxic, corrosive, or flammable.

The parallel-plate, radial-flow, plasma-assisted CVD reactor shown in Fig. 13b consists of a cylindrical glass or aluminum chamber sealed with

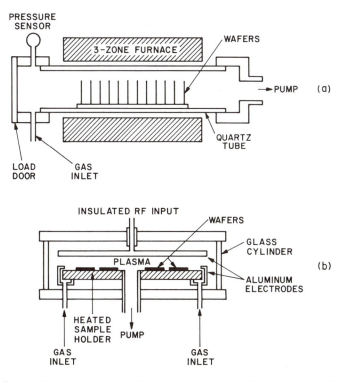

Fig. 13 Schematic diagrams of chemical-vapor deposition reactors. (a) Hot-wall, reduced-pressure reactor. (b) Parallel-plate plasma deposition reactor.[10]

aluminum endplates. Inside are two parallel aluminum electrodes. An rf voltage is applied to the upper electrode, while the lower electrode is grounded. The radio-frequency voltage causes a plasma discharge between the electrodes. Wafers are placed on the lower electrode, which is heated between 100 and 400°C by resistance heaters. Process gas flows through the discharge from outlets located along the circumference of the lower electrode. The main advantage of this reactor is its low deposition temperature. However, its capacity is limited, especially for large-diameter wafers, and the wafers may become contaminated if loosely adhering deposits fall on them.

Low-temperature depositions will become more important as device dimensions are reduced to submicron regions because we must minimize thermal broadening of dopant distribution due to diffusion (refer to Chapter 10). Depositions at very low temperatures (25 to 300°C) have been investigated. Most of the methods use *energy-enhanced CVD techniques*. Figure 14a shows a CVD setup which uses a focused electron beam to deposit dielectric films.[11] Deposition occurs only beneath the regions defined by the focused electron beam. Other energy sources such as a laser beam or an ion beam can also be used. The gases in the reaction chamber are decomposed locally by the focused energy source, and localized film deposition occurs by a pyrolysis or photolysis process. The focused energy sources are also potentially useful to deposit metal films in selected areas to repair integrated-circuit chips. Figure 14b shows another energy-enhanced CVD method. UV radiation is used to form vapor phase reactants that enhance the deposition rates.[12] Silicon dioxide films have been deposited at a rate of 150 Å/min at temperatures as low as 50°C by means of the UV radiation method.

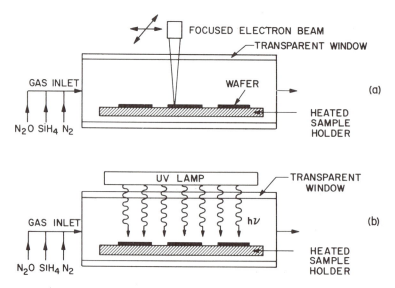

Fig. 14 Energy-enhanced CVD methods (a) using a focused energy source and (b) using a UV lamp.[12]

9.2.1 Silicon Dioxide

Chemical-vapor-deposited silicon dioxide does not replace thermally grown oxides because the best electrical properties are obtained with thermally grown films. CVD oxides are used instead to complement the thermal oxides. A layer of undoped silicon dioxide is used to insulate multilevel metallization, to mask ion implantation and diffusion, and to increase the thickness of thermally grown field oxides. Phosphorus-doped silicon dioxide is used both as an insulator between metal layers and as a final passivation layer over devices. Oxides doped with phosphorus, arsenic, or boron are used occasionally as diffusion sources.

Deposition Methods Silicon dioxide films can be deposited by several methods. For low-temperature deposition (300 to 500°C), the films are formed by reacting silane, dopant, and oxygen. The chemical reactions for phosphorus-doped oxides are

$$SiH_4 + O_2 \xrightarrow{\;450°C\;} SiO_2 + 2H_2 \tag{15}$$

$$4PH_3 + 5O_2 \xrightarrow{\;450°C\;} 2P_2O_5 + 6H_2 . \tag{16}$$

The deposition process can be performed either at atmospheric pressure in a CVD reactor or at reduced pressure in an LPCVD reactor (Fig. 13a). The low deposition temperature of the silane–oxygen reaction makes it a suitable process when films must be deposited over a layer of aluminum.

For intermediate-temperature deposition (500 to 800°C), silicon dioxide can be formed by decomposing tetraethylorthosilicate, $Si(OC_2H_5)_4$, in an LPCVD reactor. The compound, abbreviated TEOS, is vaporized from a liquid source. The TEOS compound decomposes as follows:

$$Si(OC_2H_5)_4 \xrightarrow{\;700°C\;} SiO_2 + \text{by-products} \tag{17}$$

forming both SiO_2 and a mixture of organic and organosilicon by-products. While the high temperature required for the reaction prevents its use over aluminum, it is suitable for polysilicon gates requiring a uniform insulating layer with good step coverage. The oxides can be doped by adding small amounts of the dopant hydrides (phosphines, arsine, or diborane) similar to the process in epitaxial growth.

The deposition rate as a function of temperature varies as $e^{-E_a/kT}$, where E_a is the activation energy. The E_a of the silane–oxygen reaction is quite low: about 0.6 eV for undoped oxides and almost zero for phosphorus-doped oxide. In contrast, E_a for the TEOS reaction is much higher: about 1.9 eV for undoped oxide and 1.4 eV when phosphorus doping compounds are present. Dependence of the deposition rate on TEOS partial pressure is proportional to $(1 - e^{-P/P_o})$, where P is the TEOS partial pressure and P_o is about 30 Pa. At low TEOS partial pressures the deposition rate is determined by the rate of the surface reaction. At high partial pressures, the surface becomes nearly saturated with adsorbed TEOS and the deposition rate becomes essentially independent of TEOS pressure.[10]

For high-temperature deposition (900°C), silicon dioxide is formed by reacting dichlorosilane, $SiCl_2H_2$, with nitrous oxide at reduced pressure:

$$SiCl_2H_2 + 2N_2O \xrightarrow{900°C} SiO_2 + 2N_2 + 2HCl. \tag{18}$$

This deposition gives excellent film uniformity and is sometimes used to deposit insulating layers over polysilicon.

Properties of Silicon Dioxide Deposition methods and properties of silicon dioxide films are listed[10] in Table 1. In general, there is a direct correlation between deposition temperature and film quality. At higher temperatures, deposited oxide films are structurally similar to silicon dioxide that has been thermally grown.

The lower densities occur in films deposited below 500°C. Heating deposited silicon dioxide at temperatures between 600 and 1000°C causes densification, during which the oxide thickness decreases while the density increases to 2.2 g/cm^3. The refractive index of silicon dioxide is 1.46 at a wavelength of 0.6328 μm. Oxides with lower indices are porous, such as the oxide from the silane–oxygen deposition, which has a refractive index of 1.44. The porous nature of the oxide also is responsible for the lower dielectric strength, which is the applied electric field that will cause large amounts of current to flow in the oxide film. The etch rates of oxides in a hydrofluoric acid solution depend on deposition temperature, annealing history, and dopant concentration. Usually higher-quality oxides are etched at lower rates.

Step Coverage Step coverage relates the surface topography of a deposited film to the various steps on the semiconductor substrate. In the illustration of ideal, or conformal, step coverage shown in Fig. 15a, film thicknesses are

Table 1 Properties of Silicon Dioxide

Property	Method			
	Thermally Grown at 1000°C	$SiH_4 + O_2$ at 450°C	TEOS at 700°C	$SiCl_2H_2 + N_2O$ at 900°C
Composition	SiO_2	$SiO_2(H)$	SiO_2	$SiO_2(Cl)$
Density (g/cm^3)	2.2	2.1	2.2	2.2
Refractive index	1.46	1.44	1.46	1.46
Dielectric strength (10^6 V/cm)	>10	8	10	10
Etch rate (Å/min) (100:1 H_2O:HF)	30	60	30	30
Etch rate (Å/min) (buffered HF)	440	1200	450	450
Step coverage	–	Nonconformal	Conformal	Conformal

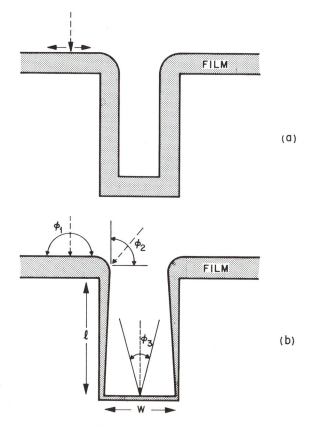

Fig. 15 Step coverage of deposited films. (*a*) Conformal step coverage. (*b*) Nonconformal step coverage.[10]

uniform along all surfaces of the step. The uniformity of film thickness, regardless of topography, is due to the rapid migration of reactants after adsorbtion on the step surfaces.[13]

Figure 15*b* shows an example of nonconformal step coverage, which results when the reactants adsorb and react without significant surface migration. In this instance the deposition rate is proportional to the arrival angle of the gas molecules. Reactants arriving along the top horizontal surface come from many different angles and ϕ_1, the arrival angle, varies in two dimensions from 0 to 180° while reactants arriving at the top of a vertical wall have an arrival angle ϕ_2 that varies from 0 to 90°. Thus, the film thickness on the top surface is double that of a wall. Further down the wall, ϕ_3 is related to the width of the opening, and the film thickness is proportional to

$$\phi_3 \cong \arctan \frac{W}{l} \tag{19}$$

where *l* is the distance from the top surface and *W* is the width of the opening.

This type of step coverage is thin along the vertical walls, with a possible crack at the bottom of the step caused by self-shadowing.

Silicon dioxide formed by TEOS decomposition at reduced pressure gives a nearly conformal coverage due to rapid surface migration. Similarly, the high-temperature dichlorosilane–nitrous oxide reaction also results in conformal coverage. However, during silane–oxygen deposition no surface migration takes place and the step coverage is determined by the arrival angle. Most evaporated or sputtered materials have a step coverage similar to that in Fig. 15b.

P-Glass Flow   A smooth topography is usually required for the deposited silicon dioxide that is used as an insulator between metal layers. If the oxide used to cover the lower metal layer is concave, circuit failure may result from an opening that may occur in the upper metal layer during deposition. Because phosphorus-doped silicon dioxide (P-glass) deposited at low temperatures becomes soft and flows upon heating, it provides a smooth surface and is often used to insulate adjacent metal layers. This process is called *P-glass flow*.

Figure 16 shows four cross sections of scanning electron microscope photographs of P-glass covering a polysilicon step.[13] All samples are heated in steam at 1100°C for 20 min. Figure 16a shows a sample of glass which contains a negligibly small amount of phosphorus and will not flow. Note the concavity of the film and that the corresponding angle θ is about 120°. Figures 16b, 16c, and 16d show samples of P-glass with progressively higher phosphorus contents up to 7.2 wt% (weight percent). In these samples the decreasing step angles of the P-glass layer indicate how flow increases with phosphorus concentration. P-glass flow depends on annealing time, temperature, phosphorus concentration, and the annealing ambient.[13]

The angle θ as a function of weight percent of phosphorus as shown in Fig. 16 can be approximated by

$$\theta \simeq 120° \left[\frac{10 - \text{wt}\%}{10} \right]. \tag{20}$$

If we want an angle smaller than 45°, we require a phosphorus concentration larger than 6 wt%. However, at concentrations above 8 wt%, the metal film (e.g., aluminum) may be corroded by the acid products formed during the reaction between the phosphorus in the oxide and atmospheric moisture. Therefore, the P-glass flow process uses phosphorus concentrations of 6 to 8 wt%.

9.2.2   Silicon Nitride

Silicon nitride films can be deposited by an intermediate-temperature (750°C) LPCVD process or a low-temperature (300°C) plasma-assisted CVD process. The LPCVD films are of stoichiometric composition (Si_3N_4) with high density (2.9 to 3.1 g/cm^3). These films can be used for passivating devices, because they serve as good barriers to the diffusion of water and sodium.

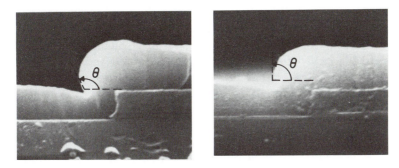

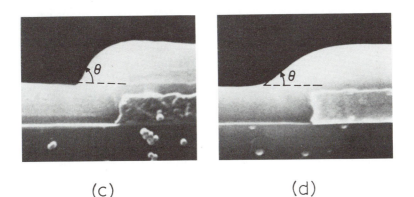

(a)   (b)

(c)   (d)

Fig. 16 Scanning-electron micrographs (10,000×) of samples annealed in steam at 1100° C for 20 min for the following weight percent of phosphorus[13]: (a) 0 wt%, (b) 2.2 wt%, (c) 4.6 wt%, (d) 7.2 wt% P.

The films also can be used as masks for the selective oxidation of silicon, because silicon nitride oxidizes very slowly and prevents the underlying silicon from oxidizing. The films deposited by plasma-assisted CVD are not stoichiometric and have a lower density (2.4 to 2.8 g/cm^3). Because of the low deposition temperature, these films can be deposited over the completely fabricated devices and serve as their final passivation. The plasma-deposited nitride provides excellent scratch protection, serves as a moisture barrier, and prevents sodium diffusion.

In the LPCVD process, dichlorosilane and ammonia react at reduced pressure to deposit silicon nitride at temperatures between 700 and 800°C. The reaction is

$$3SiCl_2H_2 + 4NH_3 \xrightarrow{\sim 750°C} Si_3N_4 + 6HCl + 6H_2 . \qquad (21)$$

Good film uniformity and high wafer throughout (i.e., number of wafers processed per hour) are advantages of the reduced-pressure process. As in the case of oxide deposition, silicon nitride deposition is controlled by temperature, pressure, and reactant concentration. The activation energy for deposition is about 1.8 eV. The deposition rate increases with increasing total pressure or dichlorosilane partial pressure, and decreases with an increasing ammonia-to-dichlorosilane ratio.

Silicon nitride deposited by LPCVD is an amorphous dielectric containing up to 8 at% hydrogen. The etch rate in buffered HF is less than 10 Å/min. The film's very high tensile stress of approximately 10^{10} dynes/cm^2 is nearly 10 times that of TEOS-deposited SiO$_2$. Films thicker than 2000 Å may crack because of the very high stress. The resistivity of silicon nitride at room temperature is about 10^{16} Ω-cm. Its dielectric constant is 6 and its dielectric strength is 10^7 V/cm.

In the plasma-assisted CVD process, silicon nitride is formed either by reacting silane and ammonia in an argon plasma or by reacting silane in a nitrogen discharge. The reactions are as follows:

$$SiH_4 + NH_3 \xrightarrow{\ 300°C\ } SiNH + 3H_2 \tag{22a}$$

$$2SiH_4 + N_2 \xrightarrow{\ 300°C\ } 2SiNH + 3H_2 . \tag{22b}$$

The products depend strongly on deposition conditions. The radial-flow parallel-plate reactor (Fig. 13b) is used to deposit the films. The deposition rate generally increases with increasing temperature, power input, and reactant gas pressure.

Large concentrations of hydrogen are contained in plasma-deposited films. The plasma nitride (also referred to as SiN) used in semiconductor processing generally contains 20 to 25 at% hydrogen. Films with low tensile stress ($\sim 2 \times 10^9$ dynes/cm^2) can be prepared by plasma deposition. Film resistivities range from 10^5 to 10^{21} Ω-cm depending on silicon-to-nitrogen ratio, while dielectric strengths are between 1×10^6 and 6×10^6 V/cm.

9.3 POLYSILICON DEPOSITION

Using polysilicon as the gate electrode in MOS devices is considered a significant development in MOS circuit technology. One important reason is that polysilicon surpasses aluminum for electrode reliability. Figure 17 shows the maximum time to breakdown for capacitors with both polysilicon and aluminum electrodes.[14] The polysilicon is clearly superior, especially for thinner gate oxides. Polysilicon is also used as diffusion source to create shallow junctions and to ensure ohmic contact to crystalline silicon. Additional uses include the manufacture of conductors and high-value resistors.

A low-pressure reactor (Fig. 13a) operated between 600 and 650°C is used to deposit polysilicon by pyrolyzing silane according to the following reaction.

$$SiH_4 \xrightarrow{\ 600°C\ } Si + 2H_2 . \tag{23}$$

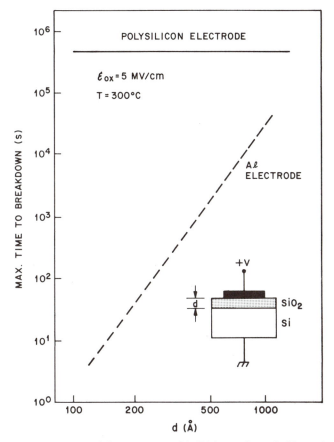

Fig. 17 Maximum time to breakdown versus oxide thickness for polysilicon electrode and an aluminum electrode.[14]

Of the two most common low-pressure processes, one operates at a pressure of 0.2 to 1.0 Torr using 100% silane, while the other process involves a diluted mixture of 20 to 30% silane in nitrogen at the same total pressure. Both processes can deposit polysilicon on hundreds of wafers per run with good uniformity (i.e., thickness within 5%).

Figure 18 shows the deposition rate at four deposition temperatures and at a partial pressure of 33 Pa (0.25 Torr).[10] At low silane partial pressure, the deposition rate is proportional to the silane pressure. At higher silane concentrations, saturation of the deposition rate occurs. Deposition at reduced pressure is generally limited to temperatures between 600 and 650°C. In this temperature range, the deposition rate varies as $\exp(-E_a/kT)$, where the activation energy E_a is 1.7 eV, which is essentially independent of the total pressure in the reactor. At higher temperatures, gas phase reactions that result in a rough, loosely adhering deposit become significant and silane depletion will

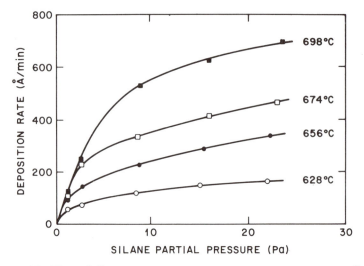

Fig. 18   Effect of silane concentration on the polysilicon deposition rate.[10]

occur, causing poor uniformity. At temperatures much lower than 600°C, the deposition rate is too slow to be practical.

Process parameters that affect the polysilicon structure are deposition temperature, dopants, and the heat cycle applied following the deposition step. A columnar structure results when polysilicon is deposited at a temperature of 600 to 650°C. This structure is comprised of polycrystalline grains ranging in size from 0.03 to 0.3 μm, at a preferred orientation of (110). When phosphorus is diffused at 950°C, the structure changes to crystallite, and grain size increases to an average of between 0.5 and 1.0 μm. When temperature is increased to 1050°C during oxidation, the grains reach a final size of 1 to 3 μm. While the initially deposited film appears amorphous when deposition occurs below 600°C, similar growth characteristics are observed after doping and heating.

Polysilicon can be doped by diffusion, ion implantation, or the addition of dopant gases during deposition, which is referred to as in-situ doping. The implantation method is most commonly used because of its lower processing temperatures. Figure 19 shows the sheet resistance of polysilicon doped with phosphorus and antimony using ion implantation[15] (refer to Chapter 10). Implant dose, annealing temperature, and annealing time all influence the sheet resistance of implanted polysilicon. Carrier traps at the grain boundaries cause a very high resistance in the highly implanted polysilicon. As Fig. 19 illustrates, resistance drops rapidly, approaching that of implanted single-crystal silicon, as the carrier traps become saturated with dopants.

9.4   METALLIZATION

Metallization refers to the formation of metal films used for interconnections, ohmic contacts, and rectifying metal–semiconductor contacts. We have

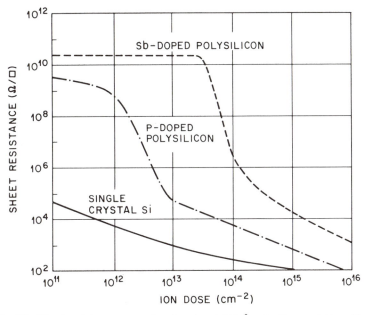

Fig. 19 Sheet resistance versus ion dose into 5000-Å polysilicon at 30 keV.[15]

considered the current–voltage characteristics of ohmic and rectifying contacts in Section 5.1. Metal films can be formed by various methods, the most important being physical vapor deposition and chemical vapor deposition, both of which shall be considered in this section. We shall also consider the two most important metal groups: aluminum and its alloys, and the silicides. These metals are used extensively for discrete devices and integrated circuits.[16]

9.4.1 Physical Vapor Deposition

In Section 8.6 we considered some basic properties of gases that are useful in physical vapor deposition. Physical vapor deposition is done under vacuum using either an evaporation or sputtering technique. The vacuum chamber is evaluated by a roughing pump from atmospheric pressure to a pressure of about 15 Pa, followed by high-vacuum pumping to reduce the pressure to 5×10^{-5} Pa (5×10^{-7} Torr) or lower. System cleanliness is required to reduce the pumpdown period. After being chemically cleaned, all chamber components are carefully dried. To remove a major source of trapped atmospheric gas, all interior film buildup must be removed. In addition, impurities such as sodium must be removed prior to MOS device deposition. Similar cleanliness precautions should be observed for sputter deposition. During film deposition, approximately 1 Pa argon pressure is needed. All connecting gas lines from source to sputtering chamber must be clean and vacuum tight to maintain high-purity gas input.

Evaporation Figure 20 shows various sources used in evaporation. A resistance-heated source is shown in Fig. 20*a* where a refractory metal (i.e.,

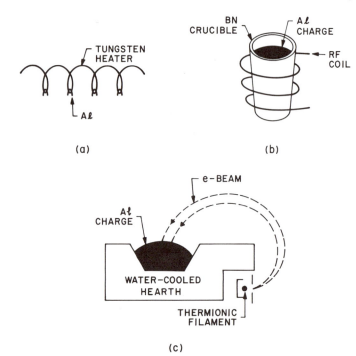

Fig. 20 Evaporation sources. (*a*) Refractory wire coils. (*b*) Inductively heated BN crucible. (*c*) Electron beam evaporation.

metal such as tungsten with a high melting temperature) is coiled into a filament and a small piece of aluminum is suspended from each coil. This approach is attractive because it is simple and inexpensive and produces no ionizing radiation. Its disadvantages are possible contamination from the heater and limited film thickness because of the small charge. Figure 20*b* shows an evaporation source that is heated by rf induction. Usually boron nitride is used to form the crucible. High deposition rates are possible through this process without ionizing radiation. However, the charge may be contaminated by the crucible.

Figure 20*c* shows a schematic view of an electron beam (e-beam) evaporation source. A thermionic filament supplies the current to the beam and the electrons are accelerated by an electric field to strike the surface of the aluminum charge to be evaporated. To prevent impurities from the filament reaching the aluminum charge in the water-cooled hearth, a magnetic field bends the e-beam path thus screening the impurities. The vacuum need not be broken to recharge the source during thick-film deposition if a large enough charge is used. Sequential deposition of different films is possible if several sources are available in the chamber. Using multiple sources also facilitates co-evaporation to produce alloy films. Very high deposition rates on the order of ~0.5 μm/min are possible in this system depending upon source-to-substrate distance. Of course, the e-beam process may be used to evaporate elements other than aluminum and aluminum alloys, for example, silicon, palladium, titanium, molybdenum, platinum, and tungsten. A disadvantage of the process is the generation of X-rays by the e-beam. This ionizing radiation can penetrate the surface layers of the devices, causing damage such as the creation of oxide-trapped charges (refer to Section 5.4), which change the device characteristics. Therefore, subsequent annealing is required to remove such damage.

We shall now derive the thickness of the film deposited from an evaporation source.[17] Consider a small sphere dS_1 evaporating material in all directions at a rate of m (g/s) as shown in Fig. 21a. Such an evaporating source is called a *point source*. The amount of material passing through a solid angle $d\omega$ in any direction per unit time is then

$$dm = \left(\frac{m}{4\pi} \right) d\omega . \tag{24}$$

A related case is the evaporation from a *plane source* of area dS_1 from which the material is evaporated from one side at a rate of m (g/s). The amount of material passing through a solid angle $d\omega$ in a direction forming an angle ϕ with the normal to the surface per unit time (Fig. 21a) is given by the cosine law:

$$dm = \frac{m}{\pi} \cos \phi \, d\omega . \tag{25}$$

If the material arrives at a small area dS_2 on a surface whose normal is inclined at an angle θ to the direction of the vapor stream as shown in Fig. 21b, we can find the thickness of the film from the solid angle:

$$d\omega = \frac{dS_2 \cdot \cos \theta}{r^2} . \tag{26}$$

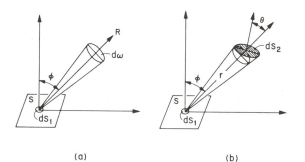

(a)   (b)

Fig. 21 (a) Evaporation from a source S through a solid angle $d\omega$. (b) Receiving surface dS_2 whose normal makes an angle θ with the direction of the vapor stream.[17]

From Eqs. 24, 25, and 26 we have

$$dm = \frac{m}{4\pi} \left[\frac{\cos \theta}{r^2} \right] dS_2 \tag{27}$$

for the point source and

$$dm = \frac{m}{\pi} \left[\frac{\cos \phi \cos \theta}{r^2} \right] dS_2 \tag{28}$$

for the plane source.

Assuming the material has a density ρ_d (g/cm^3) and the thickness of the film deposited per unit time is l (cm/s), the volume of material deposited on dS_2 must be $l dS_2$. We obtain

$$dm = \rho_d l dS_2 . \tag{29}$$

The thickness of the film at locations corresponding to the area dS_2 is then given by

$$l = \frac{m}{4\pi \rho_d} \left[\frac{\cos \theta}{r^2} \right] \tag{30}$$

and

$$l = \frac{m}{\pi \rho_d} \left[\frac{\cos \phi \cos \theta}{r^2} \right] \tag{31}$$

for the point and plane sources, respectively. We can eliminate the time element by using m (g), the total mass evaporated, to replace the mass evaporation rate m (g/s). Similarly, l (cm) becomes the total deposition thickness replacing the growth rate l (cm/s).

For discrete devices, a practical arrangement of the source and a parallel-plane receiving surface is shown in the insert of Fig. 22. The thickness at R from a point source is given by

$$l = \frac{mH}{4\pi \rho_d (H^2 + L^2)^{3/2}} = \frac{m}{4\pi \rho_d H^2} \left\{ \frac{1}{[1 + (L/H)]^{3/2}} \right\} \tag{32}$$

while the thickness from a plane source is given by

$$l = \frac{mH^2}{\pi \rho_d (H^2 + L^2)^2} = \frac{m}{\pi \rho_d H^2} \left\{ \frac{1}{[1 + (L/H)^2]^2} \right\} . \tag{33}$$

Figure 22 shows the normalized thickness distribution for each of the two evaporation sources.[17]

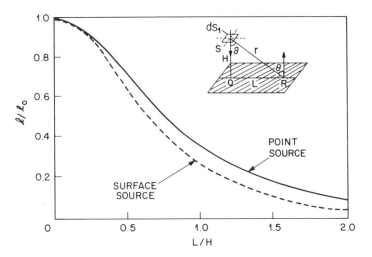

Fig. 22 Distribution of deposition on a parallel plane surface. Insert shows the evaporation arrangement.[17]

Problem

A silicon wafer is placed at a perpendicular distance of 30 cm from a plane source. If the total deposited mass is 1 g and the density is 2.7 g/cm³, what is the film thickness at $L = 0$? If the variation in film thickness must be less than 10%, how large can the wafer be?

Solution

The film thickness at $L = 0$ is given by

$$l(L = 0) = l_0 = \frac{m}{\pi \rho_d H^2} = \frac{1.0}{\pi \times 2.7 \times 30^2} = 1.3 \times 10^{-4} \text{ cm} = 1.3 \ \mu\text{m} .$$

From Eq. 33 with $l/l_0 = 0.9$ and $H = 30$ cm, we have

$$L = H[(l_0/l)^{1/2} - 1]^{1/2} = 30[(1.11)^{1/2} - 1]^{1/2} = 6.98 \text{ cm} .$$

The wafer diameter can be $6.98 \times 2 = 13.9$ cm.

For integrated-circuit processing, large numbers of wafers each requiring a uniform film thickness must be metallized per run. For such large batches we can use the planetary substrate-supporting system consisting of rotating spherical sections. The principle behind such a system is shown in Fig. 23, where the receiving surface is spherical having radius r_o while the plane source is on a stationary surface. We have

$$\cos \phi = \cos \theta = \frac{r}{2r_o} \tag{34}$$

and Eq. 31 can be written

$$l = \frac{m}{\pi \rho_d} \left[\frac{r}{2r_o} \right]^2 \frac{1}{r^2} = \frac{m}{4\pi \rho_d r_o^2} \tag{35}$$

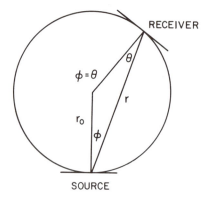

Fig. 23 Idealized view of evaporation source and film-gathering surface mounted on a sphere of radius r_o.

which is independent of the distance r from the source to the receiving surface. Therefore, the deposition rate is the same for all points on the spherical surface, and uniform film thickness can be obtained.

Sputtering Sputtering is a physical phenomenon involving the acceleration of ions, usually Ar^+, through a potential gradient and the bombardment by these ions of a "target" or cathode. Through momentum transfer, atoms near the surface of the target material become volatile and are transported as a vapor to the substrate. At the substrate the film grows through deposition.

The insert of Fig. 24 shows a cross section of a conical magnetron for sputter deposition. The concentric anode and circular symmetry are unique to this system. Electrons originating at the cathode are confined by the fields from the permanent magnets and are collected by the anode. A large fraction of the sputtered material (neutral atoms) from the target cathode is ejected forward and deposited on a substrate that need not be an electrode of the system. To ensure uniform thickness of the deposited film, appropriate mechanical motion of the substrate similar to that of the planetary system for evaporation can be employed to expose the substrate to the same average number of sputtered atoms. The magnetron operates at voltages an order of magnitude below the e-beam source voltage and thus generates less penetrating radiation. Deposition rate depends on source-to-substrate distance and can be as high as 1 μm/min for aluminum and its alloys.

The number of atoms N per unit area per unit time leaving the target is given by

$$N = \frac{J}{qZ}\, \gamma(E,\, M_1,\, M_2) \tag{36}$$

where J is the current density of the bombarding ions; q is the electronic charge; Z is the number of charges per ion; and γ is the sputter yield in atoms per incident ion, which is a function of the ion energy E, the mass of the ion species M_1, and the mass of the target material M_2. A typical sputter yield as

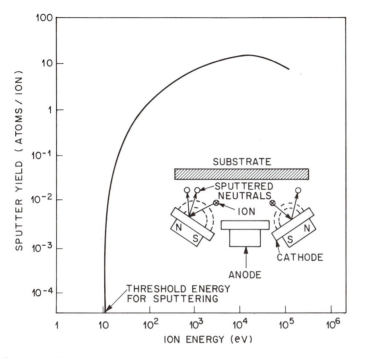

Fig. 24 Representative sputter yield characteristic. Insert shows a cross section of a conical magnetron.[16]

a function of ion energy is shown in Fig. 24. Below the threshold energy, the sputter yield is very low. The yield increases rapidly from the threshold energy level, which is usually in the range of 10 to 100 eV for common metals (e.g., 10 eV for Al, 15 eV for Pd, 21 eV for Mo, and 34 eV for Pt). To minimize the energy required to eject an atom from the target, the process uses ion energy within the range for which the sputter yield is about one atom per ion, such as 10^2 eV for the case shown in Fig. 24.

9.4.2 Chemical Vapor Deposition

Chemical vapor deposition (CVD) is attractive for metallization because it offers coatings of conformal nature with good step coverage and it can coat large numbers of wafers at a time. The basic CVD setup is the same as that used for deposition of dielectrics and polysilicon (e.g., Fig. 13a). Low-pressure CVD is capable of producing conformal step coverage over a wide range of topographical profiles often with lower electrical resistivity than that from physical vapor deposition, which also suffers from the effects of shadowing and poor step coverage.

One of the major new applications of CVD metal deposition for integrated circuit production has been in the area of refractory-metal deposition. For example, tungsten's low electrical resistivity (5.3 $\mu\Omega$-cm) and its refractory nature make it a desirable metal for use in integrated circuit fabrication. The following two chemical equations show both the pyrolysis and the reduction of tungsten[16]:

$$WF_6 \longrightarrow W + 3F_2 \tag{37}$$

and

$$WF_6 + 3H_2 \longrightarrow W + 6HF. \tag{38}$$

Many other metals such as molybdenum (Mo), tantalum (Ta), and titanium (Ti) are of interest for integrated-circuit applications. These metal films can be deposited by hydrogen reduction in an LPCVD reactor having the following overall reactions:

$$2MCl_5 + 5H_2 \xrightarrow{600-800°C} 2M + 10HCl \tag{39}$$

where M stands for the aforementioned metals. Aluminum can be deposited by using metallorganic compounds such as tri-isobutyl aluminum:

$$2[(CH_3)_2CHCH_2]_3Al \longrightarrow 2Al + 3H_2 + \text{by-products} \tag{40}$$

9.4.3 Aluminum Metallization

Aluminum and its alloy are used extensively for metallization in integrated circuits. Because aluminum and its alloys have low resistivities (i.e., 2.7 $\mu\Omega$-cm for Al and up to 3.5 $\mu\Omega$-cm for its alloys), these metals satisfy the requirements of low resistance. Aluminum also adheres well to silicon dioxide. However, the use of aluminum in integrated circuits at shallow junctions often creates problems such as *spiking* and *eletromigration*. We shall consider aluminum metallization and its problems and their various solutions in this section.

Junction Spiking   Figure 25 shows the phase digram of the Al–Si system at 1 atm.[18] The phase diagram relates these two components as a function of temperature. The Al–Si system exhibits eutectic characteristics; that is, the addition of either component lowers the system's melting point below that of either metal. Here, the minimum melting temperature, called the *eutectic temperature*, is 577°C, corresponding to a 11.3% Si–88.7% Al composition. The melting points of pure aluminum and pure silicon are 660 and 1412°C, respectively. Because of the eutectic characteristics, during aluminum deposition the temperature on the silicon substrate must be limited to less than 577°C.

The insert of Fig. 25 also shows the solid solubility of silicon in aluminum. For example, the solubility of silicon in aluminum is 0.25 wt% at 400°C, 0.5 wt% at 450°C, and 0.8 wt% at 500°C. Hence, wherever aluminum contacts silicon, the silicon will dissolve into the aluminum during annealing. The amount of silicon dissolved will depend not only on the solubility at the annealing temperature but also on the volume of aluminum to be saturated with silicon. Consider a long aluminum metal line in contact with an area ZL of silicon as shown in Fig. 26. After an annealing time t, the silicon will diffuse a distance of approximately $\sqrt{Dt}$ along the aluminum line from the edge of

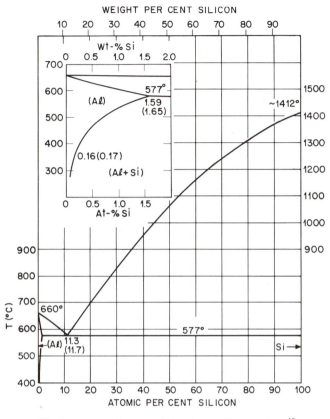

WEIGHT PER CENT SILICON

ATOMIC PER CENT SILICON

Fig. 25 Phase diagram of aluminum–silicon system.[18]

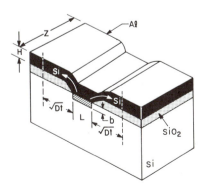

Fig. 26 Diffusion of silicon in aluminum metallization.[19]

the contact, where D is the diffusion coefficient given by $4 \times 10^{-2} \exp(-0.92/kT)$ for silicon diffusion in deposited aluminum films. Assuming that this length of aluminum is completely saturated with silicon, the volume of silicon consumed is then

$$Vol \simeq 2 \sqrt{Dt} (HZ)S \left[\frac{\rho_{Al}}{\rho_{Si}} \right] \tag{41}$$

where ρ_{Al} and ρ_{Si} are the densities of aluminum and silicon, respectively, and S is the solubility of silicon in aluminum at the annealing temperature.[19] If the consumption takes place uniformly over the contact area A (where $A = ZL$ for uniform dissolution), the depth to which silicon would be consumed is

$$b \simeq 2 \sqrt{Dt} \left[\frac{HZ}{A} \right] S \left[\frac{\rho_{Al}}{\rho_{Si}} \right]. \tag{42}$$

Problem

For $T = 500°C$, $t = 30$ min, $ZL = 16 \ \mu m^2$, $Z = 5 \ \mu m$, and $H = 1 \ \mu m$. Find the depth b, assuming uniform dissolution.

Solution

The diffusion coefficient of silicon in aluminum at $500°C$ is about $2 \times 10^{-8} \ cm^2/s$; thus, $\sqrt{Dt}$ is $60 \ \mu m$. The density ratio is $2.7/2.33 = 1.16$. At $500°C$, S is 0.8 wt%. From Eq. 42 we have

$$b = 2 \times 60 \left[\frac{1 \times 5}{16} \right] 0.8\% \times 1.16 = 0.35 \ \mu m \ .$$

Aluminum will fill to a depth of 0.35 μm from which silicon is consumed. If at the contact point there is a shallow junction whose depth is less than b, the diffusion of silicon into aluminum can short-circuit the junction.

In a practical situation, the dissolution of silicon does not take place uniformly but rather at only a few points. The effective area in Eq. 42 is less than the actual contact area, hence b is much larger. Figure 27 illustrates the actual situation in the *p–n* junction area of aluminum penetrating the silicon at only the few points where spikes are formed. One way to minimize the aluminum spiking is to add silicon to the aluminum by co-evaporation until the amount of silicon contained by the alloy satisfies the solubility requirement. Another method of satisfying the silicon requirements of the aluminum film is to deposit the film on a layer of polysilicon (Fig. 1). The polysilicon may be doped p^+ or n^+ by in-situ or post-deposition doping. A third method is to introduce a barrier metal layer between the aluminum and the silicon substrate (Fig. 28). This barrier metal layer must meet the following requirements: (1) it forms low contact resistance with silicon, (2) it will not react with aluminum, and (3) its deposition and formation are compatible with the overall process. Barrier metals such as titanium nitride (TiN) have been evaluated and found to be stable for contact annealing temperatures up to $550°C$ for 30 min.[20]

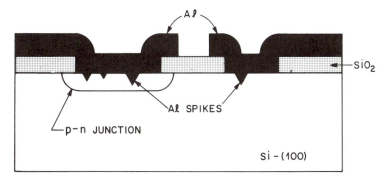

Fig. 27 Schematic view of aluminum films contacting silicon. Note that aluminum spikes in the silicon.[16]

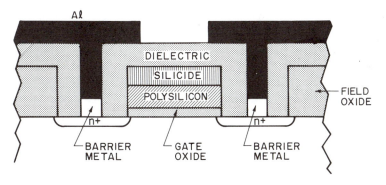

Fig. 28 Cross–sectional view of a MOSFET with barrier metal between aluminum and silicon and a composite gate electrode having silicide and polysilicon.

Electromigration In Section 5.6 we discussed scaled-down devices. As the device becomes smaller, the corresponding current density becomes larger. High current densities can cause device failure due to electromigration. The term electromigration refers to the transport of mass in metals under the influence of current. It occurs by the transfer of momentum from the electrons to the positive metal ions. When a high current passes through thin metal conductors in integrated circuits, metal ions in some regions will pile up and voids will form in other regions. The pileup can short-circuit adjacent conductors while the voids can result in an open circuit.

The mean time to failure (MTF) of a conductor due to electromigration can be related to the current density J and an activation energy E_a by

$$MTF \sim \frac{1}{J^2} \exp \left[\frac{E_a}{kT} \right]. \tag{43}$$

Experimentally, for deposited aluminum a value of $E_a \simeq 0.5$ eV is obtained and is taken to indicate that low-temperature grain–boundary diffusion is the primary vehicle of material transport, since $E_a \simeq 1.4\,eV$ would characterize the self-diffusion of single-crystal aluminum. The electromigration resistance of aluminum film conductors can be increased by using several techniques. These techniques include alloying with copper (e.g., Al with 0.5% Cu), encapsulating the conductor in a dielectric, or incorporating oxygen during film deposition.

9.4.4 Silicides

Silicides such as $TiSi_2$ and $TaSi_2$ have reasonably low resistivities ($\lesssim 50\ \mu\Omega$-cm) and are generally compatible with integrated-circuit processing. Silicides become important metallization materials as the devices become smaller. One important application of silicides is for the MOSFET gate electrode either alone or with doped polysilicon as shown in Fig. 28 above the gate oxide. These silicides remain stable during contact with polysilicon. In high-speed circuits, the RC time delay becomes an important consideration underscoring the need for higher-conductivity gates. The RC time delay per unit length at various linewidths is plotted in Fig. 29 for each of three conductive materials.[21] The materials and their respective resistivities are: polysilicon, resistivity $\gtrsim 400\ \mu\,\Omega$-cm; tantalum silicide, resistivity $\simeq 50\ \mu\,\Omega$-cm, and aluminum, resistivity $= 2.7\ \mu\,\Omega$-cm. It is assumed that two layers of silicon dioxide, each $1.5\ \mu m$ thick, enclose a 1-μm-thick layer of conductive material. Note that $TaSi_2$ gives nearly an order of magnitude reduction in the RC time constant

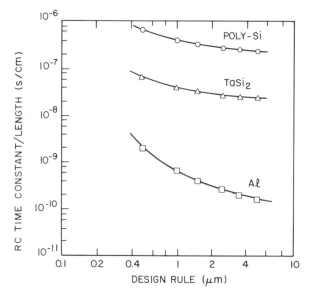

Fig. 29 *RC* time constant per unit length for three conductive materials as a function of feature size.[21]

per unit length compared with polysilicon. Thus, for a given maximum tolerable delay, the conductor length may be an order of magnitude larger if a silicide is used instead of polysilicon.

Silicides may be formed in several ways. These include (1) depositing a refractory metal film on polysilicon (or silicon) and sintering the structure to form a silicide; (2) co-depositing the metal and silicon either by sputtering or evaporating simultaneously from separate sources onto polysilicon or a dielectric layer; (3) forming the silicide by chemical vapor deposition on either an oxide or polysilicon layer.

The resistivities of various silicides on n^+-polysilicon are given[22] in Table 2. The lowest resistivity is obtained in titanium silicide ($TiSi_2$) formed by sintering a metal layer on polysilicon. For a given silicide, the resistivity of co-sputtered film is about 50 to 100% higher than that of the sintered film. This difference is due to electron mobility, which is expected to be higher in sintered film where larger crystals can be formed.

All silicides can withstand much higher processing temperatures than aluminum. Of course, the eutectic temperature will limit the maximum process temperature of the silicide in contact with silicon. Thus, palladium silicide (Pd_2Si) is limited to 700°C, platinum silicide (PtSi) to 800°C, and nickel silicide ($NiSi_2$) to 900°C. The other silicides in Table 2 are stable to temperatures above 1000°C. Another advantage of silicides is their stability in an oxidizing ambient. With the use of lower VLSI processing temperatures, silicides are expected to play a more important role in future metallization. The various aspects of metallization remain an active area of research and development, since metallization is intimately related to the ultimate capability and reliability of device operation.

Table 2 Silicide Resistivities (300 K)

Silicide	Starting Form	Sintering Temperature (°C)	Resistivity ($\mu\Omega$-cm)
$CoSi_2$	Metal on polysilicon	900	18–20
	Co-sputtered Alloy	900	25
$HfSi_2$	Metal on polysilicon	900	45–50
$MoSi_2$	Co-sputtered Alloy	1000	100
$NiSi_2$	Metal one polysilicon	900	50
	Co-sputtered Alloy	900	50–60
Pd_2Si	Metal on polysilicon	400	30–50
PtSi	Metal on polysilicon	600–800	28–35
$TaSi_2$	Metal on polysilicon	1000	35–45
	Co-sputtered Alloy	1000	50–55
$TiSi_2$	Metal on polysilicon	900	13–16
	Co-sputtered Alloy	900	25
WSi_2	Co-sputtered Alloy	1000	70
$ZrSi_2$	Metal on polysilicon	900	35–40

REFERENCES

1 For a discussion on various oxidation methods for silicon and gallium arsenide, see, for example, S. K. Ghandhi, *VLSI Fabrication Principles*, Wiley, New York, 1983.

2 For a discussion on thermal oxidation of silicon, see, for example, L. E. Katz, "Oxidation," in S. M. Sze, Ed., *VLSI Technology*, McGraw-Hill, New York, 1983.

3 E. H. Nicollian and J. R. Brews, *MOS Physics and Technology*, Wiley, New York, 1983.

4 B. E. Deal and A. S. Grove, "General Relationship for the Thermal Oxidation of Silicon," *J. Appl. Phys.*, **36**, 3770 (1965).

5 J. D. Meindl, R. W. Dutton, K. C. Saraswat, J. D. Plummer, T. I. Kamins, and B. E. Deal, "Silicon Epitaxy and Oxidation," in F. Van de Wiele, W. L. Engl, and P. O. Jespers, Eds., *Process and Device Modeling for Integrated Circuit Design*, Noorhoff, Leyden, 1977.

6 A. Fargeix, G. Ghibaudo, and G. Kamarinos, "A Revised Analysis of Dry Oxidation of Silicon," *J. Appl. Phys.*, **54**, 2878 (1983).

7 K. K. Ng, W. J. Polito, and J. R. Ligenza, "Growth Kinetics of Thin Silicon Dioxide in a Controlled Ambient Oxidation System," *J. Appl. Phys.*, **44**, 626 (1984).

8 D. W. Hess and B. E. Deal, "Kinetics of the Thermal Oxidation of Silicon in O_2/HCl Mixtures," *J. Electrochem. Soc.*, **124**, 735 (1977).

9 C. P. Ho, J. D. Plumer, J. D. Meindl, and B. E. Deal, "Thermal Oxidation of Heavily Phosphorus Doped Silicon," *J. Electrochem. Soc.*, **125**, 665 (1978).

10 For a discussion on film deposition, see, for example, A. C. Adams, "Dielectric and Polysilicon Film Deposition," in S. M. Sze, Ed., *VLSI Technology*, McGraw-Hill, New York, 1983.

11 L. R. Thompson et al., "Conformal Step Coverage of Electron-Beam-Assisted CVD of SiO_2 and Si_3N_4 Films," *J. Electrochem. Soc.*, **131**, 462 (1984).

12 J. Y. Chen, R. C. Henderson, J. T. Hall, and J. W. Peters, "Photo-CVD for VLSI Isolation," in M. D. Robinson et al., Ed., *Chemical Vapor Deposition 1984*, Electrochemical Soc., Pennington, 1984.

13 A. C. Adams and C. D. Capio, "Planarization of Phosphorus-Doped Silicon Dioxide," *J. Electrochem. Soc.*, **127**, 2222 (1980).

14 H. N. Yu, A. Reisman, C. M. Osburn, and D. L. Critchlow, "1 μm MOSFET VLSI Technology: Part I—An Overview," *IEEE Trans. Electron Devices*, **ED-26**, 318 (1979).

15 J. M. Andrews, "Electrical Conduction in Implanted Polycrystalline Silicon," *J. Electron. Mat.*, **8**, No. 3, 227 (1979).

16 D. B. Fraser, "Metallization," in S. M. Sze, Ed., *VLSI Technology*, McGraw-Hill, New York, 1983.

17 L. Holland, *Vacuum Deposition of Thin Films*, Wiley, New York, 1961.

18 M. Hansen and A. Anderko, *Constitution of Binary Alloys*, McGraw-Hill, New York, 1958.

19 D. Pramanik and A. N. Saxena, "VLSI Metallization Using Aluminum and Its Alloys," *Solid State Tech.*, **26**, No. *1*, 127 (1983); and **26**. No. *3*, 131 (1983).

20 C. Y. Ting, "TiN Formed by Evaporation as a Diffusion Barrier Between Al and Si," *J. Vac. Sci. Technol.*, **21**, 14 (1982).

21 S. K. Sinha, "Refractory Metal Silicides for VLSI Applications," *J. Vac. Sci. Technol.*, **19**, 778 (1981).

22 S. P. Murarka, "Refractory Silicides for Integrated Circuits," *J. Vac. Sci. Technol.*, **17**, 775 (1980).

PROBLEMS

1 A p-type $<100>$-oriented, silicon wafer of resistivity 10 Ω-cm is placed in a wet oxidation system to grow a field oxide of 4500 Å at 1050°C. Determine the time required to grow the oxide.

2 After the first oxidation as given in Problem 1, a window is opened in the oxide to grow a gate oxide at 1000°C for 20 minutes in dry oxidation. Find the thicknesses of the gate oxide and the total field oxide.

3 If 3% HCl is added at the start of the gate oxide growth in Problem 2, what will be the gate oxide thickness and the thickness of the field oxide?

4 Determine the diffusion coefficient D for dry oxidation of $<100>$-oriented silicon samples (*a*) at 980°C and 1 atm, (*b*) at 980°C and 0.1 atm, and (*c*) explain why the diffusion coefficients are different.

5 A silicon n^+-p lateral junction is made by the diffusion of phosphorus into a $<100>$-oriented, p-type substrate, If $N_D = 8 \times 10^{20}$ cm^{-3} and $N_A = 10^{15}$ cm^{-3}, find the oxide step at the junction when it is oxidized in dry oxygen at 900°C for 5 h.

6 In a silane–oxygen reaction to deposit undoped SiO_2 film, the deposition rate is 150 Å/min at 425°C. What temperature is required to double the deposition rate? Repeat the calculation for TEOS in which the temperature for a deposition rate of 150 Å/min is 300°C higher than that for silane.

7 The P-glass flow process requires temperatures above 1000°C. As device dimensions become smaller in VLSI, we must use lower temperatures. Suggest methods so that a smooth topography can be obtained at $< 900°$C for deposited silicon dioxide that can be used as an insulator between metal layers.

8 (*a*) In a plasma-deposited silicon nitride that contains 20 at% hydrogen and has a silicon-to-nitrogen ratio (Si/N) of 1.2, find x and y in the empirical formula of $SiN_x H_y$. (*b*) If the variation of film resistivity with Si/N ratio is given by $5 \times 10^{28} \exp(-33.3 r)$ for $2 > r > 0.8$, where r is the ratio, find the resistivity of the film in (*a*).

9 An e-beam evaporation system is used to deposit aluminum to form MOS capacitors. If the flatband voltage of the capacity is shifted by 0.5 V due

to e-beam radiation, find the number of fixed oxide charges (the silicon dioxide thickness in 500 Å). How to remove these charges?

10 To avoid electromigration problems, the maximum allowed current density in an aluminum runner is about 5×10^5 A/cm^2. If the runner is 2 mm long, 1 μm wide, and nominally 1 μm thick and 20% of the runner length passes over steps and is only 0.5 μm thick there, find the total resistance of the runner if the resistivity is 3×10^{-6} Ω-cm, and the maximum voltage that can be applied across the runner.

10

Diffusion and Ion Implantation

Diffusion and ion implantation are the two key processes we use to introduce controlled amounts of dopants into semiconductors. They are used to dope selectively the semiconductor substrate to produce either an n- or p-type region. Until the early 1970s, selective doping was done mainly by diffusion at elevated temperatures as shown in Fig. 1a. In this method the dopant atoms are placed on or near the surface of the semiconductor wafer by deposition from the gas phase of the dopant or by using doped-oxide sources. The doping concentration decreases monotonically from the surface, and the profile of the dopant distribution is determined mainly by the temperature and diffusion time. Since the early 1970s, many doping operations have been performed by ion implantation, as shown in Fig. 1b. In this process the dopant ions are

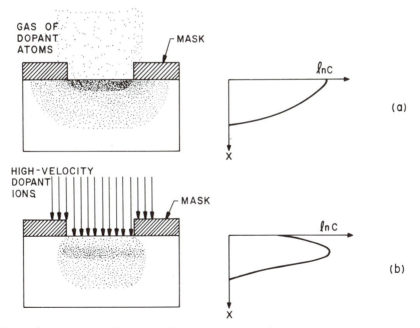

Fig. 1 Comparison of diffusion and ion implantation technique for the selective introduction of dopants into semiconductor substrate.

implanted into the semiconductor by means of a high-energy ion beam. The doping concentration has a peak inside the semiconductor; and the profile of the dopant distribution is determined mainly by the ion mass and the implanted-ion energy. The advantages of the ion implantation process are precise control of the total amount of dopants, improved reproducibility of impurity profiles, and lower-temperature processing.

Both diffusion and ion implantation are used for fabricating discrete devices and integrated circuits because these processes generally complement each other.[1, 2] For example, diffusion is used to form a deep junction (e.g., an *n*-tub in CMOS) while ion implantation is used to form a shallow junction (e.g., a source/drain junction of a MOSFET). In this chapter we consider the basic theoretical analysis and experimental results of using these two doping methods.

10.1 BASIC DIFFUSION THEORY AND PRACTICE

Diffusion of impurities is typically done by placing semiconductor wafers in a furnace and passing an inert gas that contains the desired dopant through it. The furnace and gas flow arrangements are similar to those used in thermal oxidation (Chapter 9). The temperature usually ranges between 800 and 1200°C for silicon and between 600 and 1000°C for gallium arsenide.

For diffusion in silicon, boron is the most popular dopant for introducing a *p*-type impurity, while arsenic and phosphorus are used extensively as *n*-type dopants. These three elements are highly soluble in silicon as they have solubilities above 5×10^{20} cm^{-3} in the diffusion temperature range. These dopants can be introduced in several ways, including solid sources (e.g., BN for boron, As_2O_3 for arsenic, and P_2O_5 for phosphorus), liquid sources (BBr_3, $AsCl_3$ and $POCl_3$), and gaseous sources (B_2H_6, AsH_3, and PH_3). Usually, the source material is transported to the semiconductor surface by an inert carrier gas (e.g., N_2) and is then reduced at the surface. An example of the chemical reaction for a solid source is[3]

$$2As_2O_3 + 3Si \longrightarrow 4As + 3SiO_2 . \tag{1}$$

In this reaction an oxide layer is formed on the silicon surface.

For diffusion in gallium arsenide, because of the high vapor pressure of arsenic special methods are used to prevent the loss of arsenic by decomposition or evaporation.[3] These methods include diffusion in sealed ampules with an overpressure of arsenic and diffusion in an open-tube furnace with a doped-oxide capping layer (e.g., silicon nitride). Most of the studies on *p*-type diffusion have been confined to the use of zinc in the forms of Zn–Ga–As alloys and $ZnAs_2$ for the sealed-ampule approach or ZnO–SiO_2 for the open-tube approach. The *n*-type dopants in gallium arsenide include sulfur and selenium. However, very little work has been done with these dopants.

10.1.1   Diffusion Equation

Diffusion in a semiconductor can be visualized as atomic movement of the diffusant (dopant atoms) in the crystal lattice by vacancies or interstitials. Figure 2 shows the two basic atomic diffusion models in a solid.[1] The open circles represent the host atoms occupying the equilibrium lattice positions. The solid dots represent impurity atoms. At elevated temperatures, the lattice atoms vibrate around the equilibrium lattice sites. There is a finite probability that a host atom acquires sufficient energy to leave the lattice site and to become an interstitial atom thereby creating a vacancy. When a neighboring impurity atom migrates to the vacancy site, as illustrated in Fig. 2a, the mechanism is called *vacancy diffusion*. If an interstitial atom moves from one place to another without occupying a lattice site (Fig. 2b), the mechanism is *interstitial diffusion*. An atom smaller than the host atom often moves interstitially.

The basic diffusion process of impurity atoms is similar to that of charge carriers (electrons and holes) discussed in Chapter 2. Accordingly, we define a flux F as the number of dopant atoms passing through a unit area in a unit time and C as the dopant concentration per unit volume. From Eq. 26 in Chapter 2, we have

$$F = -D\frac{\partial C}{\partial x} \qquad (2)$$

where we have substituted C for the carrier concentration and the proportionality constant D is the diffusion coefficient or diffusivity. Note that the basic driving force of the diffusion process is the concentration gradient dC/dx. The flux is proportional to the concentration gradient, and the dopant atoms will move (diffuse) away from a high-concentration region toward a lower-concentration region.

If we substitute Eq. 2 into the one-dimensional continuity equation Eq. 79

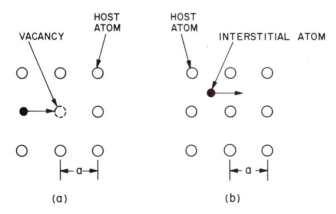

(a)   (b)

Fig. 2 Models of atomic diffusion mechanisms for a two-dimensional lattice, where *a* is the lattice constant. (a) Vacancy mechanism. (b) Interstitial mechanism.[1]

given in Chapter 2 (under the condition that no materials are formed or consumed in the host semiconductor, i.e., $G_n = R_n = 0$), we obtain

$$\frac{\partial C}{\partial t} = -\frac{\partial F}{\partial x} = \frac{\partial}{\partial x}\left[D\frac{\partial C}{\partial x}\right]. \tag{3}$$

When the concentration of the dopant atoms is low, the diffusion coefficient can be considered to be independent of doping concentration, and Eq. 3 becomes

$$\frac{\partial C}{\partial t} = D\frac{\partial^2 C}{\partial x^2}. \tag{4}$$

Equation 4 is often referred to as *Fick's diffusion equation*.

Figure 3 shows the measured diffusion coefficients for low concentrations of various dopant impurities in silicon and gallium arsenide.[4] The logarithm of the diffusion coefficient plotted against the reciprocal of the absolute tempera-

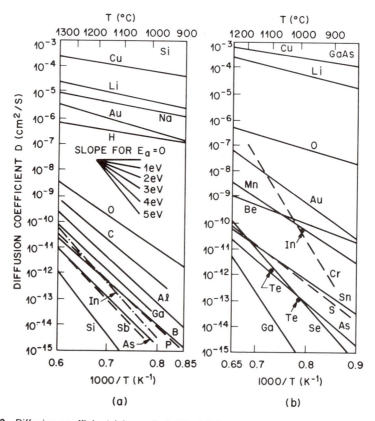

Fig. 3 Diffusion coefficient (also called diffusivity) as a function of the reciprocal of temperature for (a) silicon and (b) gallium arsenide.[4]

ture gives a straight line in most of the cases. This implies that over the temperature range, the diffusion coefficients can be expressed as

$$D \,=\, D_o \, \exp \left[\frac{-E_a}{kT} \right] \tag{5}$$

where D_o is the diffusion coefficient (in units of cm^2/s) extrapolated to infinite temperature and E_a is the activation energy (in eV).

For the interstitial diffusion model, E_a is related to the energies required to move dopant atoms from one interstitial site to another; the values of E_a are found to be between 0.5 and 1.5 eV in both silicon and gallium arsenide. For the vacancy diffusion model, E_a is related to both the energies of motion and the energies of formation of vacancies. Thus, E_a for vacancy diffusion is larger than that for interstitial diffusion usually between 3 and 5 eV. For fast diffusants as shown in the upper portion of Figs. 3a and 3b (e.g., Cu in Si and GaAs), the measured activation energies are less than 2 eV, and interstitial atomic movement is the dominant diffusion mechanism. For slow diffusants as shown in the lower portion of Figs. 3a and 3b (e.g., As in Si and GaAs), E_a is larger than 3 eV, and vacancy diffusion is the dominant mechanism.

10.1.2   Diffusion Profiles

The diffusion profile of the dopant atoms is dependent on the initial and boundary conditions. In this subsection we consider the two most important methods for diffusion, namely, constant-surface-concentration diffusion and constant-total-dopant diffusion. In the first method impurity atoms are transported from a vapor source onto the semiconductor surface and diffused into semiconductor wafers. The vapor source maintains a constant level of surface concentration during the entire diffusion period. In the second method a fixed amount of dopant is deposited onto the semiconductor surface and is subsequently diffused into the wafers.

Constant-Surface-Concentration Diffusion   The initial condition at $t = 0$ is

$$C(x, \, 0) = 0 \tag{6}$$

which states that the dopant concentration in the host semiconductor is initially zero. The boundary conditions are

$$C(0, \, t) = C_s \tag{7a}$$

and

$$C(\infty, \, t) = 0 \tag{7b}$$

where C_s is the surface concentration (at $x = 0$) which is independent of time. The second boundary condition states that at large distances from the surface there are no impurity atoms.

The solution of the diffusion equation (Eq. 4) that satisfies the initial and boundary conditions is given by[5]

$$C(x, t) = C_s \, \text{erfc} \left[\frac{x}{2\sqrt{Dt}} \right] \tag{8}$$

where erfc is the complementary error function and $\sqrt{Dt}$ is the diffusion length. Some properties of the erfc are summarized in Table 1. The diffusion profile for the constant-surface-concentration condition is shown in Fig. 4a, where we plot, on both linear and logarithmic scales, the normalized concentration as a function of depth for three values of the diffusion length $\sqrt{Dt}$ corresponding to three consecutive diffusion times (for a given diffusion temperature, D is fixed). Note that as the time progresses, the dopant penetrates deeper into the semiconductor.

The total number of dopant atoms per unit area of the semiconductor is given by

$$Q(t) = \int_0^\infty C(x, t) \, dx \ . \tag{9}$$

Substituting Eq. 8 into Eq. 9 yields

$$Q(t) = \frac{2}{\sqrt{\pi}} \, C_s \sqrt{Dt} \simeq 1.13 \, C_s \sqrt{Dt} \ . \tag{10}$$

This expression can be interpreted as follows. The quantity $Q(t)$ represents the area under one of the diffusion profiles of the linear plot in Fig. 4a. These profiles can be approximated by triangles with height C_s and base $2\sqrt{Dt}$. This leads to $Q(t) \simeq C_s \sqrt{Dt}$, which is close to the exact result obtained from Eq. 10.

Table 1 Error Function Algebra

$$\text{erf}(x) \equiv \frac{2}{\sqrt{\pi}} \int_0^x e^{-y^2} dy$$

$$\text{erfc}(x) \equiv 1 - \text{erf}(x)$$

$$\text{erf}(0) = 0$$

$$\text{erf}(\infty) = 1$$

$$\text{erf}(x) \cong \frac{2}{\sqrt{\pi}} x \quad \text{for } x \ll 1$$

$$\text{erfc}(x) \simeq \frac{1}{\sqrt{\pi}} \frac{e^{-x^2}}{x} \quad \text{for } x \gg 1$$

$$\frac{d}{dx} \text{erf}(x) = \frac{2}{\sqrt{\pi}} e^{-x^2}$$

$$\frac{d^2}{dx^2} \text{erf}(x) = -\frac{4}{\sqrt{\pi}} x e^{-x^2}$$

$$\int_0^x \text{erfc}(y') dy' = x \, \text{erfc}(x) + \frac{1}{\sqrt{\pi}} (1 - e^{-x^2})$$

$$\int_0^\infty \text{erfc}(x) \, dx = \frac{1}{\sqrt{\pi}}$$

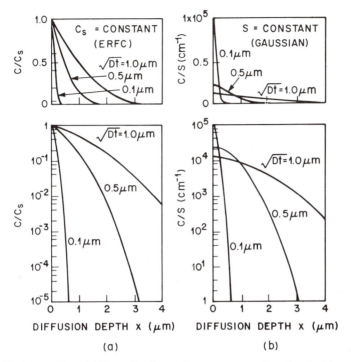

Fig. 4 Diffusion profiles. (a) Normalized complementary error function (erfc) versus distance for successive diffusion times. (b) Normalized Gaussian function versus distance for successive times.

A related quantity is the gradient of the diffusion profile dC/dx. The gradient can be obtained by differentiating Eq. 8:

$$\left.\frac{dC}{dx}\right|_{x,\,t} = -\frac{C_s}{\sqrt{\pi Dt}}\, e^{-x^2/4Dt} \tag{11}$$

Problem

For a boron diffusion in silicon at 1000°C, the surface concentration is maintained at 10^{19} cm^{-3} and the diffusion time is 1 hr. Find $Q(t)$ and the gradient at $x = 0$ and at a location where the dopant concentration reaches 10^{15} cm^{-3}.

Solution

The diffusion coefficient of boron at 1000°C, as obtained from Fig. 3, is about 2×10^{-14} cm^2/s, so that the diffusion length is

$$\sqrt{Dt} = \sqrt{2 \times 10^{-14} \times 3600} = 8.48 \times 10^{-6} \text{ cm}$$

$$Q(t) = 1.13\, C_s \sqrt{Dt} = 1.13 \times 10^{19} \times 8.48 \times 10^{-6} = 9.5 \times 10^{13} \text{ atoms/cm}^2$$

$$\left.\frac{dC}{dx}\right|_{x=0} = -\frac{C_s}{\sqrt{\pi Dt}} = \frac{-10^{19}}{\sqrt{\pi} \times 8.48 \times 10^{-6}} = -6.7 \times 10^{23} \text{ cm}^{-4}.$$

When $C = 10^{15}$ cm^{-3}, the corresponding distance x_j is given by Eq. 8, or

$$x_j = 2\sqrt{Dt}\ \mathrm{erfc}^{-1}\left[\frac{10^{15}}{10^{19}}\right] = 2\sqrt{Dt}\ (2.75) = 4.66\times10^{-5}\,\mathrm{cm} = 0.466\,\mu m$$

$$\left.\frac{dC}{dx}\right|_{x=0.466\,\mu m} = -\frac{C_s}{\sqrt{\pi Dt}}\,e^{-x_j^2/4Dt} = -3.5\times10^{20}\ \mathrm{cm}^{-4}.$$

Constant-Total-Dopant Diffusion For this case a fixed (or constant) amount of dopant is deposited onto the semiconductor surface in a thin layer, and the dopant subsequently diffuses into the semiconductor. The initial condition is the same as in Eq. 6. The boundary conditions are

$$\int_0^\infty C(x,\ t)\,dx = S \tag{12a}$$

and

$$C(\infty,\ t) = 0 \tag{12b}$$

where S is the total amount of dopant per unit area.

The solution of the diffusion equation, Eq. 4, that satisfies the above conditions is

$$C(x,\ t) = \frac{S}{\sqrt{\pi Dt}}\,\exp\left[-\frac{x^2}{4Dt}\right]. \tag{13}$$

This expression is the Gaussian distribution. Since the dopant will move into the semiconductor as time increases, to keep the total dopant S constant, the surface concentration must decrease. This is indeed the case, since the surface concentration is given by Eq. 13 with $x = 0$:

$$C_s(t) = \frac{S}{\sqrt{\pi Dt}}. \tag{14}$$

Figure 4b shows the dopant profile for a Gaussian distribution where we plot the normalized concentration (C/S) as a function of the distance for three increasing diffusion lengths. Note the reduction of the surface concentration as the diffusion time increases. The gradient of the diffusion profile is obtained by differentiating Eq. 13 and is

$$\left.\frac{dC}{dx}\right|_{x,\ t} = -\frac{xS}{2\sqrt{\pi}(Dt)^{3/2}}\,e^{-x^2/4Dt} = -\frac{x}{2Dt}\,C(x,\ t). \tag{15}$$

The gradient (or slope) is zero at $x = 0$ and at $x = \infty$, and the maximum gradient occurs at $x = \sqrt{2Dt}$.

Both the complementary error function Eq. 8 and the Gaussian distribution Eq. 13 are functions of a normalized distance $x/2\sqrt{Dt}$. Thus, if we normalize the dopant concentration with the surface concentration, we can represent

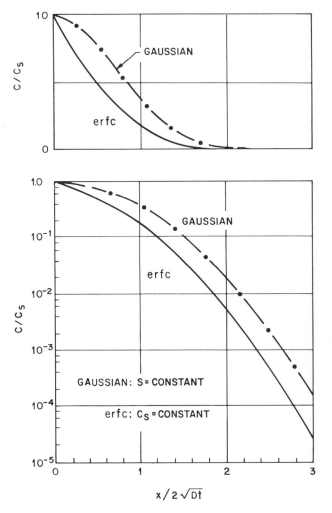

Fig. 5   Normalized concentration versus normalized distance for the erfc and the Gaussian function.

each distribution with a single curve valid for all diffusion times, as shown in Fig. 5. It is important to realize that in the case of the Gaussian distribution we have introduced a time variable into the normalizing parameter, i.e., $C_s = S/\sqrt{\pi D t}$, while for the complementary error function C_s is a constant.

In integrated-circuit processing, a two-step diffusion process is commonly used, in which a *predeposition* diffused layer is first formed under a constant-surface-concentration condition and then is followed by a *drive-in* diffusion (also called *redistribution* diffusion) under a constant-total-dopant condition. For most practical cases the diffusion length $\sqrt{D t}$ for the predeposition diffusion is much smaller than the diffusion length for the drive-in diffusion.

Therefore, we can consider the predeposition profile as a delta function at the surface, and we can regard the extent of the penetration of the predeposition profile to be negligibly small in comparison to that of the final profile that results from the drive-in step.

10.1.3 Evaluation of Diffused Layers

The results of a diffusion process can be evaluated by three measurements—the junction depth, the sheet resistance, and the dopant profile of the diffused layer. The junction depth can be delineated by cutting a groove into the semiconductor and etching the surface with a solution (e.g., 100 cm^3 HF and a few drops of HNO$_3$ for silicon) that stains the p-type region darker than the n-type region as illustrated in Fig. 6a. If R_o is the radius of the tool used to form the groove, then the junction depth x_j is given by

$$x_j = \sqrt{R_o^2 - b^2} - \sqrt{R_o^2 - a^2} \qquad (16)$$

where a and b are indicated in the figure. In addition, if R_o is much larger than a and b, then

$$x_j \simeq \frac{a^2 - b^2}{2R_o} . \qquad (17)$$

The junction depth x_j as illustrated in Fig. 6b is the position where the dopant concentration equals the substrate concentration C_B, or

$$C(x_j) = C_B. \qquad (18)$$

Thus, if the junction depth and C_B are known, the surface concentration C_s and the impurity distribution can be calculated, provided the diffusion profile follows one or the other simple equation derived in Section 10.1.2.

The resistance of a diffused layer can be measured by the four-point probe technique described in Chapter 2. The *sheet resistance R* is related to the junction depth x_j, the carrier mobility μ (which is a function of the total impurity

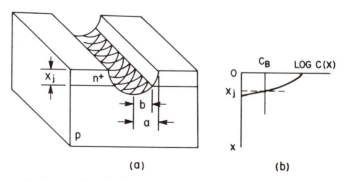

(a)   (b)

Fig. 6 Junction depth measurement by grooving and staining.

concentration), and the impurity distribution $C(x)$ by the following expression:

$$R = \frac{1}{q \int_0^{x_j} \mu C(x) \, dx}. \tag{19}$$

For a given diffusion profile, the average resistivity $\bar{\rho} = Rx_j$ is uniquely related to the surface concentration C_s and the substrate-doping concentration for an assumed diffusion profile. Design curves relating C_s and $\bar{\rho}$ have been calculated for simple diffusion profiles such as the erfc or Gaussian distribution.[6] To use these curves we must be sure that the diffusion profiles agree with the assumed profiles. For low concentration and deep diffusions, the diffusion profiles generally can be represented by the aforementioned simple functions. However, as we discuss in the next section, for high concentration and shallow diffusions, the diffusion profiles cannot be represented by these simple functions.

The diffusion profile can be measured using the capacitance–voltage technique described in Chapter 5. The majority carrier profile (which is equal to the impurity profile, if impurities are fully ionized) can be determined by measuring the reverse-bias capacitance of a p–n junction or a Schottky barrier diode as a function of the applied voltage. A more elaborate method is the secondary-ion-mass spectroscope (SIMS) technique, which measures the total impurity profile. In the SIMS technique, an ion beam sputters material off the surface of a semiconductor, and the ion component is detected and mass-analyzed. This technique has high sensitivity to many elements such as boron and arsenic, and it is an ideal tool for providing the precision needed for profile measurements in high-concentration or shallow-junction diffusions.[7]

10.2 EXTRINSIC DIFFUSION

The diffusion profiles described in Section 10.1 are for constant-diffusion coefficients. These profiles occur when the doping concentration is lower than the intrinsic-carrier concentration n_i at the diffusion temperature (e.g., at $T = 1000°C$, $n_i = 5 \times 10^{18}$ cm^{-3} for silicon and 5×10^{17} cm^{-3} for gallium arsenide). Doping profiles that have concentrations less than n_i are in the *intrinsic* diffusion region as indicated in Fig. 7. In this region the resulting dopant profiles of sequential or simultaneous diffusions of n- and p-type impurities can be determined by superposition, that is, the diffusions can be treated independently. However, for dopant concentrations above n_i, we enter the *extrinsic* diffusion region where the diffusion coefficients become concentration dependent.[8] In the extrinsic diffusion region the diffusion profiles are more complicated, and there are interactions and cooperative effects among the sequential or simultaneous diffusions.

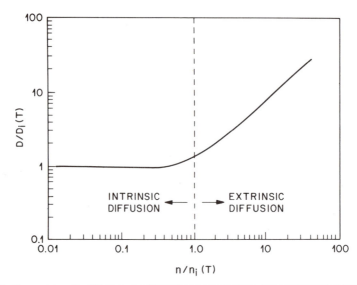

Fig. 7   Donor impurity diffusion coefficient versus electron concentration showing regions of intrinsic and extrinsic diffusion.[8]

10.2.1   Concentration-Dependent Diffusivity

As mentioned previously, when a host atom acquires sufficient energy from the lattice vibration to leave its lattice site, a vacancy is created. Depending on the charges associated with a vacancy, we can have a neutral vacancy V^0, an acceptor vacancy V^-, a double-charged acceptor vacancy V^{2-}, a donor vacancy V^+, and so forth. We expect that the vacancy density of a given charge state (i.e., the number of vacancies per unit volume, C_V) has a temperature dependence similar to that of the carrier density (refer to Eq. 32 in Chapter 1), that is,

$$C_V = C_i \exp \left[\frac{E_F - E_i}{kT} \right] \tag{20}$$

where C_i is the intrinsic vacancy density, E_F is the Fermi level, and E_i is the intrinsic Fermi level.

If the dopant diffusion is dominated by the vacancy mechanism, the diffusion coefficient is expected to be proportional to the vacancy density. At low doping concentrations ($n < n_i$), the Fermi level coincides with the intrinsic Fermi level ($E_F = E_i$). The vacancy density is equal to C_i and is independent of doping concentration. The diffusion coefficient, which is proportional to C_i, also is independent of doping concentration. At high concentrations ($n > n_i$), the Fermi level will move toward the conduction band edge (for donor-type vacancies), and the term $\exp (E_F - E_i)/kT$ becomes larger than unity. This causes C_V to increase, which in turn causes the diffusion coefficient to increase, as shown in Fig. 7.

When the diffusion coefficient varies with dopant concentration, Eq. 3 should be used as the diffusion equation instead of Eq. 4 in which D is independent of C. We shall consider the case where the diffusion coefficient can be written as

$$D = D_s \left[\frac{C}{C_s} \right]^\gamma \tag{21}$$

where C_s is the surface concentration, D_s is the diffusion coefficient at the surface, and γ is a positive integer. For such a case, we can write the diffusion equation, Eq. 3, as an ordinary differential equation and solve it numerically. The solutions for a constant-surface-concentration diffusion are shown in Fig. 8 along with the result for D = constant (or γ = 0).[9]

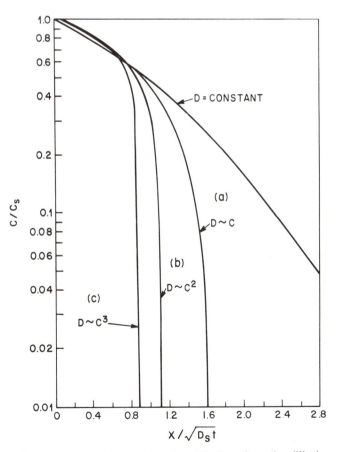

Fig. 8 Normalized diffusion profiles for extrinsic diffusion where the diffusion coefficient becomes concentration dependent.[9]

From Fig. 8 we note that for concentration-dependent diffusivities the diffusion profiles are much steeper at low concentrations ($C \ll C_s$). Therefore, highly abrupt junctions are formed when diffusions are made into a background of an opposite impurity type. The abruptness of the doping profile results in a junction depth virtually independent of the background concentration. Note that the junction depth (see Fig. 8) is given by

$$x_j = 1.6 \sqrt{D_s t} \quad \text{for } D \sim C \ (\gamma = 1)$$
$$x_j = 1.1 \sqrt{D_s t} \quad \text{for } D \sim C^2 \ (\gamma = 2) \,. \tag{22}$$
$$x_j = 0.87 \sqrt{D_s t} \quad \text{for } D \sim C^3 \ (\gamma = 3)$$

10.2.2 Diffusion Profiles

Diffusion in Silicon   The measured diffusion coefficients of arsenic and phosphorus as a function of dopant concentration are shown in Fig. 9. The diffusion of arsenic in silicon is associated with the acceptor-type vacancy V^-,

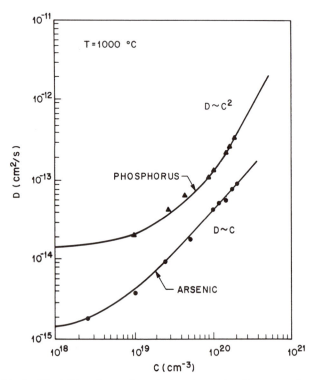

Fig. 9 Extrinsic diffusivities of arsenic and phosphorus in silicon as a function of dopant concentration.[8]

and the diffusion coefficient for $n > n_i$ can be written as:

$$D = D_o' e^{-E_a'/kT} \left[\frac{n}{n_i} \right] \qquad (23)$$

where D_o' is 45.8 cm^2/s, E_a' is 4.05 eV, and n is the carrier concentration, which equals the dopant concentration.[8]

Equation 23 has the same form as Eq. 21 for $\gamma = 1$, and the resulting doping profile is shown in curve a of Fig. 8. The junction depth is essentially independent of the background p-type concentration and is given by $x_j = 1.6 \sqrt{D_s t}$. Since the surface diffusivity D_s is given by Eq. 23 where $n = C_s$, the junction depth is then

$$x_j = 1.6 \left[D_o' e^{-E_a'/kT} \left[\frac{C_s}{n_i} \right] t \right]^{1/2} \qquad (24)$$

A closed-form solution of the diffusion equation with D given by Eq. 23 can be written in a polynomial form as

$$C = C_s(1 - 0.87Y - 0.45Y^2) \qquad (25)$$

where

$$Y \equiv \frac{x}{\sqrt{4D_s t}}. \qquad (25a)$$

The measured diffusion profile for arsenic is shown in curve a of Fig. 10 where Y_j is $x_j/\sqrt{4D_s t}$ and x_j is given by Eq. 24. Note the excellent agreement obtained between the experimental results and Eq. 25. Because of its abrupt doping profile, arsenic is used extensively in integrated circuits to form shallow junctions such as the source and drain junctions in n-channel MOSFETs.

The diffusion of boron in silicon is associated with donor-type vacancy V^+, and the diffusion coefficient varies approximately linearly with dopant concentration. The expression for D has the same form as Eq. 23, except that D_0' is 1.52 cm^2/s, E_a' is 3.46 eV, and n is replaced by p. The measured boron profile is shown in curve b of Fig. 10 and is slightly less abrupt than the arsenic profile. Note that the doping profiles for arsenic and boron are much steeper than the erfc case (curve c). The experimental data for extrinsic boron diffusion can be fitted to the curve

$$C = C_S(1 - Y^{2/3}) \qquad (26)$$

where

$$Y \equiv \left[\frac{x^2}{6D_s t} \right]^{3/2}. $$

The corresponding junction depth is also given by Eq. 24.

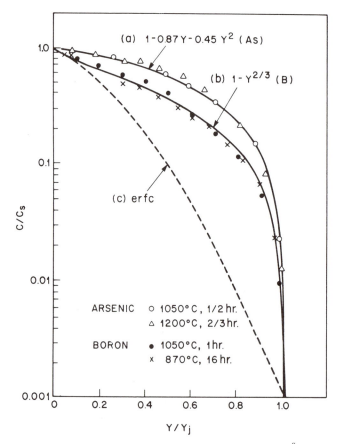

Fig. 10 Normalized diffusion profiles for arsenic and boron in silicon.[8] The erfc distribution is shown for comparison.

The diffusion of phosphorus in silicon is associated with the doubly charged acceptor vacancy V^{2-}, and the diffusion coefficient at high concentration varies as C^2 (Fig. 9). We would expect that the diffusion profile of phosphorus resembles that shown in curve b of Fig. 8. However, because of a *dissociation effect*, the diffusion profile exhibits anomalous behavior.

Figure 11 shows phosphorus diffusion profiles for various surface concentrations after diffusion into silicon for 1 hr at 1000°C.[10] When the surface concentration is low, corresponding to the intrinsic diffusion region, the diffusion profile is given by an erfc (curve a). As the concentration increases, the profile begins to deviate from the simple expression (curves b and c). At very high concentration (curve d), the profile near the surface is indeed similar to that shown in curve b of Fig. 8. However, at concentration n_e, a kink occurs and this is followed by a rapid diffusion in the tail region. The concentration n_e corresponds to a Fermi level 0.11 eV below the conduction band. At this energy level, the coupled impurity–vacancy pair (P^+V^{2-}) dissociates to P^+,

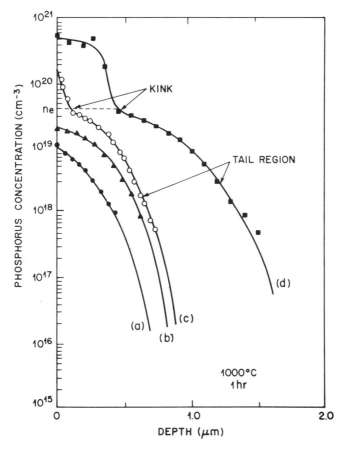

Fig. 11 Phosphorus diffusion profiles for various surface concentrations after diffusion into silicon for 1 hr at 1000° C.[10]

V^-, and an electron. Thus, the dissociation generates a large number of singly-charged acceptor vacancies V^-, which in turn enhance the diffusion in the tail region of the profile. The diffusivity in the tail region is over 10^{-12} cm²/s, which is about two orders of magnitude larger than the intrinsic diffusivity at 1000°C. Because of its high diffusivity, phosphorus is commonly used to form deep junctions such as the *n*-tubs in a CMOS.

Emitter-Push Effect In silicon *n–p–n* bipolar transistors using a phosphorus-diffused emitter and a boron-diffused base, the base region under the emitter region (inner base) is deeper by up to 0.6 μm than that outside the emitter region (outer base). This phenomena is called the *emitter push effect* and is illustrated in the insert of Fig. 12. The dissociation of phosphorus–vacancy (P^+V^{2-}) pairs at the kink region of the phosphorus profile (Fig. 11, curve *d*) provides a mechanism for the enhanced diffusion of phosphorus in the tail region. The diffusivity of boron under the emitter region (inner base)

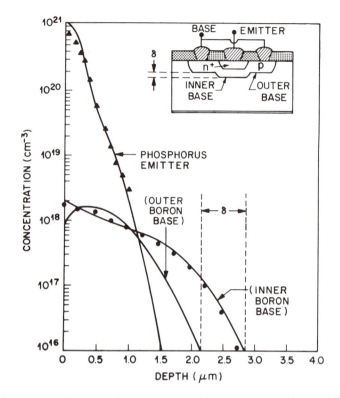

Fig. 12 Calculated and measured boron and phosphorus n–p–n transistor profile showing the emitter push effect. Emitter diffusion is at 1000° C, for 1 hr, followed by a 900° C, 45-min steam oxidation.[8]

also is expected to be enhanced by the dissociation of P^+V^{2-} pairs. Figure 12 shows close agreement between the measured profiles of boron and phosphorus and the calculated results (solid curves) based on the vacancy diffusion model.[8]

Zinc Diffusion in Gallium Arsenide   The fundamental mechanisms for the diffusion of impurities in gallium arsenide have not yet been firmly established. We expect diffusion in gallium arsenide to be more complicated than that in silicon because the diffusion of impurities may involve atomic movements on both the gallium and arsenic sublattices. Vacancies play a dominant role in diffusion processes in gallium arsenide, because both p- and n-type impurities must ultimately reside in lattice sites; however, the charge states of the vacancies have not been established.

Zinc is the most extensively studied diffusant in gallium arsenide. Its diffusion coefficient is found to vary as C^2. Therefore, the diffusion profiles are steep, as shown[11] in Fig. 13 and resembles curve b of Fig. 8. Note that even for the case of the lowest surface concentration, the diffusion is in the extrinsic-diffusion region, because n_i for GaAs at 1000°C is less than

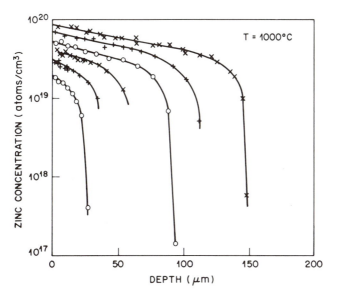

Fig. 13 Diffusion profiles of zinc in gallium arsenide after annealing at 1000° C for 2.7 hr.[11] The different surface concentrations are obtained by maintaining the Zn source at temperatures in the range 600 to 800° C.

10^{18} cm^{-3}. As can be seen in Fig. 13, the surface concentration has a profound effect on the junction depth. The diffusivity varies linearly with the partial pressure of the zinc vapor, and the surface concentration is proportional to the square root of the partial pressure. Therefore, from Eq. 22, the junction depth is linearly proportional to the surface concentration.

10.3 DIFFUSION-RELATED PROCESSES

In this section we consider briefly a few processes in which diffusion plays an important role and the impact of these processes on device performances.

10.3.1 Oxide Masking

The diffusivities of commonly used doping impurities (e.g., As, B, and P) are considerably smaller in silicon dioxide than in silicon. Therefore, silicon dioxide can be used as an effective mask against impurities. Oxide masking is important because it is the basis of present-day integrated-circuit technology. If we etch windows in the oxide and use the remaining oxide as a mask, we can incorporate dopant impurities into a silicon substrate in selective areas to form *p–n* junction regions.

For silicon dioxide, the diffusion process can be described as occurring in two steps. During the first step, the dopant impurities (e.g., phosphorus in the form of P_2O_5 vapor) react with silicon dioxide to form a glass. As the process continues, the thickness of the glass increases until the entire silicon dioxide layer is converted into a glass (e.g., phosphosilicate glass). After the glass forms, the second step begins. The dopant impurity diffuses through the glass; upon reaching the glass–silicon interface, it enters and diffuses into the silicon. During the first step, silicon dioxide is completely effective in masking the silicon against dopant impurities in the gas phase. Therefore, the thickness of silicon dioxide required for masking is determined by the rate of the formation of the glass, which in turn is determined by the diffusion of the diffusant into the silicon dioxide. Typical diffusivities in silicon dioxide at 900°C are 4×10^{-19} cm²/s for arsenic, 3×10^{-19} cm²/s for boron, and 10^{-18} cm²/s for phosphorus.

Figure 14 shows the minimum thickness d of dry oxygen-grown silicon dioxide required to mask against phosphorus and boron as a function of temperature and time.[12] Because it has a higher diffusivity in SiO₂, phosphorus

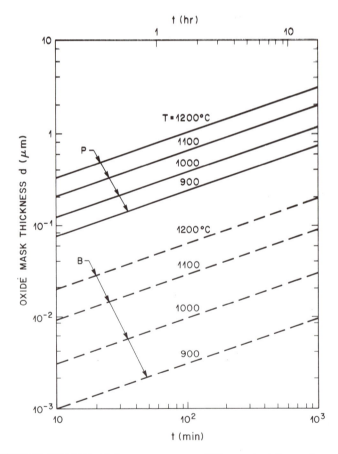

Fig. 14 Minimum thickness of dry-oxygen-grown SiO₂ required to mask against phosphorus and boron as a function of diffusion time with diffusion temperature as a parameter.[12]

requires thicker masking layers. For a given temperature, d varies as $\sqrt{t}$. This is expected since the diffusion length is given by $\sqrt{Dt}$.

10.3.2   Lateral Diffusion

The one-dimensional diffusion equation discussed previously can describe satisfactorily the diffusion process, except at the edge of the mask window. Here the impurities will diffuse downward and sideways (i.e., laterally). In this case, we must consider a two-dimensional diffusion equation and use a numerical technique to obtain the diffusion profiles under different initial and boundary conditions.

Figure 15 shows the contours of constant doping concentration for a constant-surface-concentration diffusion condition assuming that the diffusivity is independent of concentration.[13] At the far right of the figure, the variation of the dopant concentration from $0.5C_s$ to $10^{-4}C_s$ (where C_s is the surface concentration) corresponds to the erfc distribution given by Eq. 8. The contours are in effect a map of the location of the junctions created by diffusing into various background concentrations. For example, at $C/C_s = 10^{-4}$ (i.e., the background doping is 10^4 times lower than the surface concentration), we see from this constant-concentration curve that the vertical penetration is about 2.8 μm, while the lateral penetration is about 2.3 μm (i.e., the penetration along the diffusion mask–semiconductor interface). Therefore, the lateral penetration is about 80% of the penetration in the vertical direction for concentrations that are three or more orders of magnitude below the surface concentration. Similar results are obtained for a constant-total-dopant diffusion condition. The ratio of lateral to vertical penetration is about 75%. For concentration-dependent diffusivities, the ratio is found to be reduced slightly to about 65 to 70%.

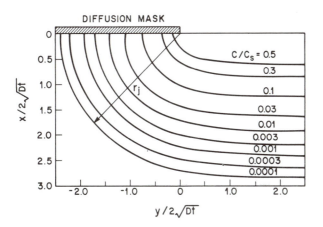

Fig. 15   Diffusion contours at the edge of an oxide window, where r_j is the radius of curvature.[13]

Because of the lateral-diffusion effect, the junction consists of a central plane (or flat) region with approximately cylindrical edges with a radius of curvature r_j as shown in Fig. 15. In addition, if the diffusion mask contains sharp corners, the shape of the junction near the corner will be roughly spherical due to lateral diffusion. Since the electric-field intensities are higher for cylindrical and spherical junction regions, the avalanche breakdown voltages of such regions can be substantially lower than that of a plane junction having the same background doping. This junction "curvature effect" has been discussed in Chapter 3.

10.3.3 Impurity Redistribution During Oxidation

Dopant impurities near the silicon surface will be redistributed during thermal oxidation. The redistribution depends on several factors. When two solid phases are brought together, an impurity in one solid will redistribute between the two solids until it reaches equilibrium. This is similar to our previous discussion on impurity redistribution in crystal growth from the melt. The ratio of the equilibrium concentration of the impurity in the silicon to that in the silicon dioxide is called the *segregation coefficient*, and is defined as

$$k = \frac{\text{equilibrium concentration of impurity in silicon}}{\text{equilibrium concentration of impurity in SiO}_2}. \tag{27}$$

A second factor that influences impurity distribution is that the impurity may diffuse rapidly through the silicon dioxide and escape to the gaseous ambient. If the diffusivity of the impurity in silicon dioxide is large, this factor will be important. A third factor in the redistribution process is that the oxide is growing, and thus the boundary between the silicon and the oxide is advancing into the silicon as a function of time. The relative rate of this advance compared to the diffusion rate of the impurity through the oxide is important in determining the extent of the redistribution. Note that even if the segregation coefficient of an impurity k equals unity, some redistribution of the impurity in the silicon will still take place. As indicated in Chapter 9, the oxide layer will be about twice as thick as the silicon layer it replaced. Therefore, the same amount of impurity will now be distributed in a larger volume resulting in a depletion of the impurity from the silicon.

Four possible redistribution processes are illustrated[5] in Fig. 16. These processes can be classified into two groups. In one group the oxide takes up the impurity (Figs. 16a and b for $k < 1$), and in the other the oxide rejects the impurity (Figs. 16c and d for $k > 1$). In each case, what happens depends on how rapidly the impurity can diffuse through the oxide. In group 1, the silicon surface is depleted of impurities; an example is boron with k approximately equal to 0.3. Rapid diffusion of the impurity through the silicon dioxide increases the amount of depletion; an example is boron-doped silicon heated in a hydrogen ambient, because hydrogen in silicon dioxide enhances the diffusivity of boron. In group 2, k is greater than unity, so that the oxide rejects the impurity. If diffusion of the impurity through the silicon dioxide is

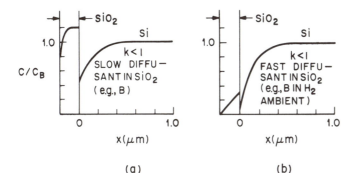

(a)   (b)

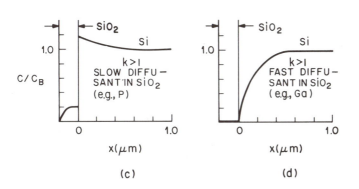

(c)   (d)

Fig. 16 Four different cases of impurity redistribution in silicon due to thermal oxidation.[5]

relatively slow, the impurity piles up near the silicon surface; an example is phosphorus, with k approximately equal to 10. When diffusion through the silicon dioxide is rapid, so much impurity may escape from the solid to the gaseous ambient that the overall effect will be a depletion of the impurity; an example is gallium, with k approximately equal to 20.

The redistributed dopant impurities in silicon dioxide are seldom electrically active. However, redistribution in silicon has an important effect on processing and device performance. For example, the oxidation rate is affected by high dopant concentration in the silicon (see Chapter 9), nonuniform dopant distribution will modify the interpretation of measurements of interface–trap properties, and the surface concentration is directly related to device contact resistance (see Chapter 5).

10.3.4 Impurity Redistribution in Epitaxial Growth

Redistribution of impurities occurs during epitaxial growth, resulting in departures from the ideal, abruptly discontinuous doping profile. The impurity redistribution is related to both the epitaxial growth process itself and to the diffusion process. In either vapor phase or liquid phase epitaxy, the process temperatures are high enough to cause appreciable impurity diffusion in both the substrate and the epitaxial layer. Because the surface of the epitaxial

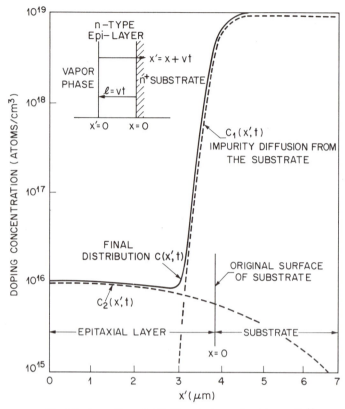

Fig. 17 Impurity redistribution during epitaxial growth.[14]

layer moves during growth, we require a solution to the problem of diffusion with a moving boundary.

The insert in Fig. 17 shows the geometry of the growth layer, where v is the epitaxial-layer growth rate (in cm/s). As the layer grows, the surface moves from the original substrate surface at the rate v. If we introduce a new variable $x' = x + vt$ into the diffusion equation (Eq. 4), we obtain

$$\frac{\partial C}{\partial t} = D \frac{\partial^2 C}{\partial x'^2} - v \frac{\partial C}{\partial x'} . \tag{28}$$

If the original concentration in the substrate is C_1 and the concentration added to the epitaxial layer during growth in C_2, the solution of Eq. 28 is

$$C(x', t) = \frac{C_2}{2} \left[\text{erfc} \left[\frac{x' - vt}{2\sqrt{D_2 t}} \right] + \exp \left[\frac{vx'}{D_2} \right] \text{erfc} \left[\frac{x' + vt}{2\sqrt{D_2 t}} \right] \right]$$

$$+ \frac{C_1}{2} \left[1 + \text{erf} \left[\frac{x' - vt}{2\sqrt{D_1 t}} \right] \right] \tag{29}$$

where D_1 and D_2 are the diffusivities of the impurities in the substrate and in the epitaxial layer, respectively.[14]

Figure 17 shows an example of the behavior of Eq. 29. Assume that in an epitaxial-growth process the temperature is 900°C and the duration is 5.5 min. The growth rate is 0.7 μm/min, and $C_1 = 10^{19}$ cm^{-3}, $C_2 = 10^{16}$ cm^{-3}, $D_1 = 6 \times 10^{-13}$ cm^2/s, and $D_2 = 4 \times 10^{-11}$ cm^2/s. The resulting dopant distribution is obtained by superposition of the two distributions of $C_1(x', t)$ and $C_2(x', t)$. Note that the distribution of $C_1(x', t)$ is displaced from the original substrate into the epitaxial layer by a significant amount. To minimize this effect, we can use higher growth rates and lower growth temperatures, which of course should be compatible with the required film qualities.

10.4 DISTRIBUTION AND RANGE OF IMPLANTED IONS

Ion implantation is the introduction of energetic, charged particles into a substrate such as silicon. The practical use of ion implantation in semiconductor technology has been mainly to change the electrical properties of the substrate. Typical ion energies are between 30 and 300 keV, and typical ion doses vary from 10^{11} to 10^{16} ions/cm^2. Note that the dose is expressed as the number of ions implanted into 1 cm^2 of the semiconductor surface area. The main advantages of ion implantation are its more precise control and reproducibility of impurity dopings and its lower processing temperature requirements compared to those of the diffusion process.

Figure 18 shows schematically an ion implantation system.[2, 15] The ion source contains the ionized dopant atoms (e.g., B$^+$ and As$^+$). The ions pass through a mass-separating analyzer magnet that eliminates unwanted ion species. The selected ions then enter the acceleration tube and are accelerated to high energies by an electric field. The high-energy ion beam passes through the vertical and horizontal scanners and is implanted into the semiconductor substrate.

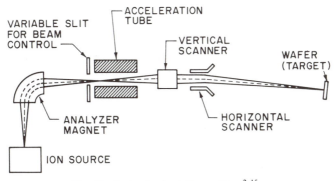

Fig. 18 An ion implantation system.[2, 15]

10.4.1 Ion Distribution

The energetic ions lose their energy through collisions with electrons and nuclei in the substrate and finally come to rest. The total distance which an ion travels in coming to rest is called its *range R*, and it is illustrated in Fig. 19a. The projection of this distance along the axis of incidence is called the *projected range R_p*. Since the number of collisions per unit distance and the energy lost per collision are random variables, there will be a spatial distribution of ions having the same mass and the same initial energy. The statistical fluctuations in the projected range are called the *projected straggle ΔR_p*. There is also a statistical fluctuation along an axis perpendicular to the axis of incidence, called the *lateral straggle $\Delta R_\perp$*.

Figure 19b shows the ion distribution. Along the axis of incidence, the implanted impurity profile can be approximated by a Gaussian distribution function:

$$n(x) = \frac{S}{\sqrt{2\pi}\,\Delta R_p} \exp\left[-\frac{(x-R_p)^2}{2\Delta R_p^2}\right] \tag{30}$$

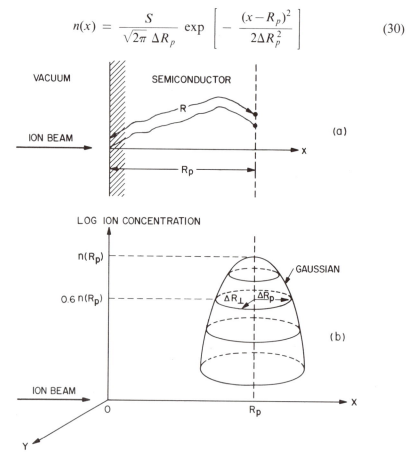

Fig. 19 (a) Schematic of the ion range R and projected range R_p.[19] (b) Two-dimensional distribution of the implanted ions.

where S is the ion dose per unit area. This equation is identical to Eq. 13 for constant-total-dopant diffusion, except that the quantity $4Dt$ is replaced by $2\Delta R_p^2$ and the distribution is shifted along the x-axis by R_p. Thus, for diffusion the maximum concentration is at $x = 0$, while for ion implantation the maximum concentration is at the projected range R_p. The ion concentration is reduced by 40% from its peak value at $(x - R_p) = \pm \Delta R_p$, by one decade at $\pm 2\Delta R_p$, by two decades at $\pm 3\Delta R_p$, and by five decades at $\pm 4.8\Delta R_p$.

Along the axis perpendicular to the axis of incidence, the distribution is also a Gaussian function of the form $\exp(-y^2/2\Delta R_\perp^2)$. Because of this distribution, there will be some lateral implantation.[16] However, the lateral penetration from the mask edge (of the order of $\Delta R_\perp$) is considerably smaller than that from the thermal diffusion process discussed in Section 10.3.

10.4.2 Ion Stopping

There are two stopping mechanisms by which an energetic ion, on entering a semiconductor substrate (also called the target), can be brought to rest. The first is by transferring its energy to the target nuclei. This causes deflection of the incident ion and also dislodges many target nuclei from their original lattice sites. If E is the energy of the ion at any point x along its path, we can define a nuclear stopping power $S_n(E) \equiv (dE/dx)_n$ to characterize this process. The second stopping mechanism is by the interaction of the incident ion with the cloud of electrons surrounding the target's atoms. The ion loses energy in collisions with electrons through Coulombic interaction. The electrons can be excited to higher energy levels (excitation), or they can be ejected from the atom (ionization). We can define an electronic stopping power $S_e(E) \equiv (dE/dx)_e$ to characterize this process.

The average rate of energy loss with distance is given by a superposition of the above two stopping mechanisms:

$$\frac{dE}{dx} = S_n(E) + S_e(E).\tag{31}$$

If the total distance traveled by the ion before coming to rest is R, then

$$R = \int_0^R dx = \int_0^{E_o} \frac{dE}{S_n(E) + S_e(E)}\tag{32}$$

where E_o is the initial ion energy. The quantity R has been defined previously as the range.

We can visualize the nuclear stopping process by considering the elastic collision between an incoming hard sphere (energy E_o and mass M_1) and a target hard sphere (initial energy zero and mass M_2) as illustrated in Fig. 20. When the spheres collide, momentum is transferred along the centers of the spheres. The deflection angle θ and the velocities v_1 and v_2 can be obtained from the requirements for conservation of momentum and energy. The maximum

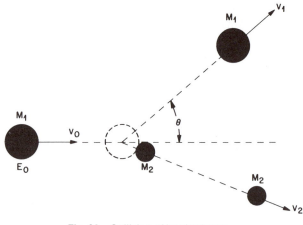

Fig. 20 Collision of hard spheres.

energy loss is in a head-on collision. For this case, the energy loss by the incident particle M_1 or the energy transferred to M_2 is

$$\frac{1}{2} M_2 v_2^2 = \left[\frac{4 M_1 M_2}{(M_1 + M_2)^2} \right] E_o. \qquad (33)$$

Since M_2 is usually of the same order of magnitude as M_1, a large amount of energy can be transferred in nuclear stopping process.

Detailed calculations show that the nuclear stopping power increases linearly with energy at low energies (similar to Eq. 33), and $S_n(E)$ reaches a maximum at some intermediate energy. At high energies, $S_n(E)$ becomes smaller because fast particles may not have sufficient interaction time with the target atoms to achieve effective energy transfer. The calculated values of $S_n(E)$ for arsenic, phosphorus, and boron in silicon at various energies are shown in Fig. 21 (solid lines, where the superscript indicates the atomic weight).[17] Note that heavier atoms (such as arsenic) have larger nuclear stopping power, that is, larger energy loss per unit distance.

The electronic stopping power is found to be proportional to the velocity of the incident ion, or

$$S_e(E) = k_e \sqrt{E} \qquad (34)$$

where the coefficient k_e is a relatively weak function of atomic mass and atomic number. The value of k_e is approximately $10^7 \, (eV)^{1/2}/cm$ for silicon and $3 \times 10^7 \, (eV)^{1/2}/cm$ for gallium arsenide. The electronic stopping power in silicon is plotted in Fig. 21 (dotted line). Also shown in the figure are the crossover energies, at which $S_e(E)$ equals $S_n(E)$. For boron, which has a relatively low ion mass compared to the target silicon atom, the crossover energy is only 10 keV. This means that over the practical implantation energy range of 30 to 300 keV, the main energy loss mechanism is due to electronic stopping.

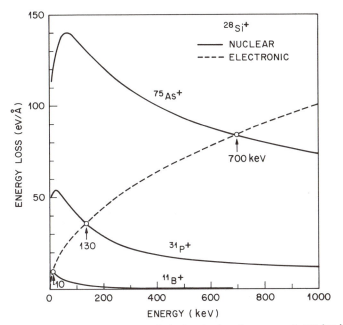

Fig. 21 Nuclear stopping power $S_n(E)$ and electronic stopping power $S_e(E)$ for As, P, and B in Si. The points of intersection of the curves correspond to the energy at which nuclear and electronic stopping are equal.[17]

On the other hand, for arsenic with relatively high ion mass, the crossover energy is 700 keV. Thus, nuclear stopping dominates over the 30- to 300-keV energy range. For phosphorus, the crossover energy is 130 keV. For an E_o less than 130 keV, nuclear stopping will dominate; for higher energies, electronic stopping will take over.

Once $S_n(E)$ and $S_e(E)$ are known, we can calculate the range from Eq. 32. This in turn can give us the projected range and projected straggle with the help of the following approximate equations[15]:

$$R_p \cong \frac{R}{1 + (M_2/3M_1)} \tag{35}$$

$$\Delta R_p \cong \frac{2}{3}\left[\frac{\sqrt{M_1 M_2}}{M_1 + M_2}\right] R_p. \tag{36}$$

The projected ranges for arsenic, boron, and phosphorus in silicon and silicon dioxide are shown[18] in Fig. 22. As expected, the larger the energy loss, the smaller the range. In a first-order approximation, the projected range increases linearly with ion energy. Figure 23 shows the projected straggle ΔR_p and the lateral straggle $\Delta R_\perp$ in silicon for the same elements shown in Fig. 22. For a given element at a given incident energy, the two straggles are comparable and usually are within $\pm 20\%$.

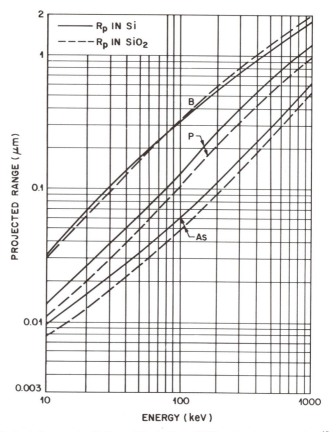

Fig. 22 Projected range for B, P, and As in Si and SiO$_2$ at various energies.[18] The results pertain to amorphous silicon target and thermal SiO$_2$.

The projected ranges for hydrogen, zinc, selenium, cadmium, and tellurium in gallium arsenide are shown in Fig. 24, and the corresponding projected and lateral straggles are shown[16] in Fig. 25. If we compare Fig. 22 with Fig. 24, we see that most of the popular dopants have larger projected ranges in silicon than they have in gallium arsenide.

Problem

Assume that a 125-mm-diameter silicon wafer is implanted uniformly with 80 keV boron ions with a dose of 3×10^{14} ions/cm^2. What are the projected range and peak concentration? If the implantation takes 10 min, what is the ion beam current?

Solution

From Figs. 22 and 23, we find that the projected range of 80 keV boron ions is 0.25 μm and the projected straggle is 0.064 μm. From Eq. 30, we obtain the peak concentration:

$$\frac{S}{\sqrt{2\pi}\,\Delta R_p} = \frac{3 \times 10^{14}}{\sqrt{2\pi}\,(0.064 \times 10^{-4})} = 1.9 \times 10^{19} \text{ ions/cm}^3 .$$

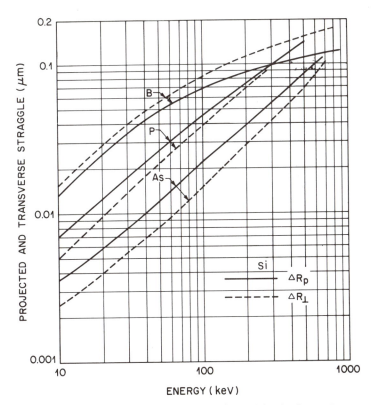

Fig. 23 Ion straggles ΔR_p (projected) and $\Delta R_\perp$ (lateral) for As, P, and B ions in silicon.[18]

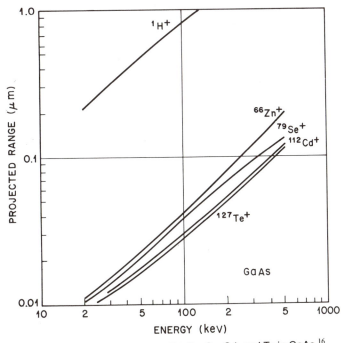

Fig. 24 Projected range of H, Zn, Se, Cd, and Te in GaAs.[16]

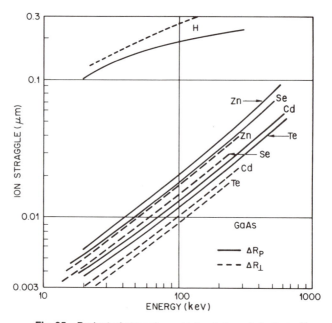

Fig. 25 Projected straggle and lateral straggle in GaAs.[16]

The total number of implanted ions is

$$Q = (3 \times 10^{14} \text{ ions/cm}^3) \left[\pi \times \left[\frac{12.5}{2} \right]^2 \right] = 3.7 \times 10^{16} \text{ ions} .$$

The required ion beam current is given by the total charges qQ divided by the implantation time:

$$I = \frac{qQ}{t} = \frac{(1.6 \times 10^{-19})(3.7 \times 10^{16})(\text{coulombs})}{10 \times 60 \ (\text{sec})} = 9.8 \times 10^{-6} \text{ A} .$$

10.4.3 Ion Channeling

The projected range and straggle of the Gaussian distribution discussed previously give a good description of the implanted ions in amorphous or fine-grain polycrystalline substrates. Both silicon and gallium arsenide behave as if they were amorphous semiconductors, provided that the ion beam is misoriented from the low-index crystallographic direction (e.g., <111>). In this situation, the doping profile described by Eq. 30 is followed closely near the peak and extended to one or two decades below the peak value. This is illustrated[19] in Fig. 26. However, even for a misorientation of 7° from the <111>-axis, there still is a tail which varies exponentially with distance as $\exp(-x/\lambda)$, where λ is typically of the order of 0.1 μm.

The exponential tail is related to the ion-channeling effect. Channeling occurs when incident ions align with a major crystallographic direction and are guided between rows of atoms in a crystal. Figure 27 illustrates a diamond

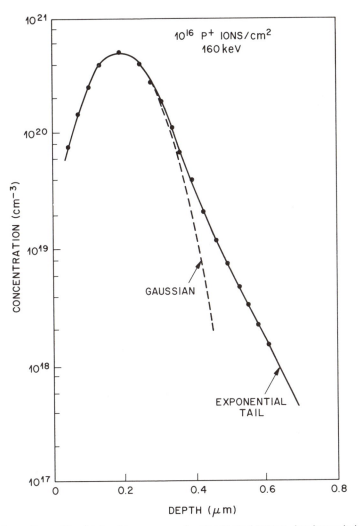

Fig. 26 Impurity profile obtained in a purposely misoriented target. Ion beam is incident 7° from the <111>-axis.[19]

lattice viewed along a <110>-direction.[20] Ions implanted in the <110>-direction will follow trajectories that will not bring them close enough to a target atom to lose significant amounts of energy in nuclear collisions. Thus, for channeled ions, the only energy loss mechanism is electronic stopping, and the range of channeled ions can be significantly larger than would be in an amorphous target.

The insert in Fig. 28 shows schematically the trajectory of a channeled ion where the path is assumed to make an angle ψ with respect to the channel axis. The largest value that the angle ψ can have before the ion is scattered out of the channel is called the *critical angle*. The critical angle is proportional to

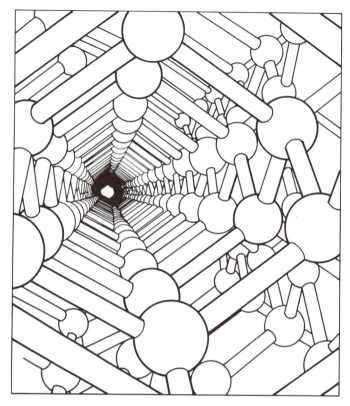

Fig. 27 Model for a diamond structure, viewed along a $<110>$-axis.[20]

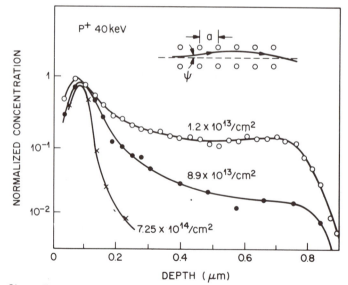

Fig. 28 Channeling of phosphorus ions in silicon. Insert shows the schematic trajectory of a channeled particle.[21]

$1/\sqrt{E_o}$, that is, the larger the incident-ion energy, the smaller the critical angle. For common dopants in silicon and gallium arsenide, the critical angle is in the range of 7 to 2° for ion energies between 30 and 300 keV.

Figure 28 shows an example of the normalized doping profiles for three doses of phosphorus in silicon for a beam aligned in the <100>-direction.[21] In all cases, the position of the Gaussian peak is near 0.1 μm, independent of the dose. The extent of the channeled region is a sensitive function of the dose and results in a penetration depth that is about a factor of 10 larger than would be obtained without channeling. For higher doses, there is more lattice disorder (to be discussed in the next section), and the semiconductor becomes amorphous. Consequently, the channeling effect is reduced correspondingly.

The sensitivity of the channeling process to the critical control of the crystal orientation makes it an undesirable process for most practical applications. To avoid channeling, the substrate is usually misoriented by an angle of 7 to 10° so that the ion beam is incident along a *random direction* in the crystal. Another way to avoid channeling is to create an amorphous surface layer using insert ions (e.g., argon ions) prior to the dopant implantation.

10.5   DISORDER AND ANNEALING

10.5.1   Disorder

When energetic ions enter a semiconductor substrate, they lose their energy in a series of nuclear and electronic collisions and finally come to rest. The electronic-energy loss can be accounted for in terms of electronic excitations to higher energy levels or in the generation of electron–hole pairs. However, electronic collisions do not displace semiconductor atoms from their lattice positions. Only nuclear collisions can transfer sufficient energy to the lattice so that host atoms are displaced resulting in lattice disorder (also called damage).[22] These displaced atoms may possess large fractions of the incident energy, and they in turn cause cascades of secondary displacements of nearby atoms to form a *tree of disorder* along the ion path. When the displaced atoms per unit volume approach the atomic density of the semiconductor, the material becomes amorphous.

The tree of disorder for light ions is quite different from that for heavy ions. Much of the energy loss for light ions (e.g., $^{11}B^+$ in silicon) is due to electronic collisions (see Fig. 21), which do not cause lattice damage. The ions lose their energies as they penetrate deeper into the substrate. Eventually, the ion energy is reduced below the crossover energy ($\cong$ 10 keV for boron) where nuclear stopping becomes dominant. Therefore, most of the lattice disorder occurs near the final ion position. This is illustrated in Fig. 29a.

We can estimate the damage by considering a 100-keV boron ion. Its projected range is 0.31 μm, or 3100 Å (Fig. 22), and its initial nuclear energy loss is only 3 eV/Å (Fig. 21). Since the spacing between lattice planes in silicon is about 2.5 Å, this means that the boron ion will lose 7.5 eV for each lattice plane due to nuclear stopping. The energy required to displace a silicon atom

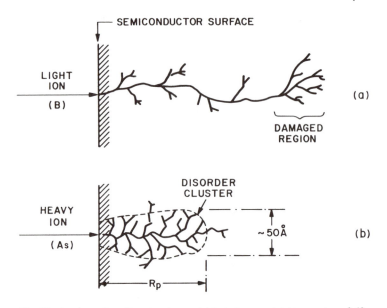

Fig. 29 Implantation disorder due to (a) light ions and (b) heavy ions.[3, 15]

from its lattice position is about 15 eV. Therefore, the incident boron ion does not release enough energy from nuclear stopping to displace a silicon atom when it first enters the silicon substrate. When the ion energy is reduced to about 50 keV (at a depth of 1500 Å), the energy loss due to nuclear stopping increases to 15 eV for each lattice plane (i.e., 6 eV/Å), sufficient to create a lattice disorder. Assuming that one atom is displaced per lattice plane for the remaining ion range, we have 600 lattice atoms displaced (i.e., 1500 Å/2.5 Å). If each displaced atom moves roughly 25 Å from its original position, the damage volume is given by $V_D \simeq \pi(25 \text{ Å})^2(1500 \text{ Å}) = 3 \times 10^{-18} \text{ cm}^3$. The damage density is $600/V_D \cong 2 \times 10^{20} \text{ cm}^{-3}$, which is only 0.4% of the atoms. Thus, very high doses of light ions are needed to create an amorphous layer.

For heavy ions, the energy loss is primarily due to nuclear collisions; therefore, we expect substantial damage. Consider a 100-keV arsenic ion with a projected range of 0.06 μm, or 600 Å. The average nuclear energy loss over the entire energy range is about 120 eV/Å (Fig. 21). This means that the arsenic ion loses about 300 eV for each lattice plane on the average. Most of the energy is given to one primary silicon atom. Each primary atom will subsequently cause 20 displaced target atoms (i.e., 300 eV/15 eV). The total number of displaced atom is 4800. Assuming a range of 25 Å for the displaced atoms, the damage volume is $V_D \cong \pi(25 \text{ Å})^2(600 \text{ Å}) = 10^{-18} \text{ cm}^3$. The damage density is then $4800/V_D \cong 5 \times 10^{21} \text{ cm}^{-3}$, or about 10% of the total number of atoms in V_D. As a result of the heavy-ion implantation, the material has become essentially amorphous. Figure 29b illustrates the situation where the damage forms a disordered cluster over the entire projected range.

To estimate the dose required to convert a crystalline material to an amorphous form, we can use the criterion that the energy density is of the same order of magnitude as that needed for melting the material (i.e., 10^{21} keV/cm^3). For 100-keV arsenic ions, the dose required to make amorphous silicon is then

$$S = \frac{(10^{21} \text{ keV/cm}^3)R_p}{E_o} = 6 \times 10^{13} \text{ ions/cm}^2. \qquad (37)$$

For 100-keV boron ions, the dose required is 3×10^{14} ions/cm^2 (because R_p for boron is five times larger). However, in practice, higher doses ($> 10^{16}$ ions/cm^2) are required for boron implant into a target at room temperature because of the nonuniform distribution of the damage along the ion path.

10.5.2 Annealing

Because of the damaged region and the disorder cluster that result from ion implantation, semiconductor parameters such as mobility and lifetime are severely degraded. In addition, most of the ions as implanted are not located in substitutional sites. To activate the implanted ions and to restore mobility and other material parameters, we must anneal the semiconductor at an appropriate combination of time and temperature.

Figure 30 shows the annealing behaviors of boron and phosphorus

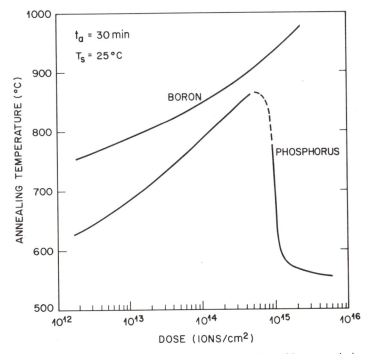

Fig. 30 Annealing temperature versus dose for 90% activation of boron and phosphorus ions.[2, 18]

implantation into silicon substrates.[2, 23] The substrate is held at room temperature during implantation. At a given ion dose, the annealing temperature corresponds to the temperature at which 90% of the implanted ions are activated by a 30-min annealing. For boron implantation, higher annealing temperatures are needed for higher doses. Note that even at 2×10^{15} boron ions/cm^2, the silicon substrate remains crystalline. For phosphorus at lower doses, the annealing behavior is similar to that for boron. However, when the dose is greater than 10^{15} cm^{-2}, the annealing temperature drops to about 600°C. This phenomenon is related to the *solid-phase epitaxy* process. At phosphorus doses greater than 10^{15} cm^{-2}, the silicon surface layer becomes amorphous. The single-crystal semiconductor underneath the amorphous layer serves as a seeding area for recrystallization of the amorphous layer. The epitaxial-growth rate along the $<100>$-direction is 100 Å/min at 550°C and 500 Å/min at 600°C, with an activation energy at 2.4 eV. Therefore, a 1000- to 5000-Å amorphous layer can be recrystallized in a few minutes. During the solid-phase epitaxial process, the impurity dopant atoms are incorporated into the lattice sites along with the host atoms; thus, full activation can be obtained at relatively low temperatures.

During the annealing, the implanted doping profile may be broadened by diffusion. The initial implanted ions are given by a Gaussian distribution, Eq. 30. The solution to a limited-source diffusion is also Gaussian.[25] Therefore a solution to the diffusion equation can be given by Eq. 30 with ΔR_p^2 replaced by ($\Delta R_p^2 + 2Dt$). For conventional furnace annealing, there is a substantial broadening of the doping profile due mainly to long annealing times (typically 30 min or longer). As device dimensions become smaller, low-temperature and short-time processing are required to minimize impurity diffusion (i.e., $\sqrt{2Dt}$ should be much less than ΔR_p). To achieve low-temperature annealing, we can implant inert ions into silicon to form an amorphous layer. The silicon can then be annealed at low temperatures ($\sim 600°$C) by the solid-phase epitaxy process discussed previously to remove the implantation damage from the dopant ions.

Recently, short-time annealing processes have been studied using a variety of energy sources with a wide range of times, 10^{-9} to 10^2 s — all short compared to standard furnace annealing. Figure 31 shows the profiles for the original implant with boron (10^{15} cm^{-2}, 40 keV) and then electron beam-annealed and furnace-annealed at 950°C for 30 min.[24] The e-beam annealing is done with a power density of 18 W/cm^2 for 8 s. The results show that e-beam annealing can activate the dopant fully with minimal redistribution, while the furnace annealing substantially broadens the profile.

Figure 32 shows various short-time annealing techniques where the power density (in W/cm^2) is plotted against the annealing time.[26] Also indicated are the average energy densities (in parentheses). For high-power techniques (such as the pulsed laser), the silicon surface is melted, and full dopant activation is achieved via liquid-phase epitaxy growth. For lower-power techniques (such as the CW laser), the damage is removed by the solid-phase epitaxy pro-

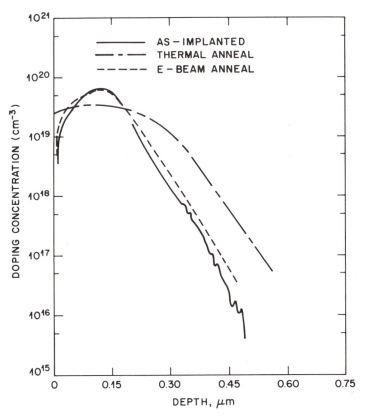

Fig. 31 Implanted boron concentration profile after an e-beam anneal and a thermal anneal.[24]

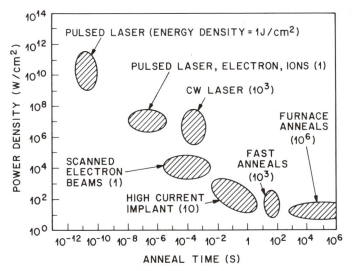

Fig. 32 Power density versus anneal time (pulse duration) for various short-time annealing techniques.[26]

cess. Among these techniques, the most attractive are the *fast anneal* technique using broad-band spectral sources (e.g., incoherent light sources) and heating with an electron beam or ion beam, because the power densities required are lower and they produce no optical-interference effects. We expect that short-time annealing will eventually replace furnace annealing, especially for very-large-scale integrated circuits where dimensional controls of dopant distributions are critical.

10.6 IMPLANTATION-RELATED PROCESSES

In this section we consider a few implantation-related process such as multiple implantation, masking, predeposition, threshold control, and pattern generation.

10.6.1 Multiple Implantation and Masking

In many applications doping profiles other than the simple Gaussian distribution are required. One such case is the pre-implantation of silicon with an inert ion to make the silicon surface region amorphous. This technique allows close control of the doping profile and permits nearly 100% dopant activation at low temperatures as discussed previously. In such a case, a deep amorphous region may be required. To obtain this type of region, we must make a series of implants at varying ion energies and doses.

Multiple implantation can also be used to form a flat doping profile as shown in Fig. 33. Here, four boron implants into silicon are used to provide a composite doping profile.[27] The measured carrier concentration and that predicted using range theory are shown in the figure. Other doping profiles,

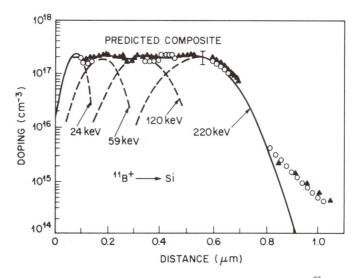

Fig. 33 Composite doping profile using multiple implants.[27]

unavailable from diffusion techniques, can be obtained by using various combinations of impurity dose and implantation energy.

To form p–n junctions in selected areas of the semiconductor substrate, an appropriate mask should be used for the implantation. Because implantation is a low-temperature process, a large variety of masking materials can be used. The minimum thickness of masking material required to stop a given percentage of incident ions can be estimated from the range parameters for ions. The insert of Fig. 34 shows a profile of an implant in a masking material. The dose implanted in the region beyond a depth d (shown shaded) is given by integration of Eq. 30 as

$$S_d = \frac{S}{\sqrt{2\pi}\,\Delta R_p} \int_d^\infty \exp\left[-\left(\frac{x - R_p}{\sqrt{2}\,\Delta R_p}\right)^2\right] dx \; . \tag{38}$$

From Table 1 we can derive the expression

$$\int_x^\infty e^{-y^2}\, dy = \frac{\sqrt{\pi}}{2}\,\mathrm{erfc}\,(x) \; . \tag{39}$$

Therefore, the fraction of the dose that has "transmitted" beyond a depth d is

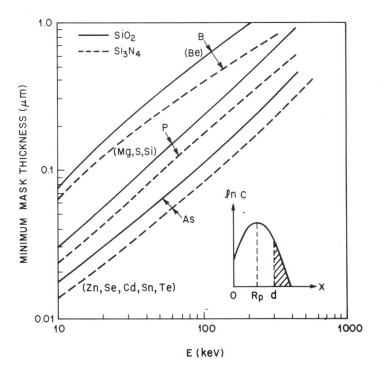

Fig. 34   Minimum thickness of SiO_2 and Si_3N_4 for a masking effectiveness of 99.99%. Insert shows ion penetration beyond a depth d.[28]

given by the transmission coefficient T:

$$T \equiv \frac{S_d}{S} = \frac{1}{2} \, \text{erfc} \left[\frac{d - R_p}{\sqrt{2} \, \Delta R_p} \right]. \tag{40}$$

The complementary error function can be approximated if the argument is large:

$$T \simeq \frac{\exp(-u^2)}{2 \sqrt{\pi} u} \tag{41}$$

where the parameter u is given by $(d - R_p)/\sqrt{2} \Delta R_p$.

In order to stop 99.99% of the incident ions, T must equal 10^{-4}. Equation 41 can be solved to give $u = 2.8$. Thus,

$$d_{\min} = R_p + 3.96 \, \Delta R_p. \tag{42}$$

The values for $d_{\min}$ are shown in Fig. 34 for SiO_2 and Si_3N_4.[18, 28] Photoresists such as KTFR can also be used as masking materials. The results for KTFR are essentially the same as for SiO_2. Mask thicknesses given in this figure are for boron, phosphorus, and arsenic implanted into silicon. These mask thicknesses also can be used as guidelines for impurity masking in gallium arsenide. The dopants are shown in the parentheses. Since both R_p and ΔR_p vary approximately linearly with energy, the minimum thickness of the masking material also increases linearly with energy. In certain applications, instead of totally stopping the beam, the masks can be used as attenuators which can provide an amorphous surface layer to the incident ion beam to minimize the channeling effect.

10.6.2 Predeposition and Threshold Control

In bipolar devices, high-dose implantations are used for buried layers, emitters, and base contact dopings; and low-dose or moderate-dose implantations are used to dope base layers and to make resistors. Implantation is used extensively for predeposition, since the amount of total dopant can be controlled precisely. Figure 35a shows a high-dose implant made in a window surrounded by oxide masking. After the predeposition, the dopant impurities can be driven in (Fig. 35b) by a high-temperature diffusion step; at the same time, the implantation damage at the surface region is annealed out.

One of the most important applications of ion implantation is the threshold voltage adjustment in MOS devices. Figure 36a shows a precisely controlled amount of dopant (in this case boron ions) implanted through the gate oxide to the channel region.[29] Because the projected ranges of boron in silicon and silicon oxide are comparable, if we choose a suitable incident energy, the ions will penetrate just the thin gate oxide but not the thicker field oxide. The threshold voltage will vary approximately linearly with the implanted dose.

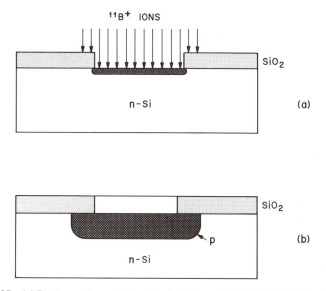

Fig. 35 (*a*) Predeposition using ion implantation. (*b*) After drive-in diffusion.

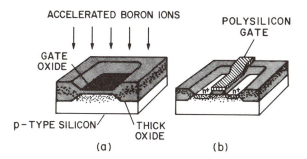

Fig. 36 Threshold voltage adjustment using boron ion implantation.[29]

After the boron implantation, polysilicon is deposited and patterned to form the gate electrode of the MOSFET. The thin oxide surrounding the gate electrode is removed, and the source and drain regions of the device are formed as shown in Fig. 36*b* by another high-dose arsenic implantation.

10.6.3 Pattern Generation

Ion implantation can be used to do resistless pattern generation. One example is the etch rate enhancement in silicon dioxide when it is implanted with various ions. As shown in Fig. 37, the etch rate increases by a factor of 2 when 10^{14} cm^{-2} argon ions are implanted.[18] Thus, oxide layers can be patterned without any resist as illustrated in the insert of Fig. 37.

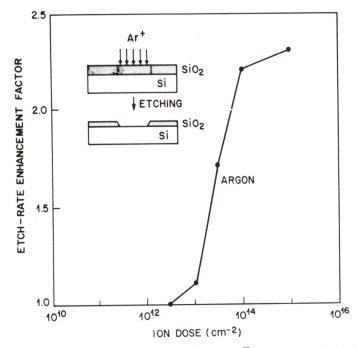

Fig. 37 Etch rate enhancement as a function of ion dose.[18] Insert shows the implantation patterning of SiO_2.

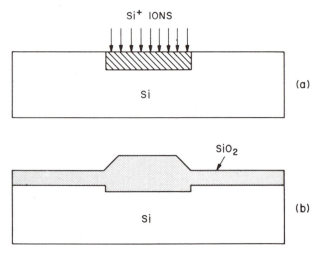

Fig. 38 Variable thickness of the silicon dioxide pattern on a silicon surface as a result of ion implantation.[15]

Another example is the modification of the oxidation rate of implanted silicon.[15] After silicon ions are implanted into the silicon substrate, followed by thermal oxidation, the thickness of silicon dioxide increases in the implanted area as shown in Fig. 38. On the other hand, nitrogen implantation into silicon retards the oxidation rate of the implanted area. These interesting effects of ion implantation are potentially very useful in the fabrication of submicron devices.

REFERENCES

1   For a discussion on diffusion in silicon, see, for example, J. C. C. Tsai, "Diffusion," in S. M. Sze, Ed., *VLSI Technology*, McGraw-Hill, New York, 1983.

2   For a discussion on ion implantation in silicon, see, for example, T. E. Seidel, "Ion Implantation," in S. M. Sze, Ed., *VLSI Technology*, McGraw-Hill, New York, 1983.

3   S. K. Ghandhi, *VLSI Fabrication Principles*, Wiley, New York, 1983.

4   H. C. Casey, Jr., and G. L. Pearson, "Diffusion in Semiconductors," in J. H. Crawford, Jr., and L. M. Slifkin, Eds., *Point Defects in Solids*, Vol. 2, Plenum, New York, 1975.

5   A. S. Grove, *Physics and Technology of Semiconductor Devices*, Wiley, New York, 1967.

6   J. C. Irvin, "Evaluation of Diffused Layers in Silicon," *Bell Syst. Tech. J.*, **41**, 2 (1962).

7   R. B. Marcus, "Diagnostic Techniques," in S. M. Sze, Ed., *VLSI Technology*, McGraw-Hill, New York, 1983.

8   R. B. Fair, "Concentration Profiles of Diffused Dopants" in F. F. Y. Wang, Ed., *Impurity Doping Processes in Silicon*, North-Holland, Amsterdam, 1981.

9   L. R. Weisberg and J. Blanc, "Diffusion with Interstitial-Substitutional Equilibrium, Zinc in GaAs," *Phys. Rev.*, **131**, 1548 (1963).

10   A. F. W. Willoughby, "Double-Diffusion Processes in Silicon," in F. F. Y. Wang, Ed., *Impurity Doping Processes in Silicon*, North-Holland, Amsterdam, 1981.

11   F. A. Cunnell and C. H. Gooch, *J. Phys. Chem. Solid*, **15**, 127 (1960).

12   E. H. Nicollian and J. R. Brews, *MOS Physics and Technology*, Wiley, New York, 1982.

13   D. P. Kennedy and R. R. O'Brien, "Analysis of the Impurity Atom Distribution Near the Diffusion Mask for a Planar *p–n* Junction," *IBM J. Res. Dev.*, **9**, 179 (1965).

14   W. R. Runyan, *Silicon Semiconductor Technology*, McGraw-Hill, New York, 1965.

15   I. Brodie and J. J. Muray, *The Physics of Microfabrication*, Plenum, New York, 1982.

16   S. Furukawa, H. Matsumura, and H. Ishiwara, "Theoretical Consideration on Lateral Spread of Implanted Ions," *Jpn. J. Appl. Phys.*, **11**, 134 (1972).

17   B. Smith, *Ion Implantation Range Data for Silicon and Germanium Device Technologies*, Research Studies, Forest Grove, Oregon, 1977.

18   K. A. Pickar, "Ion Implantation in Silicon," in R. Wolfe, Ed., *Applied Solid State Science*, Vol. 5, Academic, New York 1975.

19   J. F. Gibbons, "Ion Implantation," in S. P. Keller, Ed., *Handbook on Semiconductors*, Vol. 3, North-Holland, Amsterdam, 1980.

20   L. Pauling and R. Hayward, *The Architecture of Molecules*, W. H. Freeman, San Francisco, 1964.

21   G. Dearnaley, J. M. Freeman, G. A. Gard, and M. A. Wilkins, "Implantation Profiles of P Channeled into Silicon Crystals," *Can. J. Phys.*, **46**, 587 (1968).

22   D. K. Brice, "Recoil Contribution to Ion Implantation Energy Deposition Distribution," *J. Appl. Phys.*, **46**, 3385 (1975).

23   B. L. Crowder and F. F. Morehead, Jr., "Annealing Characteristics of *n*-type Dopants in Ion Implanted Silicon," *Appl. Phys. Lett.*, **14**, 313 (1969).

24   D. B. Rensch and J. Y. Chen, "Rapid Isothermal Healing of VLSI MOS Devices using a Scanning E-Beam System" in C. J. Dell'Oca and W. M. Bullis, Eds., *VLSI Science and Technology*, Electrochemical Soc., Pennington, 1982.

25   T. E. Seidel and A. U. MacRae, "Some Properties of Ion Implanted Boron in Silicon," *Trans. Metall. Soc. AIME*, **245**, 491 (1969).

26   M. I. Current and K. A. Pickar, "Ion Implantation Processing," Electrochemical Society Fall Meeting, Vol. 81-1, May 1982.

27   D. H. Lee and J. W. Mayer, "Ion-Implanted Semiconductor Devices," *Proc. IEEE*, **62**, 1241 (1974).

28   G. Dearnaley, J. H. Freeman, R. S. Nelson, and J. Stephen, *Ion Implantation*, North-Holland, Amsterdam, 1973.

29   W. G. Oldham, "The Fabrication of Microelectronic Circuit," in *Microelectronics*, W. H. Freeman, San Francisco, 1977.

PROBLEMS

1   For a low-concentration phosphorus drive-in diffusion in silicon at 1000°C, find the percentage change of surface concentration for a 1% variation in time or temperature.

2   Find the doping profile for a silicon epitaxial layer grown on silicon at 1100°C for 5 min at 1 μm/min with a doping concentration of 2×10^{16} phosphorus atoms/cm^3 on top of an arsenic buried layer with a concentration of 1.5×10^{21} cm^{-3}.

3   A boron predeposition into silicon substrate containing 10^{16} phosphorus atoms/cm^3 results in a sheet resistance of 5 $\Omega/\square$. The junction depth was measured by the method shown in Fig. 6 with $a = 0.1050$ cm, $b = 0.10433$ cm, and $R_o = 1$ cm. Evaluate the boron surface concentration assuming an erfc distribution ($\sqrt{Dt} = 0.1$ μm) or a constant concentration (i.e., the boron concentration remains constant from the surface to the metallurgical junction). The mobility in the *p*-region is assumed to be constant ($\mu_p = 60$ cm^2/V-s). What distribution is more realistic?

(Hint: the sheet resistance for the *p*-region is given by $(q \int_o^{x_j} \mu_p p \, dx)^{-1}$ where x_j is the junction depth).

4 (*a*) If arsenic is diffused into a thick slice of silicon doped with 10^{15} boron atoms/cm^3 at a temperature of $1100°C$ for 3 hr, what is the final distribution of arsenic if the surface concentration is held fixed at 4×10^{18} atoms/cm^3? What are the diffusion length and junction depth? (*b*) If the diffusion is performed under the same conditions as in (*a*) except that the temperature is lowered to $900°C$, what are the doping distribution and the junction depth?

5 To avoid wafer warpage due to sudden reduction in temperature, the temperature in a diffusion furnace is decreased linearly from $1000°C$ to $500°C$ in 20 min. What is the effective diffusion time at the initial diffusion temperature for a phosphorus diffusion in silicon?

6 Derive Eq. 33 for an elastic collision, that is, the total momentum and total energy are conserved after the collision.

7 Assume that a 100-mm-diameter GaAs wafer is uniformly implanted with 100-keV selenium ions for 5 min with a constant ion beam current of $10 \, \mu A$. What are the ion dose per unit area and the peak ion concentration?

8 A silicon *p–n* junction is formed by implanting boron ions at 80-keV through a window in an oxide. If the boron dose is 2×10^{15} cm^{-2} and the *n*-type substrate concentration is 10^{15} cm^{-3}, find the location of the metallurgical junction.

9 (*a*) If we approximate the average rate of energy loss with distance (Fig. 21) as a constant, find the ranges at 50- and 250-keV for As, P, and B in silicon. Compare these ranges with the corresponding projected ranges from Fig. 22 and Eq. 35. (*b*) Calculate the sheet resistance for 50-keV 5×10^{15} arsenic ions/cm^2. Assume a fully activated Gaussian profile by a short-time annealing process. The mobility in the implanted layer is assumed to be a constant ($\mu_n \simeq 100$ cm^2/V-s).

10 A threshold voltage adjustment implantation is made through a 250-Å gate oxide. The substrate is a $<100>$-oriented *p*-type silicon of resistivity $10 \, \Omega$-cm. If the incremental threshold voltage due to a 40-keV boron implantation is 1 V, (*a*) what is the total implanted dose per unit area? (*b*) Estimate the location of the peak boron concentration. (*c*) What percentage of the total dose is in the silicon?

11

Lithography and Etching

Lithography is the process of transferring patterns of geometric shapes on a mask to a thin layer of radiation-sensitive material (called resist) covering the surface of a semiconductor wafer. These patterns define the various regions in an integrated circuit such as the implantation regions, the contact windows, and the bonding-pad areas. The resist patterns defined by the lithographic process are not permanent elements of the final device but only replicas of circuit features. To produce circuit features, these resist patterns must be transferred once more into the underlying layers comprising the device. The pattern transfer is accomplished by an etching process which selectively removes unmasked portions of a layer. In this chapter we consider the lithographic process and the various lithographic methods.[1] We shall also consider both the wet chemical etching and dry etching techniques.[2, 3]

11.1   OPTICAL LITHOGRAPHY

The vast majority of lithographic equipment for integrated circuit (IC) fabrication is optical equipment using ultraviolet light ($\lambda \simeq 0.2$ to $0.4\ \mu$m). In this section we consider the exposure tools, the masks, and the resists used for optical lithography. We also consider the pattern transfer process which serves as a basis for other lithographic systems. We shall first briefly consider the *clean room*, because all lithographic processes must be performed in an ultra-clean environment.

11.1.1   The Clean Room

An IC fabrication facility requires a clean processing room, especially in the area used for lithography. The need for such a clean room arises because dust particles in the air can settle on semiconductor wafers and lithographic masks and can cause defects in the devices that result in circuit failure. For example, a dust particle on a semiconductor surface can disrupt the single-crystal growth of an epitaxial film, causing the formation of dislocations. A dust particle incorporated into the gate oxide can result in enhanced conductivity and cause device failure due to low breakdown voltage. The situation is even more critical in the lithographic area. When dust particles adhere to the surface of a photomask, they behave as opaque patterns on the mask, and these patterns will be transferred to the underlaying layers along with the circuit patterns on the mask. Figure 1 shows three dust particles on a photomask.[4] Particle 1 may result in the formation of a pinhole in the underlying layer. Particle 2 is located near a pattern edge and may cause a constriction of current flow in a metal runner. Particle 3 can lead to a short circuit between the two conducting regions and render the circuit useless.

In a clean room, the total number of dust particles per unit volume must be tightly controlled along with the temperature and humidity. Figure 2 shows the particle-size distribution curves for various *classes* of clean rooms. A *class 100* clean room is one which has a dust count of 100 particles (with particles diameters of 0.5 μm or larger) per cubic foot. This corresponds to 3500 particles (with particle sizes 0.5 μm or larger) per cubic meter, as indicated in parentheses in Fig. 2. Since the number of dust particles increases as particle size decreases, a more stringent control of the clean room environment is required when the minimum feature lengths of ICs are reduced to the 1-μm range. For most IC fabrication areas, a class 100 clean room is required, that is, the dust count must be about four orders of magnitude lower than that of ordinary room air. However, for the lithographic area, a class 10 clean room or one with a lower dust count is required.

Problem

If we expose a 125-mm wafer for 1 min to an air stream under a laminar-flow condition at 30 m/min, how many dust particles will land on the wafer in a class 10 clean room?

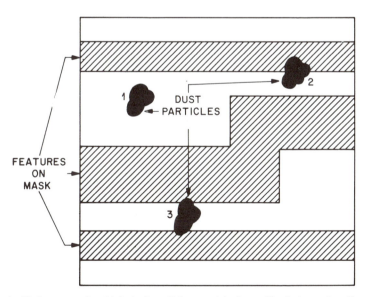

Fig. 1 Various ways in which dust particles can interfere with photomask patterns.[4]

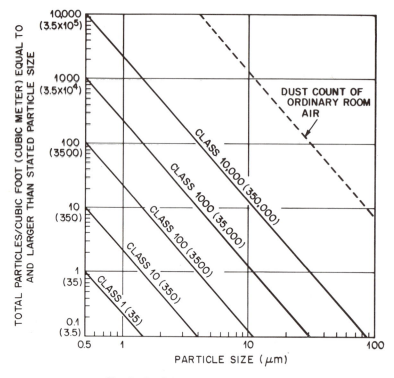

Fig. 2   Particle-size distribution curve.

Solution

For a class 10 clean room, there are 350 particles (0.5 μm or larger) per cubic meter. The air volume that goes over the wafer in 1 min is

$$(30 \text{ m/min}) \times \pi \left[\frac{0.125 \text{ m}}{2} \right]^2 \times 1 \text{ min} = 0.368 \text{ m}^3 .$$

The number of dust particles (0.5 μm or larger) contained in the air volume is

$$350 \times 0.368 = 128 \text{ particles} .$$

Therefore, if there are 200 IC chips on the wafer, the particle count amounts to one particle on each of 64% of the chips. Fortunately, only a fraction of the particles that land adhere to the wafer surface, and of those only a fraction are at a circuit location critical enough to cause a failure. However, the calculation indicates the importance of the clean room.

11.1.2 Exposure Methods

The pattern transfer process is accomplished by using a lithographic exposure tool. The performance of an exposure tool is determined by three parameters: resolution, registration, and throughput. The *resolution* is the minimum feature dimension that can be transferred with high fidelity to a resist film on a semiconductor wafer; *registration* is a measure of how accurately patterns on successive masks can be aligned (or overlaid) with respect to previously defined patterns on the wafer; and *throughput* is the number of wafers that can be exposed per hour for a given mask level.

There are basically two optical exposure methods: shadow printing and projection printing.[5, 6] Shadow printing may have the mask and wafer in direct contact with one another as in *contact printing*, or in close proximity as in *proximity printing*. Figure 3a shows a basic setup for contact printing where a resist-coated wafer is brought into physical contact with a mask, and the resist is exposed by a nearly collimated beam of ultraviolet light through the back of the mask for a fixed time. The intimate contact between resists and mask provides very high resolution ($\sim 1\ \mu$m). However, contact printing suffers a major drawback caused by dust particles. A dust particle or a speck of silicon dust on the wafer can be imbedded into the mask when the mask makes contact with the wafer. The imbedded particle causes permanent damage to the mask and results in defects in the wafer with each succeeding exposure.

To minimize mask damage, the proximity exposure method is used. Figure 3b shows the basic setup. It is similar to the contact printing method except that there is a small gap, 10 to 50 μm, between the wafer and the mask during exposure. The small gap however results in optical diffraction at feature edges on the photomask (i.e., when light passes by the edges of an opaque mask feature, fringes are formed, and some light penetrates into the shadow region) and the resolution is degraded to the 2- to 5-μm range.

In shadow printing, the minimum linewidth that can be printed is roughly

$$l_m \cong \sqrt{\lambda g} \tag{1}$$

when λ is the wavelength of the exposure radiation and g is the gap between the mask and the wafer and includes the thickness of the resist. For

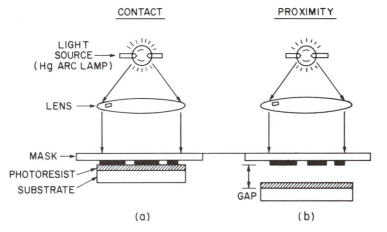

Fig. 3 Schematics of optical shadow printing techniques.[1] (a) Contact printing. (b) Proximity printing.

$\lambda = 0.4 \, \mu m$ and $g = 50 \, \mu m$, the minimum linewidth is 4.5 μm. If we reduce λ to 0.25 μm (wavelength of 0.2 to 0.3 μm are the deep-UV spectral region) and g to 15 μm, l_m becomes 2 μm. Thus, there is an advantage in reducing both λ and g. However, for a given distance g, any dust particle with a diameter larger than g potentially can cause mask damage.

To avoid the mask damage problem associated with shadow printing, projection printing exposure tools have been developed to project an image of the mask patterns onto a resist-coated wafer many centimeters away from the mask. To increase resolution, only a small portion of the mask is exposed at a time. The small image area is scanned or stepped over the wafer to cover the entire wafer surface. Figure 4a shows a 1:1 wafer scan projection system.[6, 7] A narrow, arc-shaped image field ~ 1 mm in width serially transfers the slit image of the mask onto the wafer. The image size on the wafer is the same as that on the mask. This scanning concept can be extended to two dimensions whereby a small symmetrically shaped image field is scanned in an overlapping raster fashion, as shown in Fig. 4b. These techniques are called *scanning projection*. The small image field can also be stepped over the surface of the

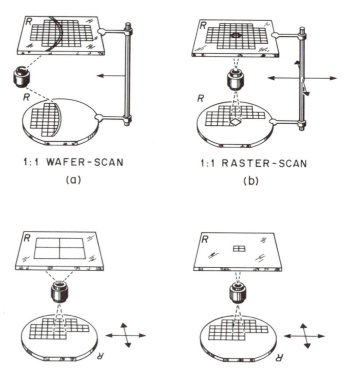

1:1 WAFER-SCAN
(a)

1:1 RASTER-SCAN
(b)

M:1 STEP-AND-REPEAT
(c)

1:1 STEP-AND-REPEAT
(d)

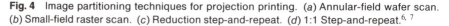

Fig. 4 Image partitioning techniques for projection printing. (*a*) Annular-field wafer scan. (*b*) Small-field raster scan. (*c*) Reduction step-and-repeat. (*d*) 1:1 Step-and-repeat.[6, 7]

wafer by two-dimensional translations of the wafer only, while the mark remains stationary. After the exposure of one chip site, the wafer is moved to the next chip site and the process is repeated. Figures 4c and 4d show the partitioning of the wafer image by *step-and-repeat projection* with the demagnification ratio M:1 (e.g., 10:1 for a 10 times reduction on the wafer) or at 1:1, respectively. The demagnification ratio is an important factor in our ability to produce both the lens and the mask from which we wish to print. The 1:1 optical systems are easier to design and fabricate than a 10:1 or a 5:1 reduction systems, but it is much more difficult to produce defect-free masks at 1:1 than it is at a 10:1 or a 5:1 demagnification ratio.

The resolution of a projection system is given by

$$l_m \cong \frac{\lambda}{NA} \tag{2}$$

where λ is again the exposure wavelength and NA is the numerical aperture, which is given by

$$NA = \bar{n} \sin \theta \tag{3}$$

with $\bar{n}$ the index of refraction in the image medium (usually air, where $\bar{n} = 1$), and θ is the half-angle of the cone of light converging to a point image at the wafer as shown[5] in Fig. 5. Also shown in the figure is the depth of focus, Δz, which can be expressed as

$$\Delta z = \frac{\pm l_m/2}{\tan \theta} \approx \frac{\pm l_m/2}{\sin \theta} = \pm \frac{\bar{n}\lambda}{2(NA)^2} . \tag{4}$$

Equation 2 indicates that resolution can be improved (i.e., smaller l_m) by either reducing the wavelength or increasing NA or both. However, Eq. 4 indicates that the depth of focus degrades much more rapidly by increasing NA than by decreasing λ. This explains the trend toward shorter wavelengths in optical lithography. Typically, scanning projection systems can achieve 1.5-μm resolution, while step-and-repeat projection systems are capable of 1-μm or better resolution.

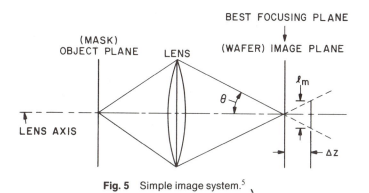

Fig. 5 Simple image system.[5]

11.1.3 Masks

For discrete devices or small-scale (up to 100 components/chip, SSI) to medium-scale (up to 1000 components/chip, MSI) integrated circuits, the first step in the generation of patterns is to draw a large composite layout of the mask set. This is typically $100\times$ (i.e., 100 times) to $2000\times$ the final size. The composite layout is broken into mask levels that correspond to the IC process sequence such as the isolation region on one level, the gate electrode on another, and so on. Artwork is drawn for each masking level. The artwork is reduced to a $10\times$ glass reticle by using a reduction camera. The final mask is made from the $10\times$ reticle using a system similar to the 10:1 step-and-repeat projection printing system described previously. All chip sites on the mask are exposed with identical patterns in the form of a matrix, and each of the sites contains a complete device or circuit pattern for that mask level. Figure 6 shows a mask on which patterns of geometric shapes have been formed. A few secondary-chip sites are also included in the mask; these sites are used for process evaluation.

For large-scale (10^3 to 10^5 components/chip, LSI) or very-large-scale ($> 10^5$ components/chip, VLSI) integrated circuits, this approach is not practical because of the large amount of time needed and the unavoidable human error involved in generating such artwork. Instead, we use computer-aided design (CAD) systems in which designers can completely describe the circuit layout electrically. The digital data produced by the CAD system then drives a pattern generator (e.g., an electron beam machine to be considered in Section 11.2) that transfers the patterns directly to photosensitized masks.

MASK AS SEEN BY NAKED EYE

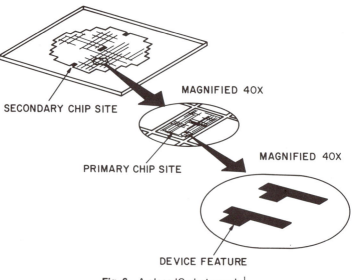

Fig. 6 A glass IC photomask.[1]

For feature sizes of the order of 5 μm or larger, masks are made from glass plates covered with a soft-surface material such as an emulsion. For smaller feature sizes, masks are made from glass covered with hard-surface materials such as chromium or iron oxide. One of the major concerns on masks is the defect density. Mask defects can be introduced during the manufacture of the mask or during subsequent lithographic processes. Even a small mask defect density has a profound effect on the final IC yield. The *yield* is defined as the ratio of good chips per wafer to the total number of chips per wafer. As a first-order approximation, the yield Y for a given masking level can be expressed as

$$Y \simeq e^{-DA} \tag{5}$$

where D is the average number of "fatal" defects per unit area and A is the area of an IC chip. If D remains the same for all mask levels (e.g., $N = 10$ levels), then the final yield becomes

$$Y \simeq e^{-NDA}. \tag{6}$$

Figure 7 shows the mask-limited yield for a 10-level lithographic process as a function of chip size for various values of defect densities. For example, for

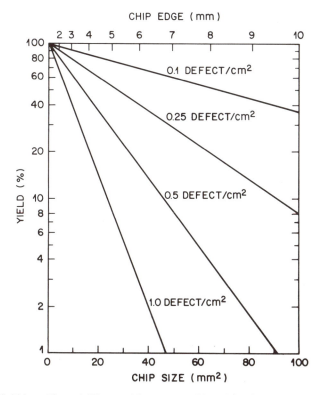

Fig. 7 Yield for a 10-mask lithographic process with various defect densities per level.

$D = 0.5$ defect/cm^2, the yield is 22% for a chip size of 30 mm^2, and it drops to about 1% for a larger chip of 90 mm^2. Therefore, inspection and cleaning of masks are important to achieve high yields on large chips. Of course, an ultra-clean processing area is mandatory for lithographic processing.

11.1.4 Photoresist

The photoresist is a radiation-sensitive compound. Photoresists can be classified as positive and negative, depending on how they respond to radiation. For positive resists, the exposed regions become more soluble and thus more easily removed in the development process. The net result is that the patterns formed (also called images) in the positive resist are the same as those on the mask. For negative resists, the exposed regions become less soluble, and the patterns formed in the negative resist are the reverse of the mask patterns.

Positive photoresists consist of three components: a photosensitive compound, a base resin, and an organic solvent. Prior to exposure, the photosensitive compound is insoluble in the developer solution. After exposure, the photosensitive compound absorbs radiation in the exposed pattern areas, changes its chemical structure, and becomes soluble in the developer solution. Upon development, the exposed areas are removed.

Negative photoresists are polymers combined with a photosensitive compound. After exposure, the photosensitive compound absorbs the optical energy and converts it into chemical energy to initiate a polymer linking reaction. This reaction causes crosslinking of the polymer molecules. The crosslinked polymer has a higher molecular weight and becomes insoluble in the developer solution. Upon development, the unexposed areas are removed. One major drawback of a negative photoresist is that in the development process the whole resist mass swells by absorbing developer solvent. This swelling action limits the resolution of negative photoresists.

Figure 8a shows a typical exposure response curve and image cross section for a positive resist.[1] The response curve describes the percent resist remaining after exposure and development versus the exposure energy. Note that the resist has a finite solubility in its developer, even without exposure to radiation. As the exposure energy increases, the solubility gradually increases until, at a threshold energy E_T, the resist becomes completely soluble. The sensitivity of a positive resist is defined as the energy required to produce complete solubility in the exposed region. Thus, E_T corresponds to the sensitivity. In addition to E_T, a parameter γ, the contrast ratio, is defined to characterize the resist:

$$\gamma \equiv \left[\ln \left(\frac{E_T}{E_1} \right) \right]^{-1} \tag{7}$$

where E_1 is the energy obtained by drawing the tangent at E_T to reach 100% resist thickness as shown in Fig. 8a. A larger γ implies a more rapid solubility

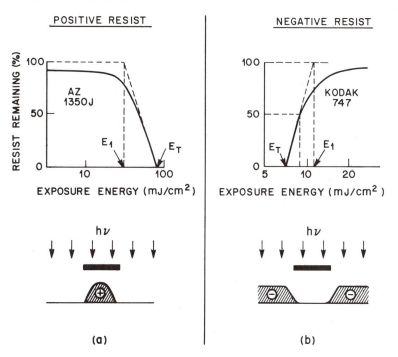

Fig. 8 Exposure response curve and cross section of the resist image after development.[1] (a) Positive photoresist. (b) Negative photoresist.

of the resist with an incremental increase of exposure energy and results in sharper images.

The image cross section in Fig. 8a illustrates the relationship between the edges of a photomask image and the corresponding edges of the resist images after development. The edges of the resist image are generally not at the vertically projected positions of the mask edges due to diffraction. The edge of the resist image corresponds to the position where the total absorbed optical energy equals the threshold energy E_T.

Figure 8b shows the exposure response curve and image cross section for a negative resist. The negative resist remains completely soluble in the developer solution for exposure energies lower than the threshold energy E_T. Above E_T, more of the resist film remains after development. At exposure energies twice the threshold energy, the resist film becomes essentially insoluble in the developer. The sensitivity of a negative resist is defined as the energy required to retain 50% of the original resist film thickness in the exposed region. The parameter γ is defined similarly as in Eq. 7 except that E_1 and E_T are interchanged. The image cross section for the negative resist (Fig. 8b) is also influenced by the diffraction effect.

Problem

Find the parameter γ for the photoresists shown in Fig. 8.

Solution

For the negative resist (Kodak 747), we have $E_T = 7\,\text{mJ/cm}^2$ and $E_1 = 12\,\text{mJ/cm}^2$:

$$\gamma = \left[\ln\left(\frac{E_1}{E_T}\right)\right]^{-1} = \left[\ln\left(\frac{12}{7}\right)\right]^{-1} = 1.9\ .$$

For the positive resist (AZ 1350J), $E_T = 90\,\text{mJ/cm}^2$ and $E_1 = 45\,\text{mJ/cm}^2$:

$$\gamma = \left[\ln\left(\frac{E_T}{E_1}\right)\right]^{-1} = \left[\ln\left(\frac{90}{45}\right)\right]^{-1} = 1.4\ .$$

Table 1 lists a few commercially available negative and positive photoresists. (The resists for electron beam and X-ray lithographies are also listed.) Note that although they have comparable γ values, the negative resist has much lower threshold energy. Therefore, for an optical lithographic system with a given exposure power (in mW/cm^2), much shorter times are needed to expose a wafer using a negative resist, permitting a large wafer exposure throughput per hour. However, as mentioned before, negative resists usually swell after development, resulting in poor resolution. Positive resists, on the other hand, do not swell so that better resolution can be obtained; but they require much larger exposure energy and consequently longer exposure time, resulting in lower throughput.

To improve resolution of a resist image, thinner resist films are generally preferred. However, because of the topography of the underlying layers in ICs, thicker resist films are required to cover the steps, thus degrading the resolution. The multilayer resist systems were developed to separate the functions of image formation and step coverage. Figure 9 shows a trilayer resist system.[8] A very thick resist layer 2 to 3 μm thick is applied to planarize the topography. A thin layer of silicon dioxide is then deposited to a thickness of 1000 Å.

Table 1 Negative and Positive Resists

Lithography	Name	Type	Sensitivity	γ
Optical	Kodak 747	Negative	$9\,\text{mJ/cm}^2$	1.9
	AZ-1350J	Positive	$90\,\text{mJ/cm}^2$	1.4
	PR102	Positive	$140\,\text{mJ/cm}^2$	1.9
e-Beam	COP	Negative	$0.3\,\mu\text{C/cm}^2$	0.45
	GeSe	Negative	$80\,\mu\text{C/cm}^2$	3.5
	PBS	Positive	$1\,\mu\text{C/cm}^2$	0.35
	PMMA	Positive	$50\,\mu\text{C/cm}^2$	1.0
X-Ray	COP	Negative	$175\,\text{mJ/cm}^2$	0.45
	DCOPA	Negative	$10\,\text{mJ/cm}^2$	0.65
	PBS	Positive	$95\,\text{mJ/cm}^2$	0.5
	PMMA	Positive	$1000\,\text{mJ/cm}^2$	1.0

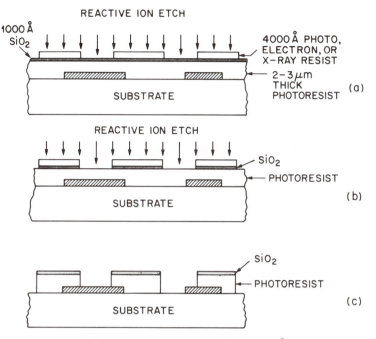

Fig. 9 Processing steps for a trilayer resist.[8]

Finally, a thin resist layer ($\sim$ 4000 Å) is applied (Fig. 9a). Because of the planarization and the opacity (nonreflecting) of the lower layer, the top photoresist layer is optimized for high-resolution imaging. After the top photoresist is developed, it serves as a mask to etch the thin silicon dioxide layer (Fig. 9b), which in turn serves as a mask for etching of the thick bottom resist layer (Fig. 9c).

Inorganic materials such as germanium selenide ($Ge_x Se_{1-x}$) can be used as a negative resist. When $Ge_x Se_{1-x}$ film is coated with silver and exposed to light, lateral diffusion of silver occurs which results in an edge-sharpening effect to give a very large contrast ratio ($\gamma = 3.5$). Lines and spaces as small as 0.5 μm have been printed in $Ge_x Se_{1-x}$ resist using step-and-repeat optical-exposure equipment.[9]

11.1.5 Pattern Transfer

Figure 10 illustrates the steps of transfer of IC patterns from a mask to a semiconductor wafer that has an insulating layer formed on its surface (e.g., thermally grown SiO_2 on Si with a thickness of 0.1 to 1 μm).[10] The wafer is placed in a clean room which typically is illuminated with yellow light, since photoresists are not sensitive to wavelengths greater than 0.5 μm. The wafer is held on a vacuum spindle, and $\sim$ 1 cm^3 of liquid resist is applied to the center of wafer. The wafer is then rapidly accelerated up to a constant rotational

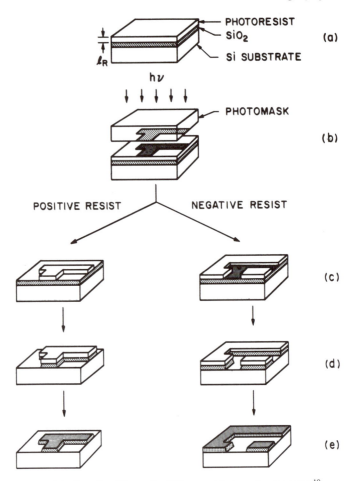

Fig. 10 Details of the optical lithographic transfer process.[10]

speed, which is maintained for about 30 s. The thickness of the resulting resist film, l_R, is given by

$$l_R \sim \frac{(\text{viscosity})(\text{percent solid content in resist})}{(\text{spin speed})^{1/2}}. \tag{8}$$

Spin speed is generally in the range 1000 to 10,000 rpm to give a uniform film of about 1 to 0.5 μm thick as shown in Fig. 10a. This procedure is also applicable to multilayer resist systems; however, considerably more processing steps are required than with the single-layer resist system.

After the spinning step, the wafer is given a pre-exposure baking (typically at 80 to 100°C) to remove the solvent from the photoresist film and to improve resist adhesion to the wafer. The wafer is aligned with respect to the mask in an optical lithographic system and the resist is exposed to UV light as shown in Fig. 10b. We shall first consider the positive photoresist. The exposed resist is dissolved in the developer, as shown in the left side of Fig. 10c. The photoresist development is usually done by spraying the wafer with the developer solution. The wafer is then rinsed and dried. After development, a postbaking (at $\sim$ 100 to 180°C) may be required to increase the adhesion of the resist to the substrate. The wafer is then put in an ambient that etches the exposed insulating layer but does not attack the resist (e.g., buffered hydrofluoric acid is a typical SiO_2 etchant), as shown in Fig. 10d. Finally, the resist is stripped (e.g., using solvent, or plasma oxidation), leaving behind an insulator image (or pattern) that is the same as the opaque image on the mask (left side of Fig. 10e).

For the negative photoresist, the procedures described are also applicable, except that the unexposed areas are removed. The final insulator image (right side of Fig. 10e) is the reverse of the opaque image on the mask.

The insulator image can be used as a mask for subsequent processing. For example, an ion implantation can be done to dope the exposed semiconductor region, but not the area covered by the insulator. The dopant pattern is a duplicate of the design pattern on the photomask (for a negative photoresist) or its complementary pattern (for a positive photoresist). The complete circuit is fabricated by aligning the next mask in the sequence to the previous pattern and repeating the lithographic transfer process. Typically, to fabricate an IC requires 5 to 10 separate masks and lithographic transfer steps.

A related pattern transfer process is the lift-off technique, shown in Fig. 11. A positive resist is used to form the resist pattern on the substrate (Fig. 11a and 11b). The film (e.g., aluminum) is deposited over the resist and the substrate (Fig. 11c); the film thickness must be smaller than that of the resist. Those portions of the film on the resist are removed by selectively dissolving

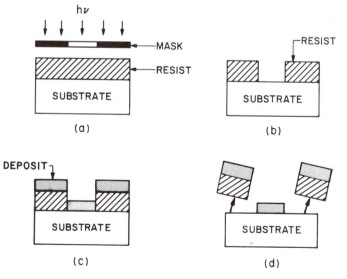

Fig. 11 Lift-off process for pattern transfer.

the resist layer in an appropriate liquid etchant so that the overlying film is lifted off and removed (Fig. 11*d*). The lift-off technique is capable of high resolution and is used extensively for discrete devices (e.g., high-power MES-FET). However, it is not as widely applicable for very-large-scale integration, in which dry etching is the preferred technique.

11.2 ELECTRON BEAM, X-RAY, AND ION BEAM LITHOGRAPHIES

Various types of advanced lithographies for IC fabrication are shown[11] in Fig. 12. As we have discussed in Section 11.1, the majority of IC exposure tools are optical systems using ultraviolet light (Fig. 12*a*). Optical exposure tools are capable of approximately 1-μm resolution, 0.5-μm registration, and a throughput of 50 to 100 wafers per hour. Electron beam lithography, shown in Fig. 12*b*, is primarily used to produce photomasks; relatively few are dedicated to direct wafer exposure (i.e., exposure of the resist directly by focused electron beam without a mask). Because of electron backscattering, electron beam exposure systems are limited to a practical minimum feature length of about 0.5 μm with 0.2 μm registration. X-Ray lithography, shown in Fig. 12*c*, has 0.5-μm or better resolution and 0.5-μm registration. However, it requires a complicated mask and is not yet used to produce ICs in volume. Ion beam lithography, shown in Fig. 12*d*, offers patterned-doping capability and very high resolution ($\sim$ 100 Å); this method is still in its initial development stage.

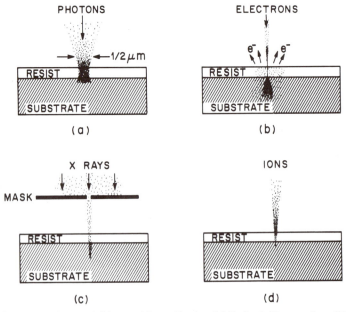

Fig. 12 Types of advanced lithographic methods. (*a*) Optical lithography. (*b*) Electron beam lithography. (*c*) X-ray lithography. (*d*) Ion beam lithography.[11]

11.2.1 Electron Beam Lithography

Figure 13 shows a schematic of an electron beam lithography system.[12] The electron gun is a device that can generate a beam of electrons with a suitable current density; a tungsten thermionic-emission cathode or single-crystal lanthanum boride (LaB_6) can be used for the electron gun. Condenser lenses are used to focus the electron beam to a spot size of 0.01 to 0.1 μm in diameter. Beam-blanking plates (to turn the electron beam on and off) and beam deflection coils are computer-controlled and operated at high rates (in MHz or higher rates) to direct the focused electron beam to any location in the scan field on the substrate. Because the scan field (typically 1 cm) is much smaller than the substrate diameter, a precision mechanical stage is used to position the substrate to be patterned.

The advantages of electron beam lithography include the generation of micron and submicron resist geometries, highly automated and precisely controlled operation, greater depth of focus than that available from optical lithography, and direct patterning on a semiconductor wafer without using a mask. The disadvantage is that electron beam lithographic machines have low throughput—approximately five wafers per hour at less than 1-μm resolution. This throughput is adequate for the production of photomasks, in situations that require small numbers of custom circuits, or for design verification. However, for maskless direct writing, the machine must have the highest possible

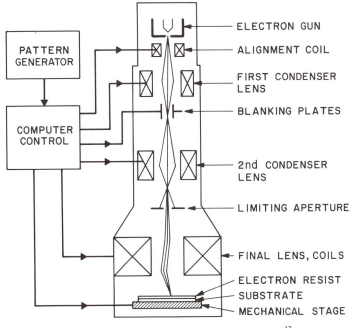

Fig. 13 Schematic of an electron beam machine.[12]

throughput and therefore the largest beam diameter possible consistent with the minimum device dimensions.

There are basically two ways to scan the focused electron beam: raster scan and vector scan. In a raster scan system, the patterns are written by a beam that moves through a regular pattern (vertically oriented, as shown in Fig. 14a).[13] The beam scans sequentially over every possible location on the mask and is blanked (turned off) where no exposure is required. All patterns on the area to be written must be subdivided into individual addresses, and a given pattern must have as a minimum increment an interval that is evenly divisible by the beam address size.

In the vector scan system as shown in Fig. 14b, the beam is directed only to the requested pattern features and jumps from feature to feature, rather than scanning the whole chip, as in raster scan. For many chips, the average exposed region is only 20% of the chip area, so we can save time by using a vector scan system. To further increase the writing speed, a variable-beam-shape system was developed in which the patterning beam has a rectangular cross section of variable size and aspect ratio.

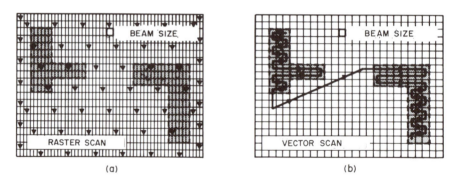

Fig. 14 Comparison of raster scan and vector scan writing schemes.[13]

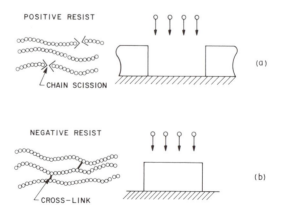

Fig. 15 Schematic of positive and negative resists used in electron beam lithography.[14]

Electron Resist   Electron resists are polymers. The behavior of an electron resist is similar to that of a photoresist, that is, a chemical or physical change is induced in the resist by radiation. This change allows the resist to be patterned. For a positive electron resist, the polymer-electron interaction causes chemical bonds to be broken (chain scission) to form shorter molecular fragments, as shown[14] in Fig. 15a. As a result, the molecular weight is reduced in the irradiated area. The irradiated area can be dissolved in a developer solution that attacks the low-molecular-weight material. Common positive electron resists are poly(methyl methacrylate), called PMMA, and poly(butene-1 sulfone), called PBS. Positive electron resists have resolutions of 0.1 μm or better.

For a negative electron resist, the irradiation causes radiation-induced polymer linking as shown in Fig. 15b. The crosslinking creates a complex three-dimensional structure with a higher molecular weight than that of the nonirradiated polymer. The nonirradiated resist can be dissolved in a developer solution that does not attack the high-molecular-weight material. Poly(glycidyl methacrylate-co-ethyl acrylate), called COP, is a common negative electron resist. COP, like a negative photoresist, also swells during development so the resolutions are limited to about 1 μm.

The characteristic response curves of electron resists are similar to that of photoresists. The sensitivity is defined as the electron dose required per unit area to give complete development of a positive resist or the electron dose required at which 50% of the original resist has been retained for a negative resist. The resist sensitivities and γ values are listed in Table 1. Note that the negative resist COP is much more sensitive than the positive resist PMMA, and that a much shorter exposure time is required for COP. However, slow resists (e.g., PMMA) have a higher resolution than fast resists (e.g., COP); so there is a trade-off between sensitivity and resolution.

The Proximity Effect   In optical lithography, the resolution is limited by diffraction of light. In electron beam lithography, the resolution is not limited by diffraction (because the wavelengths associated with electrons of a few keV and higher energies are less than 1 Å) but by electron scattering. When electrons penetrate the resist film and underlying substrate, they undergo collisions. These collisions lead to energy losses and path changes. Thus, the incident electrons spread out as they travel through the material until either all of their energy is lost or they leave the material due to backscattering.

Figure 16a shows computed electron trajectories of 100 electrons with initial energy of 20 keV incident at the origin of a 0.4-μm PMMA film on a thick silicon substrate.[15] The electron beam is incident along the z-axis and all trajectories have been projected onto the xz plane. This figure shows qualitatively that the electrons are distributed in an oblong pear-shaped volume with a diameter of the same order of magnitude as the electron penetration depth (~ 3.5 μm). Also, there are many electrons that undergo backscattering collisions and travel backward from the silicon substrate into the PMMA resist film and leave the material.

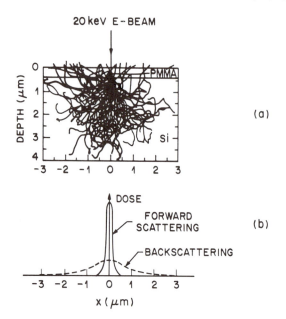

Fig. 16 (a) Simulated trajectories of 100 electrons in PMMA for a 20-keV electron beam.[15] (b) Dose distribution for forward scattering and backscattering at the resist–substrate interface.

Figure 16b shows the normalized distributions of the forward scattering and backscattering electrons at the resist-substrate interface. Because of the back-scattering, electrons effectively can irradiate several micrometers away from the center of the exposure beam. Since the dose of a resist is given by the sum of the irradiations from all surrounding areas, the electron beam irradiation at one location will affect the irradiation in neighboring locations. This phenomenon is called the *proximity effect*. The proximity effect places a limit on the minimum spacings between pattern features. To correct for the prox-imity effect, patterns are divided into smaller segments. The incident electron dose in each segment is adjusted so that the integrated dose from all its neigh-boring segments is the correct exposure dose. This approach further decreases the throughput of the electron beam system, because of the additional com-puter time required to expose the subdivided resist patterns.

11.2.2 X-Ray Lithography

X-Ray lithography uses a shadow printing method similar to optical prox-imity printing. The X-ray wavelength (4 to 50 Å) is much shorter than the wavelength of the ultraviolet light (2000 to 4000 Å) used for optical lithogra-phy. Therefore, diffraction effects are reduced and higher resolutions can be obtained. For an X-ray wavelength of 5 Å and a gap of 40 μm, the minimum line width obtained from Eq. 1 is less than 0.2 μm. Compared to the electron

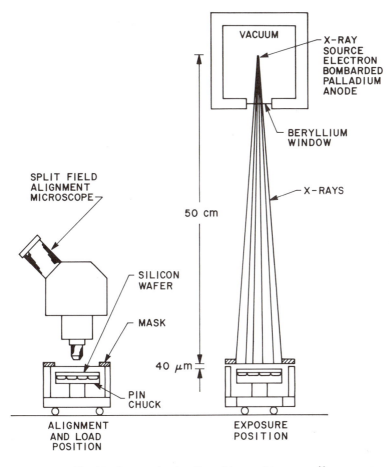

Fig. 17 Schematic of an X-ray lithographic system.[16]

beam system, X-ray lithography can have a higher throughput because it uses parallel exposure, as opposed to the serial exposure approach of electron beam lithography.

Figure 17 shows an X-ray system.[16] A 25-keV, 5-kW electron beam generated by an electron gun is incident upon a palladium target that emits X-rays with a wavelength of 4.4 Å. The X-rays pass through a beryllium window into a helium-filled chamber to the mask and wafer. Helium is used in the chamber because air is a strong absorber of X-rays. As shown on the left side of Fig. 17, the X-ray mask and semiconductor wafer are first aligned with each other. They are then moved to the exposure position as shown on the right side of the figure.

Figure 18 shows the geometric effects on X-ray lithography. Because of the finite size of the X-ray source (with diameter a) and the finite mask-to-wafer gap g, a penumbral effect results. The penumbral blur δ on the edge of the

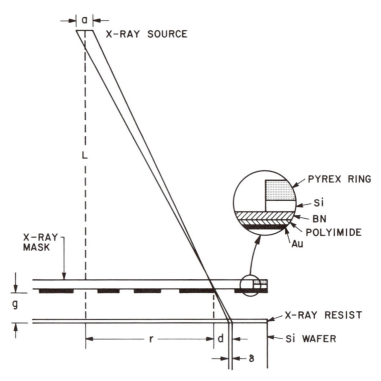

Fig. 18 Geometric effects in X-ray lithography. Insert shows the X-ray mask structure.[16]

resist image is

$$\delta = a \ \frac{g}{L} \qquad (9)$$

where L is the distance from the source to the X-ray mask. If $a = 3$ mm, $g = 40 \ \mu m$, and $L = 50$ cm, the penumbral blur is of the order of 0.2 μm. The blurring at the edge of a feature causes a loss resolution in the subsequent processing. Another geometric effect is the lateral magnification error, due to the finite gap g and the nonvertical incidence of the X-ray flux. The projected images of the mask are shifted laterally by an amount d, called runout:

$$d = r \ \frac{g}{L} \qquad (10)$$

where r is the radial distance from the center of the wafer. The runout is zero at the center of the wafer but increases linearly toward the wafer edge. For a 125-mm wafer, the runout error can be as large as 5 μm (assuming $g = 40 \ \mu m$ and $L = 50$ cm). This runout error must be compensated during the mask making process.

The insert of Fig. 18 shows the cross section of an X-ray mask. The construction of an X-ray mask is much more complicated than that of a photomask. A silicon wafer is used as the substrate. A boron nitride film ($\sim 6\ \mu$m) is deposited on the silicon substrate, followed by a spun-on polyimide film (also $\sim 6\ \mu$m). Because gold has a relatively high absorption coefficient at 4.4 Å, a gold film ($\sim 0.6\ \mu$m) is deposited on the polyimide to serve as an X-ray absorber. Mask patterns are defined in the gold film using electron beam lithography. The patterned wafer is bonded to a Pyrex ring, and the silicon substrate is removed (except underneath the Pyrex ring) by etching it from the back to form the membrane structure shown in Fig. 18.

We can use electron beam resists as X-ray resists. This is because when an X-ray is absorbed by an atom, the atom goes to an excited state with the emission of an electron. The excited atom returns to its ground state by emitting an X-ray having a different wavelength than the incident X-ray. This X-ray is absorbed by another atom, and the process repeats. Since all the processes result in the emission of electrons, a resist film under X-ray irradiation is equivalent to one being irradiated by a large number of secondary electrons from any of the processes. Once the resist film is irradiated, chain crosslinking or chain scission will occur, depending on the type of resist. Table 1 lists some X-ray resists along with their sensitivities and γ values. One of the most attractive X-ray resist is DCOPA (dichloropropyl acrylate and glycidyl methacrylate-co-ethyl acrylate), because it has a very low threshold ($\sim 10\ \mathrm{mJ/cm^2}$).

11.2.3 Ion Beam Lithography

Ion beam lithography can achieve higher resolution than optical, X-ray, or electron beam lithographic techniques because ions have a higher mass and therefore scatter less than electrons. Ion beam lithography can be operated in a shallow printing mode using PMMA as an ion beam resist. Ion beams can also be used in a focused-beam direct-writing mode. We have considered a few resistless-pattern generation approaches using an ion beam in Section 10.6.

Figure 19 shows the computer-simulated trajectories of 50 H^+ ions implanted at 60 keV into PMMA and various substrates.[17] Note that the

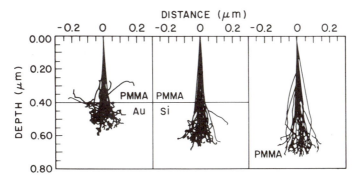

Fig. 19 Trajectories of 60-keV H^+ ions traversing through PMMA into Au, Si, and PMMA.[17]

spread of the ion beam at a depth of 0.4 μm is only 0.1 μm in all cases (compare with Fig. 16a for electrons). The backscattering is completely absent for silicon substrate, and there is only a small amount of backscattering for the gold substrate.

Ion beam lithography is still in its initial development stage. Its major advantage is high spatial resolution due to absence of proximity effect. Ion beam lithography will become important when the minimum feature dimension is reduced to about 0.2 μm and below.

11.2.4 Comparison of Lithographic Methods

The lithographic methods discussed above all have 1-μm or better resolution. However, each method has its limitations: diffraction effects in optical lithography, proximity effects in electron beam lithography, mask fabrication complexities in X-ray lithography, and beam deflection difficulties in ion beam lithography.

For IC fabrication, many mask levels are involved. However, it is not necessary to use the same lithographic method for all levels. A *hybrid lithography* approach can take advantage of the unique features of each lithographic process to improve resolution and to maximize throughput. For example, a 5:1 step-and-repeat or electron beam method can be used for the most critical mask levels, while 1:1 scanning projection can be used for the other levels.

Figure 20 shows the estimated resolution for various lithographic systems as a function of 125-mm wafer throughput.[6] For feature sizes of 1 μm or larger,

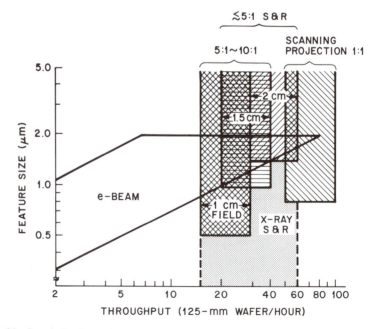

Fig. 20 Resolution for various exposure systems as a function of wafer throughput.[6]

the 1:1 scanning projection system is most attractive, because of its high throughput. For 1- to 0.5-μm feature sizes, we expect the M:1 step-and-repeat lithographic system with an image field of about 1 cm to dominate IC production. The throughput of electron beam systems varies as $1/l_m^2$, where l_m is the minimum feature length; this is due to the serial approach required to cover a two-dimensional area. For example, an electron beam system writing a feature size of 2 μm can have a throughput of 20 wafers per hour. The throughput for the same system drops to about one wafer per hour for a feature size of 0.5 μm. Of course, electron beam lithography will be the dominant mask-making process in the foreseeable future. When the step-and-repeat X-ray system is developed, X-ray lithography will serve as a submicron-feature-size, high-throughput technique filling the gap between electron beam and optical lithography.

11.3   WET CHEMICAL ETCHING

Wet chemical etching is used extensively in semiconductor processing. Starting from the sawed semiconductor wafers, chemical etchants are used for lapping and polishing to give an optically flat, damage-free surface. Prior to thermal oxidation or epitaxial growth, the semiconductor wafers are chemically cleaned and scrubbed to remove contamination that results from handling and storing. For many discrete devices and integrated circuits of relatively large dimensions ($\gtrsim 3$ μm), chemical etching is used to delineate patterns and to open windows in insulating materials. The mechanisms for wet chemical etching involve three essential steps, as illustrated in Fig. 21: (1) the reactants are transported (e.g., by diffusion) to the reacting surface, (2) chemical reactions occur at the surface, and (3) the products from the surface are transported away (e.g., by diffusion). Both agitation and the temperature of the etchant solution will influence the etch rate. In IC processing, most wet chemical etchings proceed by dissolution of a material in a solvent or by

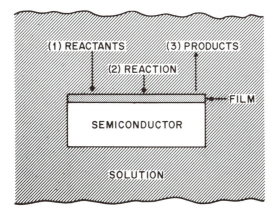

Fig. 21   Basic mechanisms in wet chemical etching.

conversion of a material into a soluble compound which subsequently dissolves in the etching medium.

11.3.1 Isotropic Etches

For semiconductor materials, wet chemical etching usually proceeds by oxidation, followed by the dissolution of the oxide by a chemical reaction. For silicon, the most commonly used etchants are mixtures of nitric acid (HNO_3) and hydrofluoric acid (HF) in water or acetic acid (CH_3COOH). The reaction is initiated by promoting silicon from its initial oxidation state to some higher oxidation state and is given by

$$Si + 2h^+ \longrightarrow Si^{2+} . \tag{11}$$

In this oxidation reaction, holes (h^+) are required. The primary oxidizing species in semiconductor etching is OH^-, which is formed by the dissociation of water:

$$H_2O \rightleftharpoons OH^- + H^+ . \tag{12}$$

The Si^{2+} in Eq. 11 combines with OH^- to give

$$Si^{2+} + 2OH^- \longrightarrow Si(OH)_2 \tag{13}$$

which subsequently liberates hydrogen to form SiO_2:

$$Si(OH)_2 \longrightarrow SiO_2 + H_2 . \tag{14}$$

Hydrofluoric acid (HF) is used to dissolve SiO_2. The reaction is

$$SiO_2 + 6HF \longrightarrow H_2SiF_6 + H_2O \tag{15}$$

where H_2SiF_6 is soluble in water.

The holes in Eq. 11 are produced by an autocatalytic process described as follows: In the presence of HNO_2 in HNO_3 solution, we have

$$HNO_2 + HNO_3 \longrightarrow 2NO_2^- + 2h^+ + H_2O \tag{16}$$

$$2NO_2^- + 2H^+ \longrightarrow 2HNO_2 . \tag{17}$$

The HNO_2 generated in the reaction, Eq. 17, reenters into reaction with HNO_3, Eq. 16. Thus, the products of the reaction promote the reaction itself.

The overall reaction is

$$Si + HNO_3 + 6HF \longrightarrow H_2SiF_6 + HNO_2 + H_2O + H_2 . \tag{18}$$

Water can be used as a diluent for this etchant. However, acetic acid is preferred because its use results in less dissociation of the nitric acid, hence a higher concentration of the undissociated species. Extensive studies of the HF–HNO_3 system have been made. Figure 22 shows the results in the form of isoetch curves for the various constituents by weight.[18] The normally available concentrated acids are 49.2 wt% HF and 69.5 wt% of HNO_3. We observe from these curves that at high HF and low HNO_3 concentrations, corresponding to

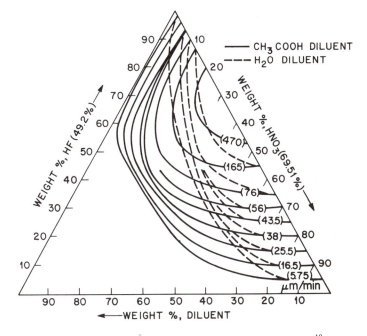

Fig. 22 Isoetch curves for silicon (HF:HNO$_3$:diluent system).[18]

the region near the upper vertex, the etch rate is controlled by the HNO$_3$ concentration (i.e., the etch rate is essentially independent of HF concentration for a given CH$_3$COOH diluent). This is because there is an excess of HF to dissolve the SiO$_2$ formed during the reaction. At low HF and high HNO$_3$ concentrations, corresponding to the region near the lower right vertex, the etch rate is controlled by the ability of HF to remove the SiO$_2$ as it is formed; etchant solutions in this region are isotropic, that is, not sensitive to crystallographic orientation, and are used as polishing etches.

From Fig. 22 we see that there are infinite choices of formulations for silicon etching. A number of these are well characterized. Two of the common etchants, CP-4A and CP-8, are listed in Table 2. Both etchants are used for polishing of silicon wafers. Also listed are a junction staining etch to reveal the p–n junction depth and an orientation-dependent etch to be considered in the next subsection.

A wide variety of etches has been investigated for gallium arsenide; however, very few of them are truly isotropic. This is because the surface activity of the (111)-Ga and (111)-As faces are very different. Most etches give a polished surface on the arsenic face, but the gallium face tends to show crystallographic defects and etches much more slowly. Figure 23 shows the isoetch curves for gallium arsenide in the H$_2$SO$_4$–H$_2$O$_2$–H$_2$O system.[19] The system can be used in a wide variety of formulations. Formulations with high hydrogen peroxide (H$_2$O$_2$) or high sulfuric acid (H$_2$SO$_4$) content fall into regions C or D, respectively, in this figure and result in surfaces with a mirror finish.

Table 2 Etchants for Silicon and Gallium Arsenide

Semi-Conductor	Etchant	Purpose	Composition	Etch Rate (μm/min)
Si	CP-4A	Polishing or lapping	3 ml HF 5 ml HNO$_3$ 3 ml CH$_3$COOH	34.8
	CP-8	Polishing	1 ml HF 5 ml HNO$_3$ 2 ml CH$_3$COOH 0.3 g I$_2$/250 ml solution	7.4
	Junction-staining etch	Measurement of Junction depth	HF + 0.1% HNO$_3$	—
	Orientation-dependent etch	Groove etching	23.4 wt% KOH 13.3 wt% Propyl alcohol 63.3 wt% H$_2$O	0.6 for <100> 6×10^{-3} for <111>
GaAs	H$_2$SO$_4$–H$_2$O$_2$–H$_2$O System	Polishing	8 ml H$_2$SO$_4$ 1 ml H$_2$O$_2$ 1 ml H$_2$O	0.8 for <111>-Ga 1.5 for all other
	H$_3$PO$_4$–H$_2$O$_2$–H$_2$O System	Polishing	3 ml H$_3$PO$_4$ 1 ml H$_2$O$_2$ 50 ml H$_2$O	0.4 for <111>-Ga 0.8 for all other

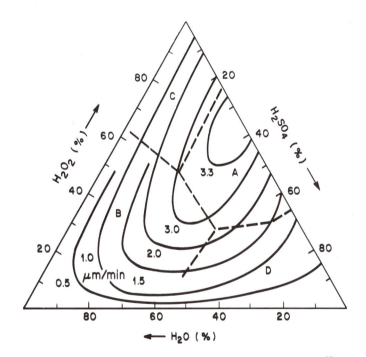

Fig. 23 Isoetch curves for GaAs (H$_2$SO$_4$:H$_2$O$_2$:H$_2$O system).[19]

Formulations in region A, having large amounts of both of these chemicals, etch relatively rapidly (~ 3 $\mu m/min$) and result in a cloudy or frosted appearance. Region B gives slower etch rates and can be used for delineating surface defects.

For an etchant with an 8:1:1 volume ratio of $H_2SO_4:H_2O_2:H_2O$, the etch rate is 0.8 $\mu m/min$ for the $<111>$-Ga face and 1.5 $\mu m/min$ for all other faces. This result and the result for an etchant of the $H_3PO_4-H_2O_2-H_2O$ system are listed in Table 2.

Etching of insulating and metal films is usually done with the same chemicals that dissolve these materials in bulk form and involves their conversion into soluble salts or complexes. Generally, film materials will etch more rapidly than their bulk counterparts. Also, the etch rates are higher for films that have poor microstructure, built-in stress, or departures from stoichiometry

Table 3   Etchants for Insulators and Conductors

Material	Etchant Composition		Etch Rate
SiO_2	28 ml HF 170 ml H_2O 113 g NH_4F	Buffered HF	1000 Å/min
	15 ml HF 10 ml HNO_3 300 ml H_2O	P-Etch	120 Å/min
Si_3N_4	Buffered HF		5 Å/min
	H_3PO_4		100 Å/min
Al	1 ml HNO_3 4 ml CH_3COOH 4 ml H_3PO_4 1 ml H_2O		350 Å/min
Au	4 g KI 1 g I_2 40 ml H_2O		1 $\mu m/min$
Mo	5 ml H_3PO_4 2 ml HNO_3 4 ml CH_3COOH 150 ml H_2O		0.5 $\mu m/min$
Pt	1 ml HNO_3 7 ml HCl 8 ml H_2O		500 Å/min
W	34 g KH_2PO_4 13.4 g KOH 33 g $K_3Fe(CN)_6$ H_2O to make 1 liter		1600 Å/min

or have been irradiated. Some useful etchants for insulating and metal films are listed in Table 3.

11.3.2 Orientation-Dependent Etching

Some etchants dissolve a given crystal plane of a semiconductor much faster than other planes; this results in orientation-dependent etching. In diamond and zincblend lattices, the (111)-plane is more closely packed than the (100)-plane; therefore, the etch rate is expected to be slower for the (111)-plane. A commonly used orientation-dependent etch for silicon consists of a mixture of KOH in water and isopropyl alcohol. The etch rate is 0.6 μm/min for the (100)-plane, 0.1 μm/min for the (110)-plane, and only 0.006 μm/min (60 Å/min) for the (111)-plane at about 80°C; thus, the ratio of the etch rates for the (100)-, (110)-, and (111)-planes is 100:16:1.

Orientation-dependent etching of <100>-oriented silicon through a patterned silicon dioxide mask creates precise V-shaped grooves, the edges being (111)-planes at an angle of 54.7° from the (100)-surface, as shown[20] in the left of Fig. 24a. If the window in the mask is sufficiently large or if the etching time is short, a U-shaped groove will be formed, as shown in the right of

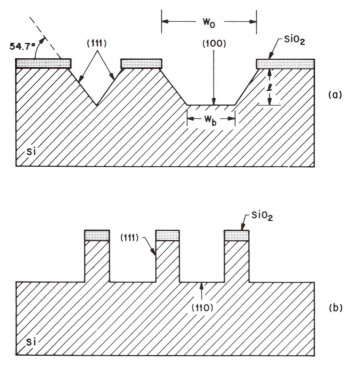

Fig. 24 Orientation-dependent etching. (a) Through window patterns on <100>-oriented silicon. (b) Through window patterns on <110>-oriented silicon.[20]

Fig. 24a. The width of the bottom surface is given by

$$W_b = W_o - 2l \cot 54.7°$$
$$= W_o - \sqrt{2}\, l \qquad (19)$$

where W_o is the width of the window on the wafer surface and l is the etched depth. If <110>-oriented silicon is used, essentially straight-walled grooves with sides of (111)-planes can be formed, as shown in Fig. 24b. We can use the large orientation dependence in the etch rates to fabricate device structures with submicron feature lengths.

Orientation-dependent etching in gallium arsenide is quite different from that in silicon because the (111)-gallium planes usually have the slowest etch rate, while the (111)-arsenic planes have the fastest. As a result, when the mask window is aligned with the <110>-axis, the etch profile is trapezoidal in one direction and dovetailed in the other, as shown in Fig. 25a. If the mask window is turned 45° with respect to the <110>-direction, as shown in Fig. 25b, we obtain a straight-walled groove.[21]

11.4 DRY ETCHING

In pattern transfer operations, a resist pattern is defined by a lithographic process to serve as a mask for etching of its underlying layer (Fig. 26a).[22] Most of the layer materials (e.g., SiO_2, Si_3N_4, and deposited metals) are amorphous or polycrystalline thin films. If they are etched in a wet chemical etchant, the etch rate is generally isotropic (i.e., the lateral and vertical etch rates are the same), as illustrated in Fig. 26b. Let h_f be the thickness of the layer material and l the lateral distance etched underneath the resist mask. We can define the degree of anisotropy A_f by

$$A_f \equiv 1 - \frac{l}{h_f} = 1 - \frac{v_l t}{v_v t} = 1 - \frac{v_l}{v_v} \qquad (20)$$

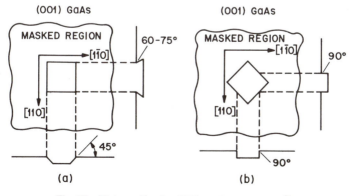

Fig. 25 Etch profiles for (100)-gallium arsenide.[21]

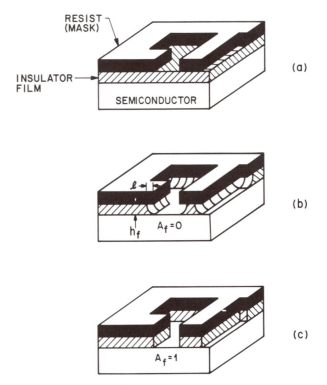

RESIST (MASK)

INSULATOR FILM

SEMICONDUCTOR

(a)

ℓ

h_f $A_f = 0$

(b)

$A_f = 1$

(c)

Fig. 26 Comparison of wet chemical etching and dry etching for pattern transfer.[22]

where t is the time and v_l and v_v are the lateral and vertical etch rates, respectively.[3] For isotropic etching, $v_l = v_v$ and $A_f = 0$.

The major disadvantage of wet chemical etching for pattern transfer is the undercutting of the layer underneath the mask, resulting in a loss of resolution in the etched pattern. In practice, for isotropic etching the film thickness should be about one third or less of the resolution required. If patterns are required with resolutions much smaller than the film thickness, anisotropic etching (i.e., $1 \geqslant A_f > 0$) must be used. In practice, the value of A_f is chosen to be close to unity. Figure 26c shows the limiting case where $A_f = 1$, corresponding to $l = 0$ (or $v_l = 0$).

To achieve $A_f = 1$, dry etching methods have been developed. Dry etching is synonymous with plasma-assisted etching, which denotes several techniques that use plasma in the form of low-pressure discharges. These techniques are commonly used in very-large-scale IC processing because of their capability for high-fidelity transfer of the resist patterns.

11.4.1 Plasma-Assisted Etching Techniques

A plasma is a fully or partially ionized gas composed of ions, electrons, and neutrons. A plasma is produced when an electric field of sufficient magnitude is applied to a gas, causing the gas to break down and become ionized. The plasma is initiated by free electrons that are released by some means such as field emission from a negatively biased electrode. The free electrons gain kinetic energy from the electric field. In the course of their travel through the gas, the electrons collide with gas molecules and lose their energy. The energy transferred in the collisions causes the gas molecules to be ionized (i.e., to free electrons). The freed electrons gain kinetic energy from the field, and the process continues. Therefore, when the applied voltage is larger than the breakdown potential, a sustained plasma is formed throughout the reaction chamber.

The electron concentrations in the plasma for dry etchings are relatively low, typically on the order of 10^9 to 10^{12} cm^{-3}. At a pressure of 1 Torr, the concentrations of gas molecules are 10^4 to 10^7 times higher than the electron concentrations. This results in an average gas temperature in the range 50 to 100°C. Therefore, the plasma-assisted dry etching is a low-temperature process.

Figure 27 shows two dry-etching systems.[3] Figure 27a is a sputter-etching system which uses relatively high-energy ($\gtrsim 500$ eV) noble gas ions, such as

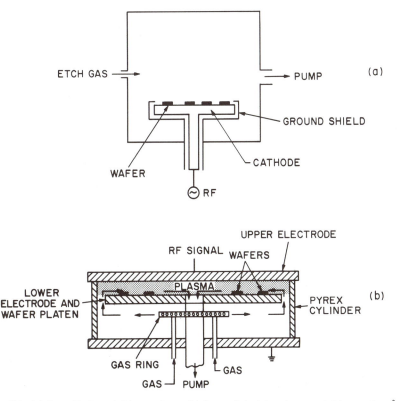

Fig. 27 (a) A sputtering-etching system. (b) A parallel-plate, plasma-etching system.[3]

argon ions (Ar^+). This process is essentially the reverse of sputtering deposition as discussed in Section 9.4. The wafer to be etched (also called the target) is placed on the powered electrode, and argon ions are accelerated by the applied field to bombard the target surface. Through the transfer of a momentum, atoms near the surface become volatile and are removed (or etched). The typical operating pressure for sputter etching is 0.01 to 0.1 Torr. The direction of the electric field is normal to the target surface so that under the operating pressure, argon ions arrive predominantly normal to the surface. There are essentially no ion bombardments on the sidewalls of the etched features. As a result, the lateral etch rate v_l is much less than the vertical etch rate v_v and a high degree of anisotropy can be obtained. However, there is one major drawback of the sputter-etching process: poor selectivity, that is, the etch rates for most materials are very close, and we cannot etch only one layer and stop at the underlying material.

Figure 27b shows a parallel-plate plasma-etching system. The plasma is confined between the two closely spaced electrodes. Molecular gases containing one or more halogen atoms (e.g., CCl_4 and Cl_2) are fed through the gas ring. The typical operating pressure is relatively high, from 0.1 to 10 Torr. Another plasma-assisted etching method is reactive-ion etching (RIE), which employs apparatus similar to that for sputter etching (Fig. 27a). However, in RIE the noble-gas plasma is replaced by molecular-gas plasma similar to that in plasma etching. Under appropriate conditions, both RIE and plasma etching can give high selectivity and a high degree of anisotropy.

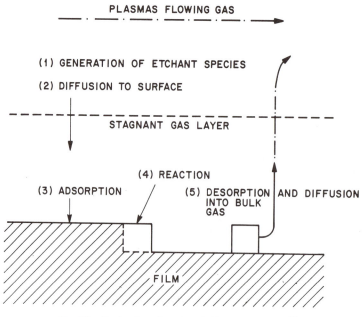

Fig. 28 Basic steps in a dry-etching processing.[23]

11.4.2 Etch Rate and Selectivity

The plasma-assisted etching process proceeds in five steps as illustrated in Fig. 28. (1) The process begins with the generation of the etchant species in the plasma. (2) The reactant is then transported by diffusion through a stagnant gas layer (refer to Chapter 8 on epitaxial growth) to the surface. (3) The reactant is adsorbed on the surface. (4) This is followed by chemical reaction (along with physical effects such as ion bombardment) to form volatile compounds. (5) These compounds are desorbed from the surface, diffused into the bulk gas, and pumped out by the vacuum system.[23]

Ion-Assisted Reaction Ions from the plasma collide with the surfaces in both plasma and reactive-ion etching processes. The pure sputtering process

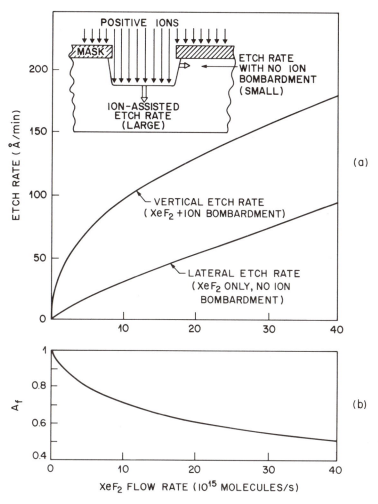

Fig. 29 Silicon etch rate versus XeF_2 flow rate with and without 1-keV Ne^+ bombardment. Insert shows the ion-assisted reaction in the vertical direction.[24]

produces only a small contribution to the etch rate. The etch rate however is substantially increased by the ion-assisted chemical reactions. Figure 29*a* shows an example in which the silicon etch rate is plotted as a function of the flow rate of XeF_2 molecules with and without 1 keV Ne^+ bombardment.[24] The insert shows the situation in which silicon is etched through a mask window under ion bombardment. The lateral etch rate depends only on the ability of the XeF_2 molecules to etch silicon in the absence of energetic ion bombardment, while the vertical etch rate is a synergistic effect due to both the Ne^+ bombardment and the XeF_2 molecules. The degree of anisotropy, which depends on the ratio of the etch rates, decreases as the flow rate increases (Fig. 29*b*). For example, a value of 0.9 can be obtained for A_f, if the XeF_2 flow rate is 2×10^{15} molecules per second. The corresponding vertical etch rate is about 50 Å/min. We can increase the energy of the ions to increase the degree of anisotropy.

Gas-Additive Effect   Another important factor in determining etch rate and selectivity is the gas composition. When a gas is mixed with one or more additive gases, the multicomponent mixture can produce the desired results in the etch rate or the selectivity, or in both. Figure 30 shows the effect on the

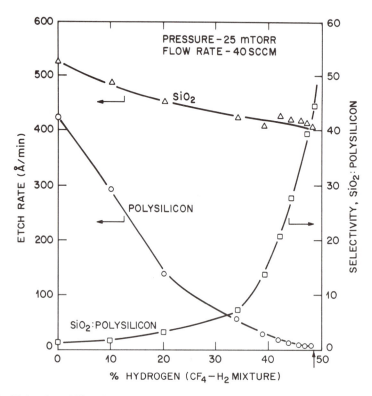

Fig. 30 Etch rates of Si and SiO_2 and the corresponding selectivity as a function of percent H_2 in CF_4.[25]

etch rates of silicon and silicon dioxide by adding hydrogen (H_2) to the carbon tetrafluoride (CF_4) gas.[25] The etch rate of silicon dioxide is approximately constant for additions of up to 40% hydrogen, while the etch rate for silicon drops monotonically to almost zero at 40% hydrogen. Also shown is the selectivity, that is, the ratio of the etch rates for silicon dioxide and silicon. Selectivities over 45:1 can be obtained with CF_4–H_2 reactive ion etching. This process is useful, for example, in selectively etching the silicon dioxide layer that covers the polysilicon gate.

In certain IC fabrication steps, we want to etch silicon instead of silicon dioxide and stop at the underlying oxide (e.g., to remove polysilicon deposited on oxide). Figure 31 shows a possible choice using the gas additive effect.[26] By varying the gas composition of sulfur hexafluoride (SF_6) and chlorine (Cl_2), we can adjust the selectivity to etch silicon 10 to 80 times faster than to etch silicon dioxide.

Table 4 shows typical etch rates and some selectivities for the dry-etching processes.[3, 27] For aluminum etching, the native oxide (~ 30 Å) must be removed first by sputtering or chemical reduction before dry etching can proceed. Figure 32 shows scanning-electron-microscope pictures that illustrate the results of highly anisotropic etching in several materials.[3, 28] Figure 32a shows a plasma-etched pattern in a 1-μm-thick phosphorous-doped polycrystalline silicon film; the substrate is silicon dioxide. Figure 32b shows

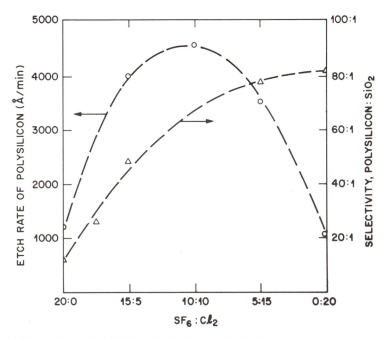

Fig. 31 Dependence of etch rates of polysilicon and selectivity to SiO_2 as gas composition in SF_6–Cl_2 (at 5.5 Pa).[26]

Table 4   Etch Rates and Selectivities for Dry Etching

Material (M)	Gas	Etch Rate ($\text{Å}/\text{min}$)	Selectivity		
			M/Resist	M/Si	M/SiO$_2$
Si	SF$_6$ + Cl$_2$	1000–4500	5	—	80
SiO$_2$	CF$_4$ + H$_2$	400–500	5	40	—
Al, Al–Si, Al–Cu	BCl$_3$ + Cl$_2$	500	5	5	25
GaAs	CCl$_4$ + O$_2$	6000	—	—	—

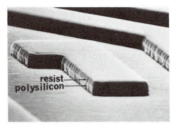

resist
polysilicon

(a)

resist
aluminum

(b)

1 μm

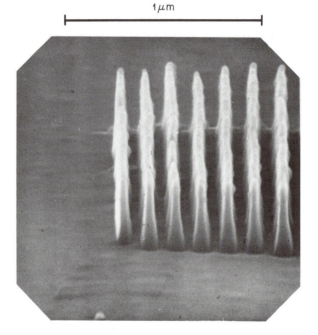

(c)

Fig. 32   SEM micrographs illustrating the results of highly anisotropic etching in reactive plasmas. (a) Plasma-etched pattern in polysilicon film. (b) Plasma-etched pattern in Al-0.7% Cu film. (c) Reactive-ion-etched pattern in a silicon substrate.[3, 28]

the essentially vertical edge profile of a 1.5-μm-thick Al (containing 0.7% Cu) film; the substrate is silicon dioxide. Figure 32c shows the silicon patterns fabricated using multilayer electron beam resist and reactive-ion etching. Note that micron- and submicron-pattern features can be transferred with very high fidelity and high selectivity by using the dry-etching processes.

REFERENCES

1 For a more detailed discussion on lithography, see D. A. McGillis, "Lithography," in S. M. Sze, Ed., *VLSI Technology*, McGraw-Hill, New York, 1983.

2 For a more detailed discussion on wet etchings, see S. K. Ghandhi, *VLSI Fabrication Principles*, Wiley, New York, 1983.

3 For a more detailed discussion on dry etching, see C. J. Mogab, "Dry Etching," in S. M. Sze, Ed., *VLSI Technology*, McGraw-Hill, New York, 1983.

4 J. M. Duffalo and J. R. Monkowski, "Particulate Contamination and Device Performance," *Solid State Technol.* **27**, No. 3, 109 (1984).

5 M. C. King, "Principles of Optical Lithography," in N. G. Einspruch, Ed., *VLSI Electronics*, Vol. 1, Academic, New York, 1981.

6 J. H. Bruning, "A Tutorial on Optical Lithography," in D. A. Doane, D. B. Fraser, and D. W. Hess, Eds. *Semiconductor Technology*, Electrochemical Soc., Pennington, 1982.

7 R. K. Watts and J. H. Bruning, "A Review of Fine-Line Lithographic Techniques: Present and Future," *Solid State Technol.*, **24**, No. 5, 99 (1981).

8 J. M. Moran and D. Maydan, "High Resolution, Steep Profile Resist Patterns," *J. Vac. Sci. Technol.*, **16**, 1620 (1979).

9 K. L. Tai, W. R. Sinclair, R. G. Vadimsky, and J. M. Moran, "Bilevel High Resolution Photolithographic Technique for Use with Wafers with Stepped and/or Reflecting Surfaces," *J. Vac. Sci. Technol.*, **16**, 1977 (1979).

10 W. C. Till and J. T. Luxon, *Integrated Circuits, Materials, Devices, and Fabrication*, Princeton-Hall, Englewood Cliffs, 1982.

11 E. D. Wolf and J. M. Ballantyne, "Research and Resource at the National Submicron Facility," in N. G. Einspruch, Ed., *VLSI Electronics*, Vol. 1, Academic, New York, 1981.

12 D. P. Kern, P. J. Coane, P. J. Houzego, and T. H. P. Chang, "Practical Aspects of Microfabrication in the 100 nm Region," *Solid State Technol.*, **27**, No. 2, 127 (1984).

13 J. A. Reynolds, "An Overview of e-Beam Mask-Making," *Solid State Technol.*, **22**, No. 8, 87 (1979).

14 W. L. Brown, T. Venkatesan, and A. Wagner, "Ion Beam Lithography," *Solid State Technol.*, **24**, No. 8, 60 (1981).

15 D. S. Kyser and N. W. Viswanathan, *J Vac. Sci. Technol.*, **12**, 1305 (1975).

16 M. P. Lepselter, D. S. Alles, H. J. Levinstein, G. E. Smith, and H. A. Watson, "A System Approach to 1-μm NMOS," *Proc. IEEE*, **71**, 640 (1983).

17 L. Karapiperis, I. Adesida, C. A. Lee, and E. D. Wolf, "Ion Beam Exposure Profiles in PMMA-Computer Simulation," *J. Vac. Sci. Technol.*, **19**, 1259 (1981).

18 H. Robbins and B. Schwartz, "Chemical Etching of Silicon II, The System HF, HNO_3, H_2O and $HC_2H_3O_2$," *J. Electrochem. Soc.*, **107**, 108 (1960).

19 S. Iida and K. Ito, "Selective Etching of Gallium Arsenide Crystal in H_2SO_4–H_2O_2–H_2O System," *J. Electrochem. Soc.*, **118**, 768 (1971).

20 K. E. Bean, "Anisotropic Etching in Silicon," *IEEE Trans. Electron Devices*, **ED-25**, 1185 (1978).

21 Y. Tarui, Y. Komiya, and Y. Harada, "Preferential Etching and Etched Profile of GaAs," *J. Electrochem. Soc.*, **110**, 585 (1963).

22 E. C. Douglas, "Advanced Process Technology for VLSI Circuits," *Solid State Technol.*, **24**, No. 5, 65 (1981).

23 J. A. Mucha and D. W. Hess, "Plasma Etching," in L. F. Thompson and C. G. Willson, Eds., *Microcircuit Processing: Lithography and Dry Etching*, American Chemical Society, 1984.

24 J. W. Coburn, "Plasma-Assisted Etching," in D. A. Doane, D. B. Fraser, and D. W. Hess, Eds., *Semiconductor Technology*, Electrochemical Soc., Penningston, 1982.

25 L. M. Ephrath, "Etching Needs for VLSI," *Solid State Technol.*, **25**, No. 7, 87 (1982).

26 W. Beinuogl and B. Hasler, "Reactive Ion Etching of Polysilicon and Tantalum Silicide," *Solid State Technol.*, **26**, No. 4, 125 (1983).

27 R. H. Burton, R. A. Gottscho, and G. Smolinsky, "Dry Etching of Group III–V Compound Semiconductors," in F. F. Y. Wang, Ed., *Materials Processing—Theory and Practices*, North Holland, Amsterdam, 1983.

28 R. E. Howard, E. L. Hu, and L. D. Jackel, "Multilevel Resist for Lithography Below 100 nm," *IEEE Trans. Electron Devices*, **ED-28**, 1378 (1981).

PROBLEMS

1 For a class-100 clean room, find the number of dust particles per cubic meter with particle sizes (*a*) between 0.5 and 1 μm, (*b*) between 1 and 2 μm, and (*c*) above 2 μm.

2 Find the final yield for a nine-mask-level process in which the average fatal defect density per cm^2 is 0.1 for four levels, 0.25 for four levels and 1.0 for one level. The chip area is 50 mm^2.

3 An optical lithographic system has an exposure power of 0.3 mW/cm^2. The required exposure energy for a positive photoresist is 140 mJ/cm^2, and for a negative photoresist, 9 mJ/cm^2. Assuming negligible times for loading and unloading wafers, compare the wafer throughput for positive photoresist and negative photoresist.

4 Specify the photoresist thickness in a production lithographic process based on the following conditions: the resist thickness l_R must be in the

range 0.5 to 2.0 μm to give acceptable resolution for 1-μm minimum features; each wafer has 250 chip sites and each chip has an area of 30 mm^2; five mask levels are used to fabricate the IC chips; 2000 finished wafers will be produced each day (20 hr per day). The fatal defect density is given by $D = 1.5l_R^{-3}$, where l_R is in micrometers; and the exposure tool throughput in wafers per hour is given by $125 - 50l_R$ (for $0.5 \leqslant l_R \leqslant 2.0$ μm). Justify your recommendation with tabular or graphic data.

5 (a) What resist would you use for high resolution? Why? (b) What are the advantages of e-beam lithography? (c) State two ways by which the exposure time can be reduced in an e-beam system.

6 Calculate the wavelength associated with electrons of 20-keV energy. If the gap between the mask and the wafer is 10 μm, find the minimum line-width due to diffraction effect of these electrons.

7 (a) Silicon is etched in a $HF:HNO_3:H_2O$ etchant. If the etch rate is 5.75 μm/min and the ratio of $HF:HNO_3$ is maintained at 3:7, what is the ratio of the H_2O used in the above etchant? (b) Silicon is etched in $HF:HNO_3:CH_3COOH$ with a ratio of 1:x:1. Sketch the etch rate as x is varied from 0 to 10. Describe the characteristic of etches at $x = 0.1, 1.0$, and 10.

8 A 125-mm-diameter <100>-oriented silicon wafer is 300 μm thick. The wafer has 1000 μm $\times$ 1000 μm ICs on it. The IC chips are to be separated by orientation-dependent etching. Describe two methods for doing this and calculate the fraction of the surface area that is lost in these processes.

9 If the mask and the substrate cannot be etched by a particular etchant, sketch the edge profile of an isotropically etched feature in a film of thickness h_f for (a) etching just to completion, (b) 100% overetch, and (c) 200% overetch. What shape does the profile tend toward as overetching proceeds? What would be the apparent degree of anisotropy for (b) and (c) if the masking layer is removed after etching?

10 Find the selectivity required to etch a 4500 Å polysilicon layer without removing more than 10 Å of its underlying gate oxide, assuming that the polysilicon is etched with a process having 10% etch rate uniformity. What is the required gas composition of SF_6 and Cl_2 for such a selectivity (the gas pressure is maintained at 5.5 Pa)?

12

Integrated Devices

Microwave, photonic, and power applications generally employ discrete devices. For example, an IMPATT diode is used as a microwave generator, an injection laser as an optical source, and a thyristor as a high-power switch. However, most electronic systems are built on the integrated circuit (IC), which is an ensemble of both active (e.g., transistor) and passive devices (e.g., resistor and capacitor) formed on and within a single-crystal semiconductor substrate and interconnected by a metallization pattern.[1] In this chapter we combine the basic processes described in previous chapters to fabricate active and passive components in an IC. Because the key element of an IC is the transistor, specific processing sequences are developed to optimize its performance. We shall consider three major IC technologies associated with the

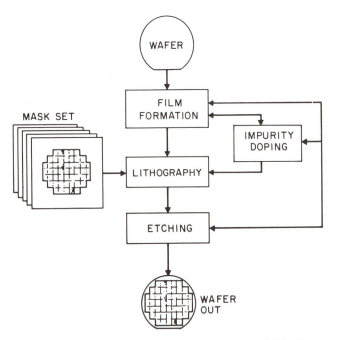

Fig. 1 Schematic flow diagrams of integrated-circuit fabrication.

three transistor families: the bipolar transistor, the MOSFET, and the MES-FET.

Figure 1 illustrates the interrelationship between the major process steps used for IC fabrication. Polished wafers with a specific resistivity and orientation are used as the starting material. The film formation steps include epitaxially grown semiconductor films (Chapter 8); thermally grown oxide films; and deposited polysilicon, dielectric, and metal films (Chapter 9). Film formation is often followed by impurity doping by methods such as diffusion and ion implantation (Chapter 10) or lithography (Chapter 11). Lithography is generally followed by etching (Chapter 11), which in turn is often followed by another impurity doping or film formation. The final IC is made by sequentially transferring the patterns from each mask, level by level, onto the surface of the semiconductor wafer.

After processing, each wafer contains hundreds of identical rectangular chips (or dice), typically between 1 and 10 mm on each side, as shown in Fig. 2a. The chips are separated by sawing or laser cutting; Fig. 2b shows a separated chip. Schematic top views of a single MOSFET and a single bipolar transistor are shown in Fig. 2c to give some perspective of the relative size of a component in an IC chip. Prior to chip separation, each chip is electrically tested. Defective chips are usually marked with a dab of black ink. Good chips are selected and packaged to provide an appropriate thermal, electrical, and interconnection environment in electronic applications.[2]

Integrated-circuit chips may contain from a few components (transistors, diodes, resistors, capacitors, etc.) to as many as a million or more. Since the invention of the integrated circuit in 1958, the number of components on a state-of-the-art IC chip has grown exponentially. We usually refer to the complexity of an IC as small-scale integration (SSI) for up to 100 components per chip, medium-scale integration (MSI) for up to 1000 components per chip,

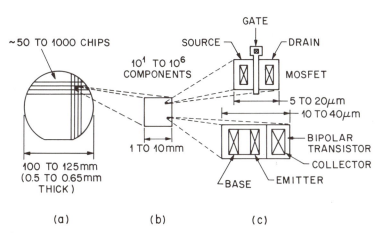

Fig. 2 Size comparison of a wafer to individual components. (a) Semiconductor wafer. (b) Chip. (c) MOSFET and bipolar transistor.

large-scale integration (LSI) for up to 100,000 components per chip, and very-large-scale integration (VLSI) for even larger numbers of components per chip. In this chapter we show two VLSI chips: a 32-bit microprocessor chip which contains over 150,000 components, and a 1-megabit dynamic-random-access-memory (DRAM) chip which contains over 2,200,000 components.

12.1 PASSIVE COMPONENTS

12.1.1 The Integrated-Circuit Resistor

To form an integrated-circuit resistor, we can define a window in a silicon dioxide layer grown thermally on a silicon substrate and then implant (or diffuse) impurities of the opposite conductivity type into the wafer. Figure 3 shows the top and cross-sectional views of two resistors; one has a meander shape and the other has a bar shape.

Consider the bar-shaped resistor first. The differential conductance dG of a thin layer of the p-type material that is of thickness dx parallel to the surface and at a depth x (as shown by the B–B cross section) is

$$dG = q \mu_p p(x) \frac{W}{L} dx \tag{1}$$

where W is the width of the bar and L is the length of the bar (we neglect the end contact areas for the time being). The total conductance of the entire implanted region of the bar is given by

$$G = \int_0^{x_j} dG = q \frac{W}{L} \int_0^{x_j} \mu_p p(x) dx \tag{2}$$

where x_j is the junction depth. If the values of μ_p (which is a function of the hole concentration) and the distribution of $p(x)$ are known, the total conductance can be evaluated from Eq. 2. We can write

$$G \equiv g \frac{W}{L} \tag{3}$$

where $g \equiv q \int_0^{x_j} \mu_p p(x) dx$ is the conductance of a square resistor pattern, that is, $G = g$ when $L = W$.

The resistance is therefore given by

$$R \equiv \frac{1}{G} = \frac{L}{W} \left[\frac{1}{g} \right] \tag{4}$$

where $1/g$ usually is denoted by the symbol $R_\square$ and is called the *sheet resistance*. The sheet resistance has units of ohms but is conventionally specified in units of ohms per square ($\Omega/\square$).

Many resistors in an integrated circuit are fabricated simultaneously by defining different geometric patterns in the mask (such as those shown in

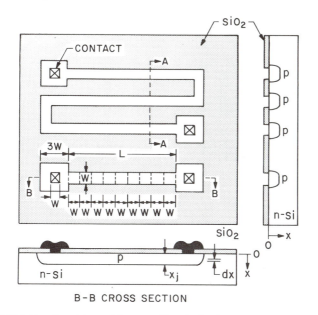

Fig. 3 Integrated-circuit resistors. All narrow lines in the large square area have the same width W, and all contacts are the same size.

Fig. 3). Since the same processing cycle is used for all these resistors, it is convenient to separate the resistance into two parts: the sheet resistance $R_\square$, determined by the implantation (or diffusion) process; and the ratio L/W, determined by the pattern dimensions. Once the value of $R_\square$ is known, the resistance is given by the ratio L/W, or the number of squares (each square has an area of $W \times W$) in the resistor pattern. The end contact areas will introduce additional resistance to the integrated-circuit resistors. For the type shown in Fig. 3, each end contact corresponds to approximately 0.65 square. For the meander-shaped resistor, the electric-field lines at the bends are not spaced uniformly across the width of the resistor but are crowded toward the inside corner. A square at the bend does not contribute exactly 1 square, but rather 0.65 square. For example, a resistor 90 μm long and 10 μm wide (such as the bar-shaped resistor in Fig. 3) contains 9 squares (9$\square$). The two end contacts correspond to 1.3$\square$. If the implanted layer has a sheet resistance of 1 k$\Omega/\square$, the value of the resistor is 10.3$\square \times$ 1 k$\Omega/\square$ = 10.3 kΩ.

12.1.2   The Integrated-Circuit Capacitor

There are basically two types of capacitors used in integrated circuits: MOS capacitors and p–n junctions. The MOS (metal–oxide–semiconductor) capacitor can be fabricated by using a heavily doped region (such as an emitter region) as one plate, the top metal electrode as the other plate, and the intervening oxide layer as the dielectric. The top and cross-sectional views of an MOS capacitor are shown in Fig. 4a. To form an MOS

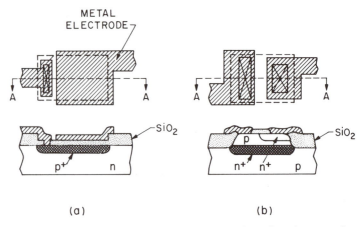

Fig. 4 (a) Integrated MOS capacitor. (b) Integrated p–n junction capacitor.

capacitor, a thick oxide layer is thermally grown on a silicon substrate. Next, a window is lithographically defined and then etched in the oxide. Diffusion or ion implantation is used to form a p^+-region in the window area, while the surrounding thick oxide serves as a mask. A thin oxide layer is then thermally grown in the window area, followed by a metallization step. The capacitance per unit area is given by

$$C = \frac{\epsilon_{ox}}{d} \quad \text{F/cm}^2 \tag{5}$$

where ϵ_{ox} is the dielectric permittivity of silicon dioxide (the dielectric constant ϵ_{ox}/ϵ_0 is 3.9) and d is the thin-oxide thickness. To increase the capacitance further, insulators with higher dielectric constants are being studied (e.g., Si_3N_4 and Ta_2O_5, with dielectric constants of 8 and 22, respectively). The MOS capacitance is essentially independent of the applied voltage, because the lower plate of the capacitor is made of heavily doped material. This also reduces the series resistance associated with it.

A p–n junction is sometimes used as a capacitor in an integrated circuit. The top and cross sectional views of an n^+-p junction capacitor are shown in Fig. 4b. The detailed fabrication process will be considered in Section 12.2, because this structure forms part of a bipolar transistor. As a capacitor, the device is usually reverse-biased, that is, the p-region is reverse-biased with respect to the n^+-region. The capacitance is not a constant but varies as $(V_R + V_{bi})^{-1/2}$, where V_R is the applied voltage and V_{bi} is the built-in potential. The series resistance is considerably higher than that of MOS capacitor because the p-region has higher resistivity than does the p^+-region.

12.2 BIPOLAR TECHNOLOGY

For IC applications, especially for VLSI, bipolar transistors must be reduced in size to meet the high-density requirement. Figure 5 illustrates the

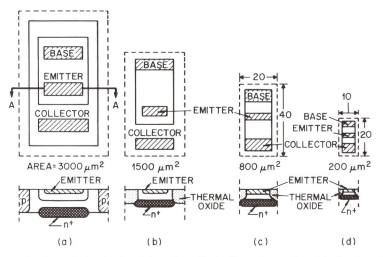

Fig. 5 Reduction of the horizontal and vertical dimensions of a bipolar transistor. (a) Junction isolation. (b) Oxide isolation. (c) and (d) Scaled oxide isolation.[3]

reduction in the size of the bipolar transistor in recent years.[3] The main differences in a bipolar transistor in an IC compared to a discrete transistor are that all electrode contacts are located on the top surface of the IC wafer, and each transistor must be electrically isolated to prevent interactions between devices. Prior to 1970, both the lateral and vertical isolations were provided by p–n junctions (Fig. 5a), and the lateral p-isolation region was always reverse-biased with respect to the n-type collector. In 1971, thermal oxide was used for lateral isolation, resulting in a substantial reduction in device size (Fig. 5b), because the base and collector contacts abut the isolation region. In the mid-1970s, the emitter was extended to the walls of the oxide, resulting in an additional reduction in area (Fig. 5c). At the present time, all the lateral and vertical dimensions have been scaled down and emitter stripe widths have dimensions in the micron region (Fig. 5d).

12.2.1 The Basic Fabrication Process

The majority of bipolar transistors used in ICs are of the n–p–n type, because the higher mobility of minority carriers (electrons) in the base region results in higher speed performance than can be obtained with p–n–p types. Figure 6 shows a perspective view of an n–p–n bipolar transistor in which lateral isolation is provided by oxide walls, and vertical isolation is provided by the n^+–p junction.[4] The lateral oxide isolation approach not only reduces the device size but also reduces parasitic capacitance because of the smaller dielectric constant of silicon dioxide (3.9, as compared to 11.7 for silicon). We shall consider the major process steps that are used to fabricate the device shown in Fig. 6.

For an n–p–n bipolar transistor, the starting material is a p-type lightly doped ($\sim 10^{15}$ cm^{-3}), $<111>$- or $<100>$- oriented, polished silicon wafer.

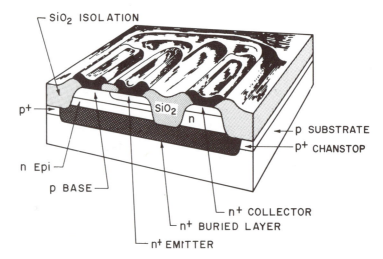

Fig. 6   Perspective view of an oxide-isolated bipolar transistor.[4, 9]

Because the junctions are formed inside the semiconductor, the choice of crystal orientation is not as critical as for MOS devices. The first step is to form a buried layer. The main purpose of this layer is to minimize the series resistance of the collector. A thick oxide (0.5 to 1 μm) is thermally grown on the wafer, and a window is then opened in the oxide. A precisely controlled amount of low-energy arsenic ions ($\sim$ 30 keV, $\sim$ 10^{15} cm^{-2}) is implanted into the window region to serve as a predeposit (Fig. 7a). Next, a high-temperature ($\sim$ 1100°C) drive-in step forms the n^+ buried layer, which has a typical sheet resistance of 20 $\Omega/\square$.

The second step is to deposit an n-type epitaxial layer. The oxide is removed and the wafer is placed in an epitaxial reactor for epitaxial growth. The thickness and the doping concentration of the epitaxial layer are determined by the ultimate use of the device. Analog circuits (with their higher voltages for amplification) require thicker layers ($\sim$ 10 μm) and lower dopings ($\sim$ 5 $\times$ 10^{15} cm^{-3}), while digital circuits (with their lower voltages for switching) require thinner layers ($\sim$ 3 μm) and higher dopings ($\sim$ 2 $\times$ 10^{16} cm^{-3}). Figure 7b shows a cross-sectional view of the device after the epitaxial process. Note that there is some outdiffusion from the buried layer into the epitaxial layer. To minimize the outdiffusion, a low-temperature epitaxial process should be employed, and low-diffusivity impurities should be used in the buried layer (e.g., As).

The third step is to form the lateral oxide isolation region. A thin-oxide pad ($\sim$ 500 Å) is thermally grown on the epitaxial layer, followed by a silicon–nitride deposition ($\sim$ 1000 Å). If nitride is deposited directly onto the silicon without the thin oxide pad, the nitride may cause damages to the silicon surface during the subsequent high-temperature steps. Next, the nitride–oxide layers and about half of the epitaxial layer are etched using a photoresist

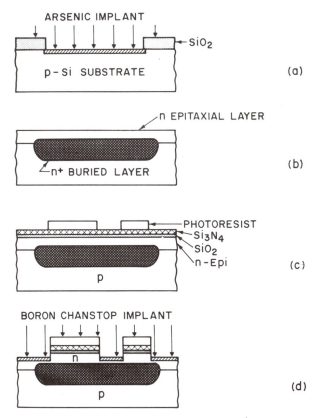

Fig. 7 Cross-sectional views of bipolar transistor fabrication.[1] (a) Buried-layer implantation. (b) Epitaxial layer. (c) Photoresist mask. (d) Chanstop implant.

as mask (Figs. 7c and 7d). Boron ions are then implanted into the exposed silicon areas (Fig. 7d).[1]

The photoresist is removed and the wafer is placed in an oxidation furnace. Since the nitride layer has a very low oxidation rate, thick oxides will be grown only in the areas not protected by the nitride layer. The isolation oxide is usually grown to a thickness such that the top of the oxide becomes coplanar with the original silicon surface to minimize the surface topograph. Figure 8a shows the cross section of the isolation oxide after the removal of the nitride layer. Due to segregation effects, most of the implanted boron ions are pushed underneath the isolation oxide to form a p^+-layer. This is called p^+ channel stop (or *chanstop*), because the high concentration of p-type semiconductor will prevent surface inversion and eliminate possible high-conductivity paths among neighboring buried layers.

The fourth step is to form the base region. A photoresist is used as a mask to protect the right half of the device; then, boron ions ($\sim 10^{12}$ cm^{-2}) are implanted to form the base region as shown in Fig. 8b. Another lithographic

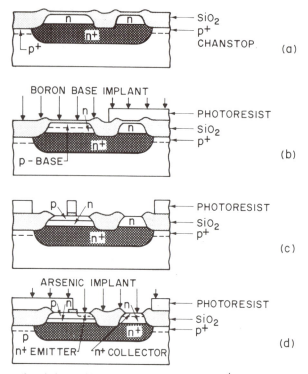

Fig. 8 Cross-sectional views of bipolar transistor fabrication.[1] (a) Oxide isolation. (b) Base implant. (c) Removal of thin oxide. (d) Emitter implant.

process removes all the thin pad oxide except a small area near the center of the base region (Fig. 8c).

The fifth step is to form the emitter region. As shown in Fig. 8d, the base contact area is protected by a photoresist mask; then a low-energy high-arsenic-dose ($\sim 10^{16}$ cm^{-2}) implantation forms the n^+-emitter and the n^+-collector contact regions.[1] The photoresist is removed; and a final metallization step forms the contacts to the base, emitter, and collector as shown in Fig. 6.

In this above basic bipolar process, there are six film formation operations, six lithographic operations, four ion implantations, and four etching operations. Each operation must be precisely controlled and monitored. Failure of any one of the operations generally will render the wafer useless.

The doping profiles of the completed transistor along a coordinate perpendicular to the surface and passing through the emitter, base, and collector are shown[5] in Fig. 9. The emitter profile is quite abrupt due to the concentration-dependent diffusivity of arsenic. The base doping profile beneath the emitter does not decrease smoothly as described by a Gaussian distribution for limited-source diffusion because of the emitter push effect discussed in Chapter 10. The collector doping is given by the epitaxial doping level

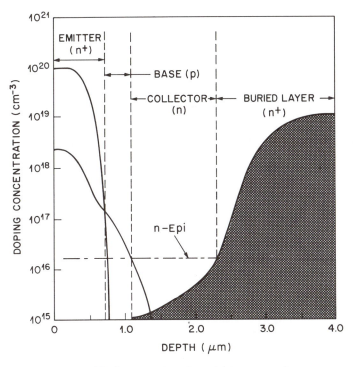

Fig. 9 *n–p–n* transistor doping profiles.[5]

($\sim 2 \times 10^{16}$ cm^{-3} for a representative switching transistor); however, at larger depths the collector doping concentration increases due to outdiffusion from the buried layer.

Dielectric Isolation In the isolation scheme described for the bipolar transistor, the device is isolated from other devices by the oxide layer around its periphery and is isolated from its common substrate by a $n^{+} - p$ junction (buried layer). In high voltage applications, a different approach, called *dielectric isolation,* is used to form insulating tubs to isolate a number of pockets of single-crystal semiconductors. In this approach the device is isolated from both its common substrate and its surrounding neighbors by a dielectric layer.

The process sequence for the dielectric isolation is shown[6] in Fig. 10. An oxide is thermally grown on a <100>-oriented silicon wafer, and windows are lithographically defined in the oxide as shown in Fig. 10*a*. An orientation-dependent etching is used to delineate the V-shaped grooves using the oxide as an etching mask (Fig. 10*b*). After removal of the mask, a thermal oxide is grown across the wafer to a thickness of typically 1 μm (Fig. 10*c*). This oxide layer serves as the dielectric isolation between the single-crystal and the polycrystalline silicon which is subsequently deposited to a thickness of about 250 μm (Fig. 10*d*). The polysilicon can be deposited by thermal reduc-

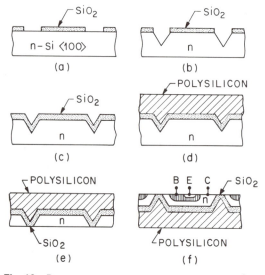

Fig. 10   Process sequence for dielectric isolation.[6]

tion of SiHCl$_3$ in hydrogen or by thermal decomposition of silane as described
in Chapter 9. Finally, the single-crystal side of the wafer is lapped down to
give the structure shown in Fig. 10e. The resulting wafer consists of the
desired tubs of single-crystal silicon, isolated from each other by a layer of
dielectric (silicon dioxide). Various types of devices can be fabricated within
these tubs, as illustrated in Fig. 10f (inverted from Figs. 10a through 10e).

The main advantage of this technique is its higher degree of isolation compared to the lateral oxide isolation technique shown in Fig. 6. The main
disadvantages of dielectric isolation are the requirement of precise mechanical
alignment of the wafer during the lapping operation and the warping of the
wafer caused by the high-temperature deposition of polysilicon. Because of
these disadvantages and the added process steps required to form the isolated
tubs, the dielectric isolation approach is mainly used for high-voltage ICs or
for circuits which must be insensitive to radiation. This approach is valuable
for such circuits since all the electron–hole pairs generated beneath the insulator (Fig. 10f) by high-energy radiations cannot participate in the operation of
the devices located in the isolated tubs.

12.2.2   The Bipolar Inverter

The processing steps for passive and active devices described in previous
sections can now be combined to show how various integrated circuits are
fabricated. As an example, we shall consider the bipolar-inverter circuit,
which is one of the basic elements of most digital systems. An inverter causes
its output to be in the logic state opposite that of its input. In other words,
when the input is at logic 0, the output is at logic 1, or vice versa. In terms of
voltage levels and positive logic, this means that with a low-voltage level at the

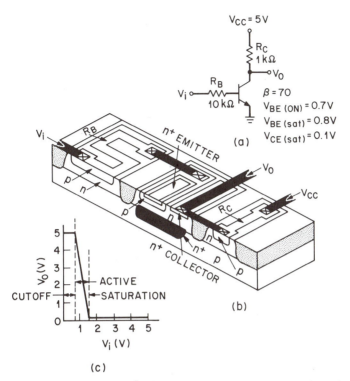

Fig. 11 Bipolar transistor inverter.[7] (a) Circuit diagram. (b) Perspective view of the inverter structure. (c) Voltage transfer characteristics.

input, the output is at a high-voltage level, or vice versa. A representative circuit of a bipolar-transistor logic inverter[7] is shown in Fig. 11a. A perspective view of an IC realization of the circuit is shown in Fig. 11b. The bipolar transistor and the two resistors are fabricated using the lateral oxide isolation technique. For clarity, the intermediate dielectric layers are not shown.

Figure 11c shows the voltage transfer characteristic of the bipolar inverter. The transfer characteristic relates the output voltage to the input voltage. When the input voltage (V_i) is less than the turn-on voltage, $V_{BE(on)}$, the transistor is in the cutoff region; and the collector current I_C is essentially zero and the output voltage V_o will be equal to $(V_{CC} - I_C R_C) \approx V_{CC} = 5$ V. When V_i is increased to a value above $V_{BE(on)}$, the transistor turns on and enters the forward active region, where the collector current is related to the base current as $I_C \simeq \beta_0 I_B$. As the input voltage increases, the base current will increase; this results in a decrease of the output voltage, which is given by $V_o = V_{CC} - \beta_0 I_B R_C$. Thus, the direction of the voltage change is inverted. With sufficient input voltage, the output voltage will decrease until the transistor enters into the saturation region. In the saturation region, the output voltage will remain at a constant low value as the input voltage is further increased.

12.2.3   Integrated Injection Logic

Integrated injection logic (I^2L) is extensively used in IC logic and memory designs. Attractive features of I^2L include compatibility with bipolar transistor processing, ease of layout, and high packing density.[1] Figures 12a and 12b show the electrical schematic diagram and cross section of an I^2L gate. The basic logic cell has a lateral $p-n-p$ transistor (Q_1) and an inverted vertical $n-p-n$ transistor (Q_2) with multiple collector contacts.

In the lateral $p-n-p$ transistor, the p-type emitter and the collector regions are formed simultaneously during the base implantation (or diffusion) step of the $n-p-n$ transistor fabrication. The n-epitaxial layer provides the base region. Since the current flow is mainly along the lateral direction, this device is called a lateral transistor. The advantage of the lateral transistor is that it does not require additional processing steps beyond those necessary for the fabrication of the standard $n-p-n$ transistor. However, the performance of the lateral transistor is inferior to that of the vertical $n-p-n$ transistor, because some of the injected carriers from the emitter will flow vertically toward the n^+-buried layer. These carriers recombine in the base and are not available at the collector; therefore, the current gain is lower.

For the inverted vertical $n-p-n$ transistor, the buried layer serves as its emitter, the collection region of the lateral $p-n-p$ transistor serves as its base,

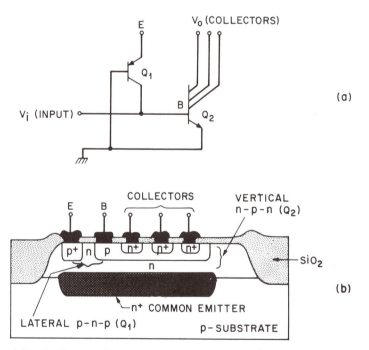

Fig. 12   (a) Circuit diagram of an integrated injection logic (I^2L). (b) Cross section of an I^2L device.

and the top n^+-regions serve as its multiple collectors. With node E (Fig. 12a) at a fixed positive bias, and with a high input voltage (logic 1) applied to V_i, additional current is injected from the emitter of the lateral p–n–p transistor (Q_1) into its collector, which is also the base of Q_2, thus driving Q_2 into a saturation condition. This, in turn, causes the outputs of Q_2 (V_o) to drop to a logic 0. Therefore, the I^{2}L can perform the functions of a logic inverter. Since I^{2}L does not require resistors or isolation regions between the multiple collectors and common emitter, its circuit density can be very high. The I^{2}L structure shown in Fig. 12b uses lateral oxide isolation and its fabrication is compatible with the bipolar processing described in Section 12.2.1.

12.3 MOSFET TECHNOLOGY

At present, the MOSFET is the dominant device used in VLSI circuits because it can be scaled to smaller dimensions than can other types of devices. MOSFET technology can be subdivided into NMOS (n-channel MOSFET) technology and CMOS (complementary MOSFET) technology, which provides n-channel and p-channel MOSFETs on the same chip. Both technologies are attractive because the NMOS circuit has fewer processing steps than does the bipolar transistor, while the CMOS circuit has substantially lower power consumption compared to both bipolar and NMOS circuits.

Figure 13 shows the reduction in the size of the MOSFET in recent years.[8] In the early 1970s, the gate length was 7.5 μm and the corresponding device area was about 6000 μm^2. As the device is scaled down, there is a drastic reduction in the device area. For a MOSFET with a gate length of 1 μm, the device area shrinks to less than 1% of the early MOSFET. We expect that device miniaturization will continue. We consider the fundamental limits of the devices in Section 12.5.

12.3.1 The Basic Fabrication Process

Figure 14 shows a perspective view of an n-channel MOSFET prior to its final metallization.[9] The top layer is a phosphorus-doped silicon dioxide (P-glass) which is used as an insulator between the polysilicon gate and the gate metallization. Compare Fig. 14 with Fig. 6 for the bipolar transistor, and note that a MOSFET is considerably simpler in its basic structure. Although both devices use lateral oxide isolation, there is no need for vertical isolation in the MOSFET, while a buried layer n^+-p junction is required in the bipolar transistor. The doping profile in a MOSFET is not as complicated as that in a bipolar transistor, and the control of the dopant distributions is also less critical. We shall consider the major process steps that are used to fabricate the device of Fig. 14.

For NMOS processing, the starting material is a p-type lightly doped ($\sim 10^{15}$ cm^{-3}), $<100>$-oriented, polished silicon wafer. The $<100>$-orientation is preferred over $<111>$ because it has an interface trap density which is about one tenth that of $<111>$. The first step is to form the oxide

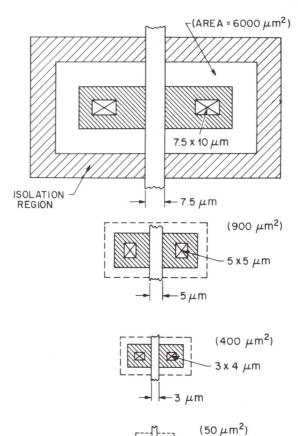

Fig. 13 Reduction in the area of MOSFET as the gate length (minimum feature length) is reduced.[8]

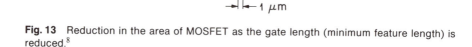

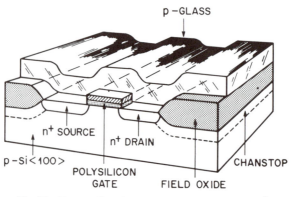

Fig. 14 Perspective view of an n-channel MOSFET.[9]

isolation region. The process sequence for this step is similar to that for bipolar transistor. A thin oxide pad ($\sim$ 500 Å) is thermally grown, followed by a silicon nitride ($\sim$ 1000 Å) deposition (Fig. 15a). The active device area is defined by a photoresist mask, and a boron chanstop layer is then implanted through the composite nitride–oxide layer (Fig. 15b). The nitride layer not covered by the photoresist mask is subsequently removed by etching. After stripping the photoresist, the wafer is placed in an oxidation furnace to grow an oxide (called the field oxide) where the nitride layer is removed, and to drive in the boron implant. The thickness of the field oxide is typically 0.5 to 1 μm.

The second step is to grow the gate oxide and to adjust the threshold voltage. The composite nitride–oxide layer over the active device area is removed, and a thin gate oxide layer (a few hundred angstroms) is grown. For an enhancement-mode n-channel device, boron ions are implanted in the channel region as shown in Fig. 15c to increase the threshold voltage to a predetermined value (e.g., $+0.5$ V). For a depletion-mode n-channel device, arsenic ions are implanted in the channel region to decrease the threshold voltage (e.g., -0.5 V).

The third step is to form the gate. A polysilicon layer is deposited and is heavily doped by diffusion or implantation of phosphorus to a typical sheet

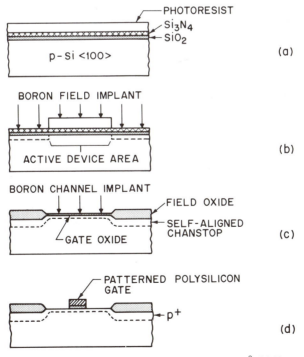

Fig. 15 Cross-sectional views of NMOS fabrication sequence.[9] (a) Formation of SiO_2, Si_3N_4 and photoresist layer. (b) Boron implant. (c) Field oxide. (d) Gate.

resistance of 20 to 30 $\Omega/\square$. This resistance is adequate for MOSFETs with gate lengths larger than 3 μm. For smaller devices, refractory metals (e.g., Mo) or polycides (i.e., a composite layer of metal silicide and polysilicon) can be used as the gate materials, to reduce the sheet resistance to about 1 $\Omega/\square$.

The fourth step is to form the source and drain. After the gate is patterned (Fig. 15d), it serves as a mask for the arsenic implantation ($\sim 30\,$keV, $\sim 10^{16}\,$cm^{-2}) to form the source and drain (Fig. 16a), which are self-aligned with respect to the gate. At this stage, the only overlapping of the gate is due to lateral straggling of the implanted ions (for 30 keV As, $\Delta R_\perp$ is only 50 Å). If low-temperature processes are used for subsequent steps to minimize lateral diffusion, the parasitic gate-drain and gate-source coupling capacitances can be much smaller than the gate-channel capacitance.

The last step is the metallization. A phosphorus-doped oxide (P-glass) is deposited over the entire wafer and is flowed by heating the wafer to give a smooth surface topography (Fig. 16b). Contact windows are defined and etched in the P-glass. A metal layer, such as aluminum, is then deposited and patterned. A cross-sectional view of the completed MOSFET is shown in Fig. 16c, and the corresponding top view is shown in Fig. 16d. The gate con-

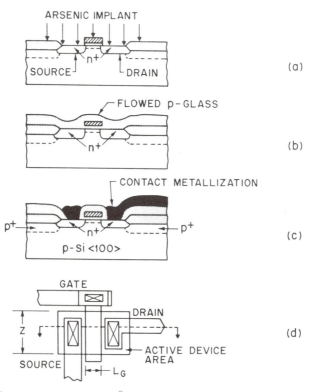

Fig. 16 NMOS fabrication sequence.[9] (a) Source and drain. (b) P-glass deposition. (c) Cross section of the MOSFET. (d) Top view of the MOSFET.

tact is usually made outside the active device area to avoid possible damage to the thin gate oxide.

In this NMOS process, there are six film-formation operations, four lithographic operations, three ion implantations, and four etching operations. There are savings of two lithographic operations and one implantation compared to the basic bipolar process.

12.3.2 The NMOS Logic Gate

Figure 17a shows the circuit of a basic logic gate, the two-input NOR gate, which has two enhancement mode MOSFETs (these devices are called *drivers*) and one depletion mode MOSFET (called the *load*).[10] The layout of the NOR gate is shown in Fig. 17b, and a cross-sectional view of the gate along the line AA' is shown in Fig. 17c. The cross-sectional view also corresponds to a depletion load inverter. When there is a logic 0 input signal (low voltage at V_i) at the driver device, the device has high impedance (with very small channel current), and the output is at logic 1 (close to V_{DD}). On the other hand, when

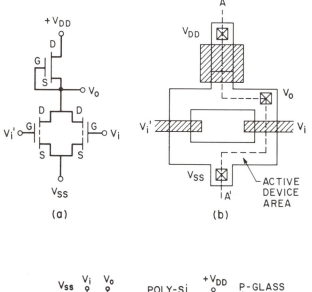

(a)   (b)

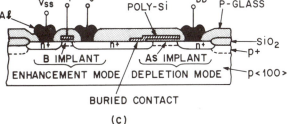

(c)

Fig. 17 An NMOS two-input NOR gate.[10] (a) Circuit diagram. (b) Circuit layout. (c) Cross section along the dotted line of (b).

the input signal is at logic 1, the driver conducts a large current with a small voltage drop across the device, and the output is at logic 0. The input and output signals are inverted, and the transfer characteristic is similar to that of Fig. 11c. For the two-input NOR gate (Fig. 17a), a logic 1 output signal can be obtained if, and only if, both inputs have logic 0 signals. For any other combinations of input signals, the output will have a logic 0 signal.

The processing steps are essentially the same as described in Section 12.3.1. The enhancement mode MOSFETs (with corresponding input voltages of V_i and V_i') are formed by implanting boron ions in the channel region to control the threshold voltage; while the depletion mode MOSFET is formed by implanting arsenic ions in the channel region to reduce the threshold voltage. The gate electrode of the depletion mode MOSFET is connected to its source by means of a buried contact as shown in Fig. 17c. Similarly, many other logic circuits can be implemented using NMOS technology.

12.3.3 NMOS Memory Devices

Memories are devices which can store digital information (or data) in terms of *bits* (*bi*nary dig*its*). Various memory chips have been designed and fabricated using NMOS technology. For most large memories, the random access memory (RAM) organization is preferred. In a RAM, memory cells are organized in a matrix structure, and they can be accessed in random order, independent of their physical locations, to store (write) or to retrieve (read) data. A static random-access memory (SRAM) can retain stored data indefinitely. The SRAM can be implemented as a flip-flop circuit to store one bit of information. A SRAM cell has four enhancement mode MOSFETs and two depletion mode MOSFETs. The depletion mode MOSFETs can be replaced by resistors formed in undoped polysilicon to minimize power consumption.[7, 11]

To further reduce area and power consumption, dynamic random-access memory (DRAM) was developed. Figure 18a shows the circuit diagram of the one-transistor DRAM cell in which the transistor serves as a switch, and one bit of information can be stored in the storage capacitor. The voltage level on the capacitor determines the state of the cell. For example, $+5$ V may be defined as logic 1 and 0 V defined as logic 0. The stored charge will be removed (typically in a few milliseconds) by the leakage currents of the capacitor; thus, dynamic memories require periodic "refreshing" of the stored charge. Because of its small cell area and low power consumption, the DRAM has the highest component density per chip.

Figure 18b shows the layout of a DRAM cell, and Fig. 18c shows the corresponding cross section through AA. The storage capacitor uses the channel region as one plate, the polysilicon gate as the other plate, and the gate oxide as the dielectric. The row line is an aluminum track to minimize RC delays. The column line is formed by n^+-diffusion. The internal drain region of the MOSFET serves as a conductive link between the inversion layers under the storage gate and the transfer gate. This drain region can be eliminated by using the double-level polysilicon approach shown in Fig. 18d. The second polysilicon electrode is separated from the first polysilicon capacitor plate by an oxide layer that is thermally grown on the first-level polysilicon before the second electrode has been defined. The charge from the column line can therefore be transmitted directly to the area under the storage gate by the continuity of inversion layers under the transfer and storage gates.

Figure 19 shows three-dimensional views of a fabrication sequence for the double-level polysilicon DRAM cell.[12] Many steps in this sequence are identical to those used to fabricate an individual MOSFET (described in Section 12.3.1). The first few process steps involve the selective oxidation of silicon using silicon nitride–pad oxide layers as the oxidation mask (Figs. 19a, 19b, and 19c). The silicon nitride–pad oxide is then removed in a selective etchant that does not attack silicon, and the first gate oxide is grown (Fig. 19d). The first-level polysilicon layer is deposited and patterned as shown in Fig.

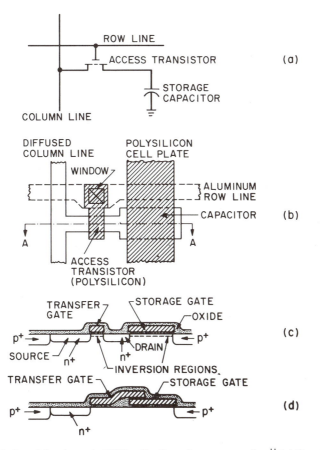

Fig. 18 Single-transistor dynamic-RAM cell with a storage capacitor.[11] (a) Circuit diagram schematic. (b) Cell layout. (c) Cross section through A–A. (d) Double-level polysilicon cell.

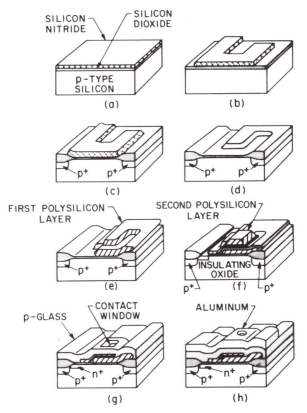

Fig. 19   Three-dimensional views of the fabrication sequence for the double-level polysilicon DRAM cell.[12]

19e. The second gate oxide is grown, followed by the deposition of the second-level polysilicon, which in turn is patterned as shown in Fig. 19f. At this stage, the exposed gate oxide regions may be implanted with an n-type dopant. A thick layer of silicon dioxide (P-glass) is deposited next, and contact windows are opened in the oxide to reach the second-level polysilicon (Fig. 19g) The contact to the first-level polysilicon is not shown. Finally, a layer of aluminum is deposited and patterned as shown in Fig. 19h. A protective coating of silicon nitride can be deposited on the wafer to seal it from contaminations.

Figure 20a shows a 100-mm silicon wafer containing fifty-seven 1-megabit DRAM chips.[13] From the identifying flats on the wafer, we recognize that it is a <100>-oriented p-type wafer. The memory cell uses 1.3-μm NMOS design rules (and its peripheral circuits are in CMOS to be considered in Section 12.3.5). The memory chip has an area of about 70 mm^2 that contains over 2,200,000 components. The active power of the chip is 160 mW, and the standby power is 2.5 mW. Figure 20b shows a 1-megabit DRAM mounted in

Fig. 20 A 1-megabit DRAM. (a) 100-mm Silicon wafer containing 1-megabit DRAM chips.[13] (b) DRAM mounted in a 18-pin cerdip.

a standard 18-pin ceramic dual-in-line package or *cerdip* (top cover removed). This package can provide adequate heat dissipation for the DRAM.

The dynamic RAMs are volatile memories, which means that stored data will be lost when power is removed from the chip. Nonvolatile memories can semipermanently retain data that have been preprogrammed into them either electrically or by other means. Figure 21*a* shows a floating-gate nonvolatile memory, which is basically a conventional MOSFET that has a modified gate electrode.[14] The composite gate has a regular (control) gate and a *floating gate* which is surrounded by insulators. When a large positive voltage is applied to the control gate, charge will be injected from the channel region through the gate oxide into the floating gate. When the applied voltage is removed, the injected charge can be stored in the floating gate for a long time. To remove this charge, a large negative voltage must be applied to the control gate, so that the charge will be injected back into the channel region.

Another version of the nonvolatile memory is the metal–insulator–oxide–semiconductor (MIOS) type shown in Fig. 21*b*. When a positive gate voltage is applied, electrons can tunnel through the thin oxide layer (~ 20 Å) and be captured by the traps at the oxide–nitride interface, and thus become stored charges there. The equivalent circuit for both types of nonvolatile memories can be represented by two capacitors in series for the gate structure as illustrated in Fig. 21*c*. The charge stored in the capacitor C_1 causes a shift in the threshold voltage, and the device remains at the higher threshold voltage state (logic 1). For a well-designed memory device, the charge retention time can be over 100 years. To erase the memory (e.g., the stored charge) and return the device to a lower threshold voltage state (logic 0), a gate voltage or other means (such a ultraviolet light) can be used.

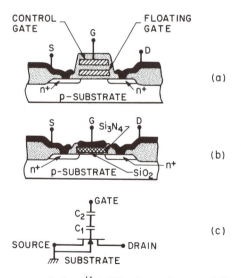

Fig. 21 Nonvolatile memory devices.[14] (*a*) Floating-gate nonvolatile memory. (*b*) MIOS nonvolatile memory. (*c*) Equivalent circuit of either type of nonvolatile memory.

12.3.4 Charge-Coupled Devices

A schematic view of a charge-coupled device (CCD) is shown in Fig. 22. The basic device consists of a closely spaced array of MOS diodes on a continuous insulator (oxide) layer that covers the semiconductor substrate.[15] Figure 22a shows a CCD to which sufficiently large, positive bias pulses have been applied to all the electrodes to produce surface depletion; a slightly higher bias has been applied to the center electrode so that the center MOS structure is under greater depletion and a potential well is formed there. If minority carriers (electrons) are introduced, they will be collected in the potential well. If the potential of the right-hand electrode is increased to exceed that of the central electrode, we obtain the potential distribution shown in Fig. 22b. In this case, the minority carriers will be transferred from the central electrode to the right-hand electrode. Subsequently, the potential on the electrodes can be readjusted so that the quiescent storage site is located at the right-hand electrode. By continuing this process, we can transfer the carriers successively along a linear array. Using this basic mechanism, CCDs can perform a wide range of electronic functions including image sensing and signal processing.

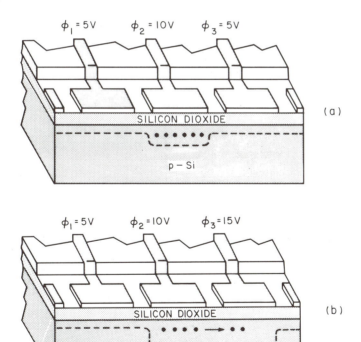

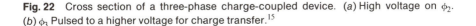

Fig. 22 Cross section of a three-phase charge-coupled device. (a) High voltage on ϕ_2. (b) ϕ_3 Pulsed to a higher voltage for charge transfer.[15]

12.3.5 CMOS Technology

Figure 23a shows a CMOS inverter.[10] The gate of the upper PMOS device is connected to the gate of the lower NMOS device. Both devices are enhancement mode MOSFETs with threshold voltages V_{Tp} less than zero for the PMOS device and V_{Tn} greater than zero for the NMOS device (typically $V_{Tp} = -0.5$ to -1.0 V, and $V_{Tn} = 0.5$ to 1.0 V). When the input voltage V_i is at ground or at small positive values, the PMOS device is turned on (the gate-to-ground potential of PMOS is $-V_{DD}$, which is more negative than V_{Tp}), and the NMOS device is off. Hence, the output voltage V_o is very close to V_{DD} (logic 1). When the input is at V_{DD}, the PMOS (with $V_{GS} = 0$) is turned off, and the NMOS is turned on ($V_i = V_{DD} > V_{Tn}$). Therefore, the output voltage V_o equals zero (logic 0). The behavior of the CMOS inverter has a transfer characteristic similar to that of other inverters described previously. However, the CMOS inverter has a unique feature: in either logic state, one device in the series path from V_{DD} to ground is nonconductive. The current that flows in either steady state is a small leakage current; and only

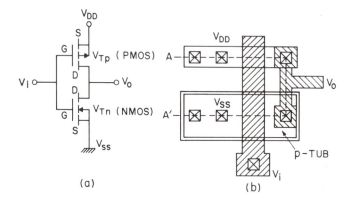

(a)   (b)

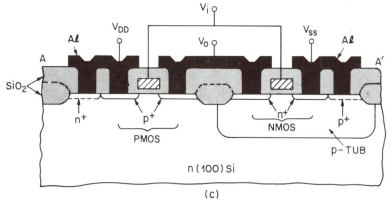

(c)

Fig. 23 CMOS inverter.[10] (a) Circuit diagram. (b) Circuit layout. (c) Cross section along dotted A — A′ line of (b).

when both devices are on during switching does a significant current flow through the CMOS inverter. Thus, the average power dissipation is small, in the order of nanowatts. As the number of components per chip increases, the power dissipation becomes a major limiting factor. The low power consumption is the most attractive feature of the CMOS circuit.

Figure 23b shows a layout of the CMOS inverter, and Fig. 23c shows the device cross section along the A−A′ line. In the processing, a p-tub (also called a p-well) is first implanted and subsequently driven into the n-substrate. The p-type dopant concentration must be high enough to overcompensate the background doping of the n-substrate. The subsequent processes for the n-channel MOSFET in the p-tub are identical to those described previously. For the p-channel MOSFET, $^{11}B^+$ or $^{49}(BF_2)^+$ ions are implanted into the n substrate to form the source and drain regions. A channel implant of $^{75}As^+$ ions may be used to adjust the threshold voltage, and an n^+-chanstop is formed underneath the field oxide around the p-channel device. Because of the p-tub and the additional steps needed to make the p-channel MOSFET, the number of steps to make a CMOS circuit is essentially double that to make an NMOS circuit. Thus, we have a trade-off between the complexity of processing and reduction in power consumption.

Instead of the p-tub described above, an alternate approach is to use an n-tub formed in p-type substrate, as shown in Fig. 24a. In this case, the n-type

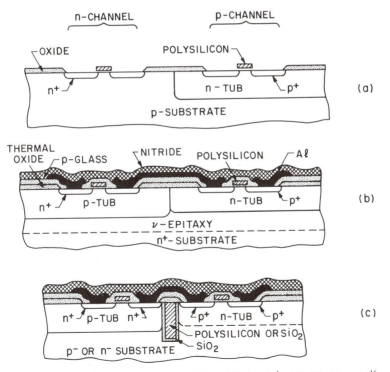

Fig. 24 Various CMOS structures. (a) n-Tub. (b) Twin tub.[1] (c) Refilled trench.[16]

(a)

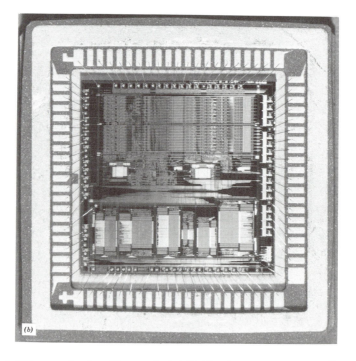

(b)

Fig. 25 (a) Micrograph of a silicon wafer of 32-bit microprocessor chips.[17] (b) A 32-bit microprocessor chip mounted in a chip carrier package.

dopant concentration must be high enough to overcompensate for the background doping of the *p*-substrate (i.e., $N_D > N_A$). In both the *p*-tub and the *n*-tub approach, the channel mobility will be degraded, because mobility is determined by the total dopant concentration $(N_A + N_D)$. A recent approach uses two separate tubs implanted into a lightly doped substrate as shown in Fig. 24*b*. This is called the *twin tub* approach.[1] Because no overcompensation is needed in either of the twin tubs, higher channel mobilities can be obtained.

All CMOS circuits have the potential for a troublesome problem called *latchup* that is associated with parasitic bipolar transistors. To see how this problem can occur, refer to Fig. 24*b*. Note that an *n–p–n* transistor can be formed with an n^+-source or drain as its emitter, the *p*-tub as its base, and the adjacent *n*-tub as its collector. Similarly, a *p–n–p* transistor can be formed with a p^+-source or drain as its emitter and with the *n*-tub and *p*-tub as its base and collector, respectively. These two transistors can be coupled together to act as a thyristor (refer to Chapter 4). If the product of the current gains of the two transistors exceeds unity, a large current can flow between V_{DD} and V_{SS}, which is the phenomenon known as latchup. Because of this high current, latchup can cause permanent damage to CMOS circuits.

To avoid latchup, we must reduce the current gain of the parasitic bipolar transistors. One method is to use gold doping or neutron irradiation to lower the minority carrier lifetimes. However, this approach is difficult to control and increases the leakage current. An effective technique is to use the trench isolation as shown[16] in Fig. 24*c*. In this technique a trench is formed in the silicon by anisotropic reactive-sputter etching. An oxide layer is thermally grown on the bottom and walls of the trench, which is then refilled by deposited polysilicon or silicon dioxide. This technique can eliminate latchup because the *n*-channel and *p*-channel devices are physically isolated by the refilled trench.

Figure 25*a* shows a 100-mm silicon wafer containing forty eight 32-bit microprocessor chips and eight test chips.[17] From the identifying flats on the wafer, we recognize that it is a <100>-oriented *n*-type wafer. The devices are fabricated using twin-tub 2-μm CMOS design rules. The area of each chip is about 100 mm^2, containing 150,000 components. The chip can perform 1 million instructions per second, and it consumes 700 mW power at full speed. Figure 25*b* shows a mounted 32-bit microprocessor chip in a 84 pin ceramic chip carrier package (top cover removed).

12.3.6 Novel CMOS Structures

Many CMOS structures are being investigated. One interesting approach is to use stacked transistors,[18] as shown in Fig. 26*a*, to achieve a multilevel layout. If we stack a *p*-channel device on top of an *n*-channel device, the two can share a common gate electrode. This structure can result in very dense static- and dynamic-RAM cells because it eliminates the conventional device isolation problems. The common output connection of transistor drains is shown at the right, and the supply connections to the transistor sources are shown at the left.

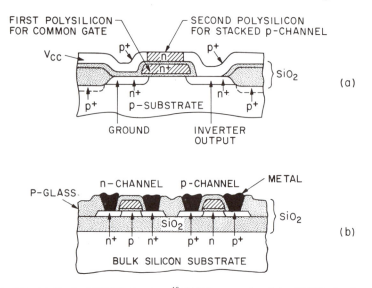

FIRST POLYSILICON
FOR COMMON GATE

SECOND POLYSILICON
FOR STACKED p-CHANNEL

(a)

GROUND

INVERTER
OUTPUT

(b)

Fig. 26 (a) Stacked CMOS structure.[18] (b) Silicon-on-insulator CMOS structure.

A pair of these devices forms a static memory cell that resembles the static RAM with the polysilicon resistor load cell. Such a static RAM substantially reduces the cell area while it offers even lower standby power.

Another potentially useful CMOS technology is the silicon-on-insulator (SOI) structure. A cross section of CMOS on SOI is illustrated in Fig. 26b. One technique to form such a structure is to grow a layer of silicon dioxide on silicon to a thickness of about 1 μm. A polysilicon layer is then deposited on the silicon dioxide and is subsequently recrystallized with laser or incoherent light sources. Semiconductor islands are patterned and defined in the recrystallized silicon. MOSFETs are then fabricated as previously described.

The SOI approach is attractive because it offers high component density and complete immunity to latchup. Circuit performance of SOI is also expected to be superior to conventional CMOS due to reduced parasitic capacitance. The mobility degradation in recrystallized materials has a minimal effect on device performance, since the device speed is determined by the saturation velocity, which is insensitive to defects or crystal imperfection.

12.4 MESFET TECHNOLOGY

Recent advances in gallium arsenide processing techniques in conjunction with new fabrication and circuit approaches have made possible the development of "silicon-like" gallium arsenide IC technology. There are three inherent advantages of gallium arsenide as compared to silicon: (1) higher electron mobility, which results in lower series resistance for a given device geometry, (2) higher drift velocity at a given electric field, which improves device speed, and (3) it can be made semi-insulating, which can provide a lattice-matched dielectric-insulated substrate. However, gallium arsenide also has three disadvantages: (1) a very short minority carrier lifetime; (2) lack of a stable, passivating native oxide; and (3) crystal defects are many orders of magnitude higher than in silicon. The short minority carrier lifetime and the lack of high-quality insulating films have prevented the development of bipolar devices and delayed MOS technology using gallium arsenide. Thus, the main emphasis of gallium arsenide IC technology is in the MESFET area in which majority carriers are transported across metal–semiconductor contacts.

A perspective view of various devices used in gallium arsenide ICs is shown[19] in Fig. 27. The major processing steps are shown in Fig. 28. The starting material is semi-insulating gallium arsenide. Insulators (e.g., SiO_2 or Si_3N_4) are deposited and masked by photoresist for a light-dose implantation (e.g., Se, S, or Si implants) to form the channel and diode regions (Fig. 28a). An implant depth of 0.1 μm and an impurity concentration of 10^{17} cm^{-3} can be obtained for a selenium implantation at 400 keV with a dose of 2.2×10^{12} cm^2 through a 1100-Å silicon nitride layer. The threshold voltage for the depletion mode MESFET is about 1 V. The next lithographic step is used to open windows for n^+-implantation to form the ohmic-contact regions (Fig. 28b). Following the implantations, additional dielectric (SiO_2) is added prior to the post implantation annealing (Fig. 28c). Ohmic contact metallization is done next to the source, drain, and diode regions (Fig. 28d). This is followed by Schottky barrier and interconnect metallization using the lift-off method described in Chapter 11 (Fig. 28e). Finally, another insulator and a second-level metal is deposited and patterned for the second-level interconnections (Fig. 28f). Note that gallium arsenide MESFET processing technology is quite similar to silicon-based MOSFET processing technology.

Gallium arsenide ICs with complexities up to the medium-scale integration level (~ 1000 components per chip) have been fabricated. Because of the higher drift velocity ($\sim 20\%$ higher than silicon), gallium arsenide ICs will

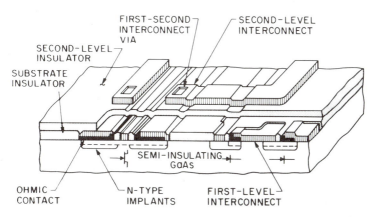

Fig. 27 Cutaway view of various planar gallium arsenide IC devices.[19]

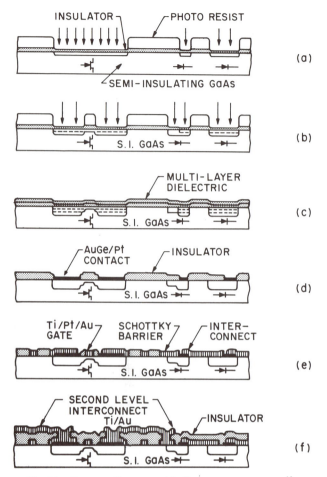

Fig. 28 Planar gallium arsenide IC process sequence.[19]

have a 20% higher speed than silicon ICs that use the same design rules. However, substantial improvements in crystal quality and processing technology are needed before gallium arsenide can seriously challenge the preeminent position of silicon in VLSI applications.

Gallium arsenide and other direct-gap semiconductors can be used to provide monolithic integration of electronic and photonic devices on a single semiconductor substrate. There are many inherent advantages of this approach compared to hybrid integration (i.e., separate substrates for electronic and photonic devices) such as lower parasitic capacitance, smaller size, and intimate optical coupling between devices. Figure 29a shows an example of a monolithically integrated structure consisting of two MESFETs and one laser.[20] For the MESFETs, the basic process sequences are similar to that described previously. For the laser, the active layer consists of gallium

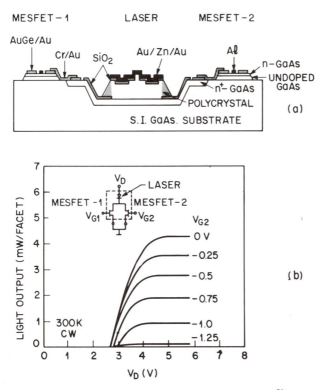

Fig. 29 Monolithic integration of photonic and electronic devices.[20] (*a*) Cross-sectional view of a structure consisting of two MESFETs and one laser. (*b*) Laser output power versus applied voltages.

arsenide and a ternary compound ($Al_x Ga_{1-x} As$). The laser has a 300-μm cavity and is formed by cleaving. The equivalent circuit of the monolithically integrated structure is shown in the insert of Fig. 29*b*. The laser current is controlled by the voltage V_D and the gate voltages V_{G1} and V_{G2}. Figure 29*b* shows the laser output power as a function of the voltage V_D and for different gate voltages V_{G2} (MESFET-2 is operated under pinch-off condition). Note that the laser output can be varied over a wide range (from zero to 4.5 mW) by varying the gate voltage of a MESFET.

12.5 FUNDAMENTAL LIMITS OF INTEGRATED DEVICES

Since the beginning of the integrated-circuit era in 1958, the minimum device dimension has been reduced at an annual rate of about 13%. At this rate, the minimum device dimension will shrink to 0.5 μm in the year 1990 and to 0.1 μm around 2000. Assuming that the advanced technology required to fabricate submicron or even smaller devices can be developed, what are the fundamental limits of device dimensions based on physical laws? We shall consider these fundamental limits in terms of intrinsic device limitations, wiring limitations, and power limitations. We shall be mainly concerned with MOSFET-type devices, since MOSFET is the dominant technology for VLSI applications.

12.5.1   Intrinsic Device Limitations

We first consider the quantum limit, which states that a physical operation performed in a time τ must involve an energy

$$E \geqslant \frac{\hbar}{\tau} \qquad (6)$$

where $\hbar$ is the reduced Planck constant. This energy is dissipated as heat. The power dissipated during the operation is

$$P = \frac{E}{\tau} \geqslant \frac{\hbar}{\tau^2} \qquad (7)$$

which is a lower limit for power dissipation per unit operation. For an operation speed of 10 ps (10^{-11} s), the minimum energy dissipated in a switching device as obtained from Eq. 6 is in the order of 10^{-23} J per operation. The actual value for a MOSFET is about 10^{-14} J, which is quite remote from this quantum limit.

Another intrinsic device limit is the thickness of the gate oxide. When the gate oxide is scaled below 50 Å, there is a finite probability that electrons will pass through the gate oxide by a quantum-mechanical tunneling process. For proper device operation, this tunneling current must be small. Therefore, the tunneling effect sets a fundamental lower limit for the thickness of the gate oxide to about 50 Å.

Semiconductor material properties such as critical electric field and saturation velocity will impose limits on device operation. Consider the minimum propagation time in a cube of silicon material of dimension Δx. The voltage ΔV and the electric field $\mathscr{E}$ are related by

$$\Delta V = \mathscr{E} \Delta x \ . \qquad (8)$$

If the electrons are traveling at their saturation velocity v_s, the time τ needed to traverse Δx is given by

$$\tau = \frac{\Delta x}{v_s} \ . \qquad (9)$$

If we assume that the minimum voltage corresponds to the thermal energy divided by the elementary charge (i.e., $\Delta V = kT/q$) and we use $\mathscr{E}_c = 5 \times 10^5$ V/cm as the critical field and $v_s = 10^7$ cm/s as the saturation velocity in silicon, the minimum transit time is

$$\tau_{\min} = \frac{kT/q}{\mathscr{E}_c v_s} \simeq 5 \times 10^{-15} \text{ s} \ . \qquad (10)$$

Therefore, the critical field and saturation velocity limit the minimum transit time in silicon to 5×10^{-15} s. Since gallium arsenide has comparable values of $\mathscr{E}_c$ and v_s, the minimum transit time due to material limitations is also about 5×10^{-15} s.

Because of thermal fluctuations, a semiconductor device may randomly switch from a logic 1 to a logic 0, or vice versa. Therefore, the thermal limit of the switching energy is kT, and the operating voltage of a device should be many times kT/q.

Figure 30 shows the propagation delay (transit time) versus the switching power required per bit of information due to limits imposed by the quantum ($\hbar/\tau$), thermal (kT), and material (5×10^{-15} s) limitations.[21] Also shown are the performances of ring oscillators having minimum feature lengths of 2 to 5 μm, 1 μm (dots), and 0.5 μm (circles). The power–delay product (this corresponds to the switching energy) for 0.5-μm devices is about 5×10^{-14} J. By scaling the device dimensions and using advanced CMOS technology to minimize power, we may reach 0.1 μm with a switching energy three orders of magnitude lower (to 5×10^{-17} J), as indicated in Fig. 30. Smaller devices might be fabricated, but at the cost of lowering the voltage or the temperature at which the devices have to be operated.

12.5.2 Wiring Limitations

For intrachip connections (interconnections of components within a chip) and interchip connections (interconnections between chips), there are three

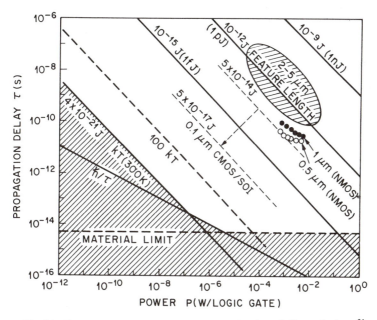

Fig. 30 Propagation delay versus power dissipation of silicon devices.[21]

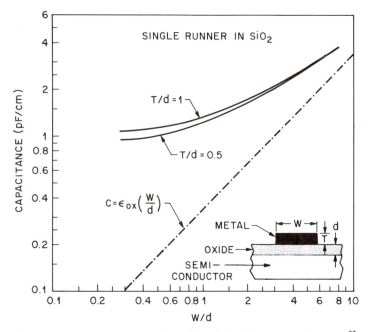

Fig. 31 Capacitance versus ratio of line width W to oxide thickness d.[22]

limitations when we shrink the interconnect wirings: electromigration limit, edge capacitance, and wire resistance.

The electromigration problem becomes worse as the minimum device dimension gets smaller. The current from a MOSFET varies as $(1/L)^n$, where L is the channel length and n is between zero and 1 depending on the carrier velocity variation with electric field. If the thickness of a conductive runner is maintained, its area varies as L, so the current density varies as $(1/L)^{n+1}$. A typical 1-μm MOSFET gate can output 1 mA, so that a 1-μm square line will carry 10^5 A/cm^2, which is close to the electromigration limit. This can of course be circumvented by making the conductors wider and reducing the circuit packing density.

The capacitance of an interconnect consists of parallel-plate and edge capacitances. The parallel-plate part varies directly with line width, but the edge component stays constant, causing a saturation in line capacitance as the line-width shrinks. At 1-μm line widths, these two components are approximately equal; and for practical field oxide thicknesses (~ 0.3 to 1 μm), the edge capacitance starts to be significant at about 2 μm. A plot of capacitance versus the ratio of line-width to oxide thickness is shown[22] in Fig. 31. The dot–dash line shows the parallel-plate component. The solid lines are the total capacitance for T/d ratios (conductor thickness to field oxide thickness) of 0.5 to 1. The field oxide is normally chosen as large as possible to keep total capacitance low. A related problem is that line-to-line capacitance will dominate line-to-substrate capacitances as dimensions shrink, causing a crosstalk problem. This can be minimized by routing ground wires next to signal wires at the expense of an increase in the area.

Since the resistivity of a conductor remains constant, the resistance of wires increases as dimensions shrink. This increased resistance coupled with the previously mentioned capacitance result in an increase of the RC time constant, which will limit device speed. All of the wiring-related limitations can be viewed as design constraints, but they are fundamental ones and they detract from the advantages of reducing device dimensions.

12.5.3 Power Limitations

The power required merely to charge and discharge circuit nodes in an integrated circuit is proportional to the number of gates and the frequency at which they are switched (clock frequency). The power can be expressed as $P \simeq \frac{1}{2} C V^2 nf$, where C is the capacitance per device, V is the applied voltage, n is the number of devices per chip, and f is the clock frequency. The temperature rise caused by this power dissipation in an IC package is limited by the thermal conductivity of the package material, unless auxiliary liquid or gas cooling is used. The maximum allowable temperature rise is limited by the bandgap of the semiconductor ($\sim 100°C$ for Si with a bandgap of 1.1 eV). For such a temperature rise, the maximum power dissipation of a typical high-performance package is about 10 W. As a result, we must limit either the maximum clock rate or the number of gates on a chip. As an example, in an IC containing 1-μm NMOS devices with $C = 5 \times 10^{-3}$ pF, running at a 2-GHz clock rate, the maximum number of gates we can have is about 10^5 if we assume a 10% duty cycle. Again, this can be looked at as a design constraint fixed by basic material parameters.

12.5.4 Ultimate Device Limits

If new device structures are allowed, several basic reasons may limit the smallest dimension to about 100 Å. One is that this dimension corresponds to the spacing between dopant atoms at the degeneracy limit ($\sim 10^{19}$ cm^{-3} for Si). Another is that our ability to define a line by any imaginable lithographic means is limited by basic particle scattering considerations to about 100 Å. Certainly, statistical considerations of device geometry and ion dose variation will prohibit any reduction beyond this point.

Figure 32 summarizes the improvements attained and those anticipated in device miniaturation as a function of time.[21, 23] As we discussed previously, the minimum channel length for MOSFETs is probably around 0.1 μm (1000 Å). Similar limits can be placed on the minimum feature size in bipolar transistors and MESFETs. Because the fabrication tools will soon be available, we expect that miniaturization will reach these limits in the next two decades.

Beam fabrication techniques such as electron beam[24] and ion beam resistless processes (refer to Section 10.6) will eventually allow devices with minimum feature lengths approaching 100 Å. However, even to enter the

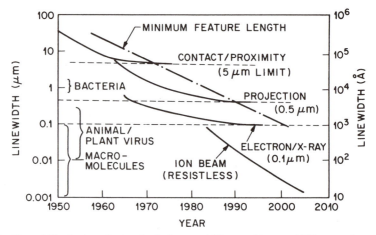

Fig. 32 Line width of microelectronic devices and lithographic capabilities as a function of time.[21] Dot–dash line shows exponential decrease of the minimum device dimensions.[23]

regions below 1000 Å will require that we use quantum theory rather than Boltzmann statistics and that we devise radically new device structures (such as molecular electronic structures).[25]

REFERENCES

1 For a detailed discussion on IC process integration, see L. C. Parrillo, "VLSI Process Integration," in S. M. Sze, Ed., *VLSI Technology*, McGraw-Hill, New York, 1983.

2 C. A. Steidel, "Assembly Techniques and Packaging," in S. M. Sze, Ed., *VLSI Technology*, McGraw-Hill, New York, 1983.

3 D. Rice, "Isoplanar-S Scales Down for New Heights in Performance," *Electronics*, **52**, 137 (1979).

4 E. F. Labuda and J. T. Clemens, "Integrated Circuit Technology," in R. E. Kirk and D. F. Othmer, Eds., *Encyclopedia of Chemical Technology*, Wiley, New York, 1980.

5 J. Agrag-Guerena, P. T. Panousis, and B. L. Morris, "OXIL, A Versatile Bipolar VLSI Technology," *IEEE Trans. Electron Devices*, **ED-27**, 1397 (1980).

6 K. E. Bean and W. R. Runyan, "Dielectric Isolation, Comprehensive, Current and Future," *J. Electrochem. Soc.*, **124**, 5C (1977).

7 For a detailed discussion on digital IC designs, see D. A. Hodges and H. G. Jackson, *Analysis and Design of Digital Integrated Circuits*, McGraw-Hill, New York, 1983.

8 E. C. Douglas, "Advanced Process Technology for VLSI Circuits," *Solid State Technol.*, **24**, No. 5, 65 (1981).

9 W. E. Beadle, J. C. C. Tsai, and R. D. Plummer, Eds., *Quick Reference Manual for Semiconductor Engineers*, Wiley, New York, 1985.

10 A. K. Sinha, S. M. Sze, and R. S. Wagner, "Silicon Devices of Integrated Circuit Processing," in M. Bever, Ed., *Encyclopedia of Materials Science and Engineering*, Pergamon, Oxford, 1985.

11 R. W. Hunt, "Memory Design and Technology," in M. J. Howes and D. V. Morgan, Eds., *Large Scale Integration*, Wiley, New York, 1981.

12 W. G. Oldham, "The Fabrication of Microelectronic Circuits," in *Microelectronics*, W. H. Freeman, San Francisco, 1977.

13 H. C. Kirsch, et al., "A One Megabit DRAM," in *Digest IEEE International Solid State Circuit Conference*, p. 256, 1985.

14 For a discussion on devices and their related references, see S. M. Sze, *Physics of Semiconductor Devices*, 2nd ed., Wiley, New York, 1981.

15 W. S. Boyle and G. E. Smith, "Charge-Coupled Devices—A New Approach to MIS Device Structures," *IEEE Spectrum*, **8**, 18 (1971).

16 R. D. Rung, H. Momose, and Y. Nagakubo, "Deep Trench Isolated CMOS Devices," in *Tech. Digest IEEE International Electronic Device Meeting*, 1982, p. 237.

17 B. T. Murphy, L. C. Thomas, and A. V. MacRae, *Electronics*, **54**, No. 20, 106 (1981).

18 J. F. Gibbons and K. F. Lee, "One-Gate-Wide CMOS Inverter with Stacked Transistors," *Solid State Electron.*, **15**, 789 (1982).

19 B. M. Welch, Y. D. Shen, R. Zucca, R. C. Eden, and S. I. Long, "LSI Processing Technology for Planar GaAs Integrated Circuits," *IEEE Trans. Electron Devices*, **ED-27**, 1116 (1980).

20 T. Sanada, S. Yamakoshi, O. Wada, T. Fujii, T. Sakurai, and M. Sasaki, "Monolithic Integration of an AlGaAs/GaAs Multiquantum Well Laser and GaAs MESFETs on a Semiinsulating GaAs Substrate by Molecular Beam Epitaxy," *Appl. Phys. Lett.*, **44**, 325 (1984).

21 I. Brodie and J. J. Muray, *The Physics of Microfabrication*, Plenum, New York, 1982.

22 M. P. Lepselter, D. S. Alles, H. J. Levinstein, G. E. Smith, and H. A. Watson, "A System Approach to 1-μm NMOS," *Proc. IEEE*, **71**, 640 (1983).

23 S. M. Sze, "VLSI Technology Overview and Trends," *Jpn. J. Appl. Phys.*, **22**, Suppl. 22-1, 3(1983).

24 H. G. Craighead, R. E. Howard, L. D. Jackel, and P. M. Mankiewich, "10 nm Linewidth Electron-Beam Lithography on GaAs," *Appl. Phys. Lett.*, **42**, 38 (1983).

25 F. L. Carter, "From Electroactive Polymers to the Molecular Electronic Device Computer," in D. J. McGreivy and K. A. Pickar, Eds., *VLSI Technologies Through the 80s and Beyond*, IEEE Computer Society, Los Angeles, 1982.

PROBLEMS

1 For a sheet resistance of 1 kΩ/$\square$, find the maximum resistance that can be fabricated on a 2.5×2.5-mm chip for 2-μm lines and 4-μm pitch (i.e., distance between the centers of the parallel lines).

2 Design a mask set for a 5-pF MOS capacitor. The oxide thickness is 300 Å. Assume that the minimum window size is 2×10 μm and the maximum registration errors are 2 μm.

3 Draw the circuit diagram and device cross section of a clamped transistor.

4 (a) Why is $<100>$-orientation preferred in NMOS fabrication? (b) What are the disadvantages if too thin a field oxide is used in NMOS devices? (c) What problems occur if a polysilicon gate is used for gate lengths less than 3 μm? Can another material be substituted for polysilicon? (d) How is a self-aligned gate obtained and what are its advantages? (e) What purpose does P-glass serve?

5 In NMOS processing, the starting material is a p-type 10-Ω-cm $<100>$-oriented silicon wafer. The source and drain are formed by arsenic implantation of 10^{16} ions/cm^2 at 80 keV; the channel is implanted with 8×10^{11} boron ions/cm^2 at 30 keV through a gate oxide of 250 Å. (a) Estimate the threshold voltage change of the device. (b) Draw the doping profile along a coordinate perpendicular to the surface and passing through the channel region or the source region.

6 Draw a complete step-by-step set of masks for CMOS inverter shown in Fig. 23. Pay particular attention to the cross section shown in Fig. 23c for your scale.

7 Draw the layout of NMOS and CMOS NAND gates. Use the same scale for both gates. Label your layouts. Estimate roughly how much more area is required for a CMOS NAND gate.

8 Plot the cross-sectional views of a twin-tub CMOS structure of the following stages of processing: (a) n-tub implant, (b) p-tub implant, (c) twin-tub drive-in, (d) nonselective p^+ source/drain implant, (e) selective n^+ source/drain implant using photoresist as mask, and (f) P-glass deposition.

9 For a floating-gate nonvolatile memory, the lower insulator has a dielectric constant of 4 and is 100 Å thick. The insulator above the floating gate has a dielectric constant of 10 and is 1000 Å thick. If the current density J in the lower insulators is given by $J = \sigma\mathscr{E}$, where $\sigma = 10^{-7}$ S/cm, and the current in the other insulator is negligibly small, find the threshold voltage shift of the device caused by a voltage of 10 V applied to the control gate for (a) 0.25 μs, and (b) sufficiently long time that J in the lower insulator becomes negligibly small.

10 (a) Calculate the RC time constant of a 0.5-μm-thick aluminum runner formed on 0.5-μm-thick thermally grown SiO$_2$. The length and width of the runner are 1 cm and 1 μm, respectively. The resistivity of the runner is 10^{-5} Ω-cm. (b) What will be the RC time constant for a polysilicon runner ($R_\square = 30$ $\Omega/\square$) of identical dimensions?

APPENDIX A

List of Symbols

Symbol	Description	Unit
a	Lattice constant	Å
B	Magnetic induction	T
c	Speed of light in vacuum	cm/s
C	Capacitance	F
D	Diffusion coefficient	cm^2/s
E	Energy	eV
E_C	Bottom of conduction band	eV
E_F	Fermi energy level	eV
E_g	Energy bandgap	eV
E_V	Top of valence band	eV
$\mathscr{E}$	Electric field	V/cm
$\mathscr{E}_c$	Critical field, maximum field at breakdown	V/cm
$\mathscr{E}_m$	Maximum field	V/cm
f	Frequency	Hz (cps)
$F(E)$	Fermi–Dirac distribution function	
h	Planck constant	J-s
$h\nu$	Photon energy	eV
I	Current	A
I_C	Collector current	A
J	Current density	A/cm^2
J_t	Threshold current density	A/cm^2
k	Boltzmann constant	J/K
kT	Thermal energy	eV
L	Length	cm or μm
m_0	Electron rest mass	kg
m_n	Electron effective mass	kg
m_p	Hole effective mass	kg
$\bar{n}$	Refractive index	
n	Density of free electrons	cm^{-3}
n_i	Intrinsic density	cm^{-3}
N	Doping concentration	cm^{-3}
N_A	Acceptor impurity density	cm^{-3}

Symbol	Description	Unit
N_C	Effective density of states in conduction band	cm^{-3}
N_D	Donor impurity density	cm^{-3}
N_V	Effective density of states in valence band	cm^{-3}
p	Density of free holes	cm^{-3}
P	Pressure	Pa
q	Magnitude of electronic charge	C
Q_{it}	Interface trap density	$charges/cm^2$
R	Resistance	Ω
t	Time	s
T	Absolute temperature	K
v_n	Carrier velocity	cm/s
v_s	Saturation velocity	cm/s
v_{th}	Thermal velocity	cm/s
V	Voltage	V
V_{bi}	Built-in potential	V
V_{EB}	Emitter–base voltage	V
V_B	Breakdown voltage	V
W	Thickness	cm or μm
W_B	Base thickness	cm or μm
ϵ_0	Permittivity in vacuum	F/cm
ϵ_s	Semiconductor permittivity	F/cm
ϵ_{ox}	Oxide permittivity	F/cm
ϵ_s/ϵ_0 or ϵ_{ox}/ϵ_0	Dielectric constant	
τ	Lifetime or decay time	s
θ	Angle	rad
λ	Wavelength	μm or Å
ν	Frequency of light	Hz
μ_0	Permeability in vacuum	H/cm
μ_n	Electron mobility	$cm^2/V\text{-}s$
μ_p	Hole mobility	$cm^2/V\text{-}s$
ρ	Resistivity	$\Omega\text{-}cm$
ρ_d	Specific density	$atoms/cm^3$
ρ_s	Space charge density	cm^{-3}
ϕ_{Bn}	Schottky barrier height on n-type semiconductor	V
ϕ_{Bp}	Schottky barrier height on p-type semiconductor	V
$q\phi_m$	Metal work function	eV
ω	Angular frequency ($2\pi f$ or $2\pi\nu$)	Hz
Ω	Ohm	Ω

APPENDIX B

International System of Units

Quantity	Unit	Symbol	Dimension
Length	meter	m	
Mass	kilogram	kg	
Time	second	s	
Temperature	kelvin	K	
Current	ampere	A	
Frequency	hertz	Hz	$1/s$
Force	newton	N	$kg\text{-}m/s^2$
Pressure	pascal	Pa	N/m^2
Energy	joule	J	$N\text{-}m$
Power	watt	W	J/s
Electric charge	coulomb	C	$A\text{-}s$
Potential	volt	V	J/C
Conductance	siemens	S	A/V
Resistance	ohm	Ω	V/A
Capacitance	farad	F	C/V
Magnetic flux	weber	Wb	$V\text{-}s$
Magnetic induction	tesla	T	Wb/m^2
Inductance	henry	H	Wb/A

APPENDIX C

Unit Prefixes [a]

Multiple	Prefix	Symbol	Multiple	Prefix	Symbol
10^{18}	exa	E	10^{-1}	deci	d
10^{15}	peta	P	10^{-2}	centi	c
10^{12}	tera	T	10^{-3}	milli	m
10^{9}	giga	G	10^{-6}	micro	μ
10^{6}	mega	M	10^{-9}	nano	n
10^{3}	kilo	k	10^{-12}	pico	p
10^{2}	hecto	h	10^{-15}	femto	f
10	deka	da	10^{-18}	atto	a

[a]Adopted by International Committee on Weights and Measures.
(Compound prefixes should not be used; e.g., not $\mu\mu$ but p.)

$1 \dot{A} = 10^{-10}$

APPENDIX D

Greek Alphabet

	Lowercase Letter	Uppercase Letter		Lowercase Letter	Uppercase Letter
Alpha	α	A	Nu	ν	N
Beta	β	B	Xi	ξ	Ξ
Gamma	γ	Γ	Omicron	o	O
Delta	δ	Δ	Pi	π	Π
Epsilon	ϵ	E	Rho	ρ	P
Zeta	ζ	Z	Sigma	σ	Σ
Eta	η	H	Tau	τ	T
Theta	θ	Θ	Upsilon	υ	Υ
Iota	ι	I	Phi	ϕ	Φ
Kappa	κ	K	Chi	χ	X
Lambda	λ	Λ	Psi	ψ	Ψ
Mu	μ	M	Omega	ω	Ω

Physical Constants

Quantity	Symbol/Unit	Value
Angstrom unit	Å	$1 \text{ Å} = 10^{-1} \text{ nm} = 10^{-4} \mu\text{m}$ $= 10^{-8} \text{ cm} = 10^{-10} \text{ m}$
Avogadro constant	N_{AVO}	$6.02204 \times 10^{23} \text{ mole}^{-1}$
Bohr radius	a_B	0.52917 Å
Boltzmann constant	k	$1.38066 \times 10^{-23} \text{ J/K } (R/N_{AVO})$
Elementary charge	q	$1.60218 \times 10^{-19} \text{ C}$
Electron rest mass	m_0	$0.91095 \times 10^{-30} \text{ kg}$
Electron volt	eV	$1 \text{ eV} = 1.60218 \times 10^{-19} \text{ J}$ $= 23.053 \text{ kcal/mole}$
Gas constant	R	$1.98719 \text{ cal/mole} - \text{K}$
Permeability in vacuum	μ_0	$1.25663 \times 10^{-8} \text{ H/cm } (4\pi \times 10^{-9})$
Permittivity in vacuum	ϵ_0	$8.85418 \times 10^{-14} \text{ F/cm } (1/\mu_0 c^2)$
Planck constant	h	$6.62617 \times 10^{-34} \text{ J-s}$
Reduced Planck constant	$\hbar$	$1.05458 \times 10^{-34} \text{ J-s } (h/2\pi)$
Proton rest mass	M_p	$1.67264 \times 10^{-27} \text{ kg}$
Speed of light in vacuum	c	$2.99792 \times 10^{10} \text{ cm/s}$
Standard atmosphere		$1.01325 \times 10^5 \text{ Pa}$
Thermal voltage at 300 K	kT/q	0.0259 V
Wavelength of 1-eV quantum	λ	$1.23977 \mu\text{m}$

$$k_B = 8.612 \times 10^{-5} \text{ eV}/k$$

APPENDIX F

Properties of Important Semiconductors at 300 K

Semiconductor		Lattice Constant (Å)	Bandgap (eV)	Band[a]	Mobility[b] (cm²/V-s)		Dielectric Constant
					μ_n'	μ_p	
Element	Ge	5.64	0.66	I	3900	1900	16.0
	Si	5.43	1.12	I	1450	450	11.9
IV–IV	SiC	3.08[c]	2.99	I	400	50	10.0
III–V	AlSb	6.13	1.58	I	200	420	14.4
	GaAs	5.63	1.42	D	8500	400	13.1
	GaP	5.45	2.26	I	110	75	11.1
	GaSb	6.09	0.72	D	5000	850	15.7
	InAs	6.05	0.36	D	33000	460	14.6
	InP	5.86	1.35	D	4600	150	12.4
	InSb	6.47	0.17	D	80000	1250	17.7
II–VI	CdS	5.83	2.42	D	340	50	5.4
	CdTe	6.48	1.56	D	1050	100	10.2
	ZnO	4.58	3.35	D	200	180	9.0
	ZnS	5.42	3.68	D	165	5	5.2
IV–VI	PbS	5.93	0.41	I	600	700	17.0
	PbTe	6.46	0.31	I	6000	4000	30.0

[a] I = Indirect, D = Direct.

[b] The values are for drift mobilities obtained in the purest and most perfect materials available to date.

[c] Silicon carbide crystallizes in the wurtzite structure (see Ref. 5 of Chapter 1).

Index